The Domestication of Humans

The Domestication of Humans explains the alternative to the African Eve model by attributing human modernity, not to a speciation event in Africa, but to the unintended self-domestication of humans.

This alternative account of human origins provides the reader with a comprehensive explanation of all features defining our species that is consistent with all the available evidence. These traits include, but are not limited to, massive neotenisation, numerous somatic changes, susceptibility to almost countless detrimental conditions and maladaptations, brain atrophy, loss of oestrus and thousands of genetic impairments. The teleological fantasy of replacement by a 'superior' species that has dominated the topic of modern human origins has never explained any of the many features that distinguish us from our robust ancestors. This book explains all of them in one consistent, elegant theory. It presents the most revolutionary proposal of human origins since Darwin.

Although primarily intended for the academic market, this book is perfectly suitable for anyone interested in how and why we became the species that we are today.

Robert G. Bednarik is a Professor at the International Centre for Rock Art Dating, Hebei Normal University, Shijiazhuang, China. He has served as Convener/CEO and Editor-in-Chief of the International Federation of Rock Art Organisations continuously since 1988.

The Domestication of Humans

Robert G. Bednarik

Routledge
Taylor & Francis Group

LONDON AND NEW YORK

First published 2020
by Routledge
2 Park Square, Milton Park, Abingdon, Oxon OX14 4RN

and by Routledge
52 Vanderbilt Avenue, New York, NY 10017

Routledge is an imprint of the Taylor & Francis Group, an informa business

© 2020 Robert G. Bednarik

The right of Robert G. Bednarik to be identified as author of this work has been asserted by him in accordance with sections 77 and 78 of the Copyright, Designs and Patents Act 1988.

All rights reserved. No part of this book may be reprinted or reproduced or utilised in any form or by any electronic, mechanical, or other means, now known or hereafter invented, including photocopying and recording, or in any information storage or retrieval system, without permission in writing from the publishers.

Trademark notice: Product or corporate names may be trademarks or registered trademarks, and are used only for identification and explanation without intent to infringe.

British Library Cataloguing-in-Publication Data
A catalogue record for this book is available from the British Library

Library of Congress Cataloging-in-Publication Data
A catalog record has been requested for this book

ISBN: 978-0-367-89787-1 (hbk)
ISBN: 978-1-003-02113-1 (ebk)

Typeset in Bembo
by Taylor & Francis Books

To Elfriede

Contents

	List of figures	ix
	Foreword	xi
1	Introduction: the empirical context	1

About palaeoanthropology 1
Pliocene and Pleistocene archaeology 9
Incommensurabilities 15
Formulating a credible synthesis 20

2	The gracilisation of humans	27

The 'African Eve' hoax 27
Palaeoart and seafaring 36
Robusticity and gracility 56
Human neoteny 63

3	Evolution and pathologies	68

Narratives of evolution 68
The brains of apes and humans 77
The rise of pathologies 84
The Keller and Miller paradox 94

4	Human self-domestication	99

Domestication 99
Human auto-domestication 104
Testing the domestication hypothesis 115
The implications 126

5	The unstoppable advance of exograms	136
	About exograms 136	
	Language and other exograms 146	
	The roles of exograms 165	
6	Effects of the domestication hypothesis	173
	Palaeoanthropology and archaeology 173	
	Reassessing human evolution 182	
	References	186
	Index	239

Figures

1.1	The Narmada calvarium from Hathnora, central India	11
2.1	Ivory ring fragments, perforated animal canines and fossil shell pendant made by Neanderthals, Grotte du Renne, Arcy-sur-Cure, France	35
2.2	Aurignacian lion-headed human ivory figurine, possibly female, from Hohlenstein-Stadel, Germany	41
2.3	Aurignacian pictograms apparently depicting the movement of a charging woolly rhinoceros, Chauvet Cave, France	42
2.4	Middle Palaeolithic combinations of circular and cross forms from (A) Tata (Hungary) and (B) Axlor (Spain)	43
2.5	Engraved bone fragment of the forest elephant and hominin skull fragment of the Lower Palaeolithic of Bilzingsleben, Germany	44
2.6	Lower Palaeolithic stone beads made from modified *Porosphaera globularis* fossils, Biddenham, England	45
2.7	Cupules at Nchwaneng, southern Kalahari, South Africa, thought to be c. 400,000 years old	47
2.8	Modified manuport of the Middle Acheulian, Tan-Tan, Morocco	48
2.9	The Makapangat jaspilite cobble, a manuport deposited in a cave between 2.4 and 2.9 Ma ago	49
2.10	Bamboo raft built with Lower Palaeolithic stone tools off the coast of Flores, April 2008	50
2.11	The first stone tool of the Middle Pleistocene found in Timor, at Motaoan	53
2.12	Weidenreich's (1946) original trellis model of hominin evolution (above) and Howell's false interpretation of it (below)	57
2.13	Male and female relative cranial robusticity/gracility in Europe during the final Pleistocene	61
3.1	The atrophy of the human brain during the last 40,000 years: volume in ml versus age in 1,000 years; the upper curve represents males, the lower is of females	83

x   List of figures

4.1	Presumed decorative objects, including a green chlorite bracelet fragment (i), made by Denisovans from Denisova Cave, Siberia	110
4.2	Calcined bone fragment in the earliest-known fireplace, 1.7 million years old, Wonderwerk Cave, South Africa	121
4.3	Depictions of female genitalia of the Aurignacian and Gravettian, from (a) Galgenberg, (b) Willendorf, (c) Hohle Fels, (d) Chauvet Cave, (e) Avdeevo, (f) Abri Cellier, and (g) La Ferrassie	132
4.4	Female Early Upper Palaeolithic depictions with typical items of body decoration, of marl and ivory from Kostenki (a, c) and fired clay from Pavlov (b)	134
5.1	The atrophy of the human brain during the last 40,000 years compared with the presumed rise in the cognitive demands on the brain during the same period	138
5.2	Greenstone bracelet from Denisova Cave, Middle Palaeolithic, but of Neolithic technology	147
5.3	Acheulian hand-axe from Swanscombe, bearing *Conulus* fossil cast	150
5.4	Stone plaque with seven incised grooves, c. 300 ka old, from Wonderwerk Cave	153
5.5	Microphotograph of the artificially enlarged orifice of one of the Bedford Acheulian beads	156
5.6	Flake scars at the closed end of the tunnel of one of the Bedford Acheulian beads and their sequence. Five scars are clearly visible, No. 2 displays rippling typical of impact fractures on silica stone	157
5.7	Six Acheulian *Porosphaera globularis* beads showing different degrees of wear at tunnel opening, including significant asymmetrical concave wear (b, d)	158
5.8	Schematic depiction of (a) the initial beads before anthropogenic action; (b) flaking to open the second tunnel entrance; (c) heavy wear from rubbing against other, fresher beads for many years; and (d) the outcome of beads of different ages on a string having been worn, some for very long periods of time	158
5.9	Carefully perforated wolf canine, probably c. 300 ka old, Repolusthöhle, Austria	160
5.10	Some of the hundreds of cupules on the walls of Daraki-Chattan Cave, central India, of the Lower Palaeolithic	161
5.11	Just before a memory is activated, fast ripples of brain activity occur simultaneously, in this example in two areas, the temporal association cortex (very dark areas in lower parts of the image) and the medial temporal lobe (very light areas)	166
6.1	Some of the millions of presumed geofacts found in nature	178

Foreword

The Domestication of Humans is a short but stimulating, highly readable volume that proceeds from a wide and very deep analysis of the current state of knowledge about human evolution, to presentation of several interrelated hypotheses suggesting how the later phases of our relatively recent human evolutionary transformations have occurred. Robert Bednarik is an acknowledged expert – arguably, the world's leading expert – on rock art, in pursuit of which expertise his studies have expanded widely into a great many contiguous aspects of archaeology and anthropology. He also is professionally conversant with human biology, including molecular genetics, as well as psychology, history and philosophy. We would have described him by the over-used term "Renaissance Scholar" were it not for fear that the mantle more comfortably worn in that earlier historical period might cause his comprehensive knowledge to be perceived as outdated in our own time. It is not. In fact, the author is a genuine twenty-first-century scholar of the highest caliber in the social and natural sciences, and broadly familiar with the literatures of both domains. His volume succeeds in genuinely delivering what most scientific authors promise to publishers, but nearly always fail to produce: A book that can be read profitably by the "general educated reader" and professional scientist alike. This outcome arises from a combination of organization and passion.

The organization is achieved partly by opening each chapter with an abstract of its contents and ending with a paragraph that provides a transition to the chapter that will follow. Chapters themselves contain several sub-heads that signal coherent blocks of text that advance the author's arguments. Illustrations complement presentation of the major ideas that are being developed. Passion can be seen in Bednarik's pervasive sense of purpose. His Renaissance breadth arises from the base of a Reformation of sorts, with scholarship that is pursued with the sense of mission reminiscent of a historical tradition that is neither the author's nor our own, but the spirit of which in some ways still lives in our shared culture: Martin Luther nailing his Ninety-five Theses on the door of All Saints Church in Wittenberg in 1517. Although the intellectual context here certainly is secular (as is our own frame of reference, as well as his) rather than religious, there is more than a little overlap in intent: to take knowledge from the hands of a narrow priestly class and make it more readily available to members of a wider audience genuinely in need of individual understanding.

The result is a coherent book, having three strands that are woven together with skill into a refreshingly original whole. The first strand comprises an informed review of the scholarship covering the last 5.3 million years of hominin evolution, with an emphasis on the later phases of the last several hundred thousand years that encompass members of our own genus, *Homo*. This most recent phase of our evolution – which the author shows to be widely misunderstood, misrepresented and intellectually confused in many other sources – is substantially clarified here. On this subject, Bednarik's knowledge is encyclopaedic and his explanations enlightening. Many authors writing about the evolution of the genus *Homo* and its antecedents largely play a "name game" that simply catalogues and describes various putative "species" of humans. This "family photo album" approach produces a static record of descriptions and reconstructed images, along with the places and dates of specific finds. Many creators of those multiple human taxa flourish their professional authority by claiming to have discovered the "oldest" or "most advanced" or "most complete" specimens. Exciting as this competitive approach may be, it does little to explain how evolution occurs, because the processes leading from one putative "species" to the next rarely are discussed and described. The author's critique of this present state of the discipline is accomplished in Chapter 1 ("Introduction: The Empirical Context").

The second strand of inquiry helps to explain just why the latest phase of human evolution is so widely misunderstood. In the quasi-commercialized world of research grants and contracts, investigators hoping to secure as much financial support for their work as possible commonly leap aboard the main bandwagon of opinion, to which the majority of the field's professionals already are clinging. Some younger authors exercise self-censorship in order to avoid criticism from senior researchers who could prevent publication in a "leading journal" or otherwise make their career progress difficult. Thus, as Professor Bednarik elaborates, while the overlapping traditions of palaeoanthropology and Pleistocene archaeology purport to inform and influence the public's perception of the human past, instead they very largely have reflected and amplified some rather hackneyed public, and particularly socio-political, opinions. As a prime example, for most of the last three decades the pattern of later human evolution has been forced into the Procrustean bed of the "African Eve" myth that uses a seemingly sophisticated molecular biological framework to paraphrase the Biblical story of Genesis. Just as the Biblical Eve is a character of historical revelation, so is the mitochondrial Eve equally an imaginary construct for different didactic purposes. The familiar Biblical referent helps with acceptance, though not genuine understanding.

Early in Chapter 3 ("Evolution and Pathologies") in a section entitled "Narratives of evolution", the author comments on the perennial battle in human palaeontology between "lumpers" and "splitters" that commonly is treated as a matter of taste, rather comparable to preference for eye of newt over toe of frog in the brew concocted by the proverbial witches. He notes that in many other sciences (such as ethology, genetics, geology), there are more objective frameworks for assessing evidence than competition for popular coverage in mass media of the newest new thing. (Sadly, in a time when media companies are booming financially, "economies" have ensured that there are no replacements with the intellectual stature of John Noble Wilford, who for decades was a voice of objectivity in his coverage of human evolution and other areas of science at the *New York Times*.) Stating explicitly that this faddishness is neither necessary nor desirable, the author comments "Palaeoanthropology, in common with most archaeology, operates like alchemy and is much in need of a theoretical framework, as Eckhardt (2007) so rightly observes." The author's quote was made in the context of the dispute over just 100 bones, including a single skull, from Liang Bua, Flores, first described in 2004. "Homo floresiensis" was invented from sparse evidence that plainly was misinterpreted from its inception yet sustained by journalistic enthusiasm in the face of blatant contradictions within and between the biological and cultural bodies of evidence. The inherent contradictions would have caused the "species" to be dismissed as theoretically impossible – if palaeoanthropology in fact had a body of theory and a tradition of hypothesis testing, which it manifestly lacks. The rest of Chapter 3 builds on the preceding two chapters and provides a transition to the third strand, which addresses the title of the book. It introduces a powerful, biomedically-based argument for the self-domestication hypothesis, by way of exploring the possible reasons why humans have a multitude of mental problems that do not appear in related hominoid primate species. Many of these disorders seem to occur in more recently evolved portions of the human brain, and survival of those who suffer from cognitive disorders is more likely in close-knit communities that practice collaborative care for children, the elderly, and disabled people of any age.

The book's title relates more directly to the remainder of the work: Chapter 4 ("Human Self-Domestication"), Chapter 5 ("The Unstoppable Advance of Exograms"), and Chapter 6 ("Effects of the Domestication Hypothesis"). In Chapter 4, only after having thoroughly criticized the current state of knowledge and providing very broad factual background of paleoanthropology, archaeology, human cognition, molecular genetics, psychology and psychiatry, does Professor Bednarik develop his thesis in persuasive detail. Being a meticulous scholar, he gives the background for this idea. Although parts of it have been presented by others before, his explication of the idea makes it thoroughly his own. Readers familiar with this subject's intellectual history will be aware that the topic of domestication has some roots that go as deep as Charles Darwin's volume, *The Variation of Animals and Plants under Domestication*, published in 1868, nearly a decade after the *Origin* and still elaborating the evidence on which that epic intellectual insight was based. Some previous authors also have argued that humans who accomplished domestication of plants and animals, eventually domesticated themselves by creating artificial living conditions that resulted in larger, settled and stratified human communities. However, Professor Bednarik's proposition regarding self-domestication of humans cuts through the intricacies of food production, organization of proto-cities, incipient social stratification and complex cultural rituals in the best scientific tradition of parsimony. His thought centers on the single essential feature that is necessary for domestication: Assortative breeding. He argues that toward the end of the Pleistocene, by some 50,000–30,000 years ago, Palaeolithic humans had developed complex social relations and cognitive processes that resulted in sexual selection for youthfulness and gracility of the body structure. Assortative mating for these characteristics produced a fairly rapid change of human morphology from "robust" humans with heavy bony structures supporting large muscle mass, to a gentler, more gracile morphology characteristic of younger individuals. Since Palaeolithic technological advances allowed people to survive by the intelligence made possible by evolved brains rather than just by sheer physical strength, this assortative mating for reduced physical development did not lower their reproductive fitness. According to the author, modern human males characteristically prefer to mate with youthful-looking, gracile females. As Professor Bednarik argues, the assortative mating for gracility that occurred some 50,000–30,000 years ago gradually but rapidly changed human morphology from robust forms (e.g. Neanderthal-looking) to gracile ones, essentially similar to the humans who are alive today. Formulated in this way, the "self-domestication hypothesis" is clear and unencumbered by a multitude of cultural and sociological details; and by this virtue, it also is easily testable and potentially falsifiable. This task of hypothesis testing we must, however, leave to future researchers. At the moment Bednarik's idea explains the origin of "anatomically modern humans" better than the flimsily constructed and now thoroughly falsified (though still widely accepted) "African Eve" hypothesis. In place of hominins with more robust anatomy being replaced wholesale by more gracile African invaders, Bednarik offers an intriguing nexus among mate selection, increasing self-awareness, plus abundant and suggestive archaeological objects depicting the female body and its decoration.

Chapter 5 follows the implications of these inter-related shifts into the realm of reduction in hominin brain volume resulting from the gracility of bodies, and improvements in cognitive functioning made possible by exograms (externalised memory traces); this is fresh material, comprehensively and compellingly assembled. Chapter 6 weaves all of these strands together, providing an explanatory framework that unifies the observed evidence of hominin neoteny and gracilisation with abundant material exograms that improve the cognitive functioning of communities. All of these elements combine to provide a compelling alternative to the "African Eve" replacement hypothesis, still widely accepted by academics who find safety in numbers rather than independent analysis, and by the "science writer" camp followers in need of the latest story line, however unrelated it might be to the problem as a whole. To help readers understand why this jarring frame of reference – scholars exposed as popular fad followers rather than as thought leaders – is important, it is worth a bit of what might seem to be a digression. In an essay on the techniques taught by the great American actor Gena Rowlands, Andrew Wood referred to the history of art criticism. To paraphrase Wood, in the earlier period prior to the end of the eighteenth century, art had been conceived as a mirror. It reproduced life; it showed you what life was like; figuratively, it reflected the surfaces of life. In the later age of Romanticism, art was thought to allow us to see things concealed beneath the surfaces of life and not normally visible; it served as source of illumination – a lamp. These contrasting metaphors served as context for Wood's larger point: the truth is not always what you think it is. In the context of these contrasting metaphors for art, Bednarik's view shows palaeoanthropology and Pleistocene archaeology to be representing a human past not as illuminated by data collection and hypothesis testing, but instead as factoids distorted in the funhouse mirrors of academic cliques, "leading" journals, and the popular science industry. Sadly, he is correct. Happily, there are alternatives provided by models from other fields of science. As just one example, Barry Marshall, along with his colleague Robin Warren, was in pursuit of a practical problem, the cause of stomach ulcers. Marshall's Nobel Prize lecture explicates a model of curiosity-driven research, accomplished with little financial support. Especially notable is his introductory quote, which he attributed to the author and librarian Daniel J. Boorstin: "The greatest obstacle to discovery is not ignorance – it is the illusion of knowledge." Bednarik exposes the "African Eve" illusion, replacing it with testable hypotheses, including the self-domestication of humans.

We have little doubt that some reviewers of this book will fault the author for his departure from what for several decades has been a mainstream of sorts, with its enthusiasm for "African Eve" and the "hobbits" of Flores. Before taking comfort in the security seemingly afforded by large numbers of like-minded people, however, such critics would do well to read *Extraordinary Popular Delusions and the Madness of Crowds*, written by the Scottish journalist Charles Mackay. First published in 1841, Mackay's work is a classic better known among economic historians than students of the far deeper human past. He chronicled some of the major extreme financial escapades up through his time, including the Dutch tulip mania of the early seventeenth century, the Mississippi Company Bubble of 1719–1720, and the nearly decade-long South Sea Company bubble of 1711–1720 that it overlapped. In all of these phenomena some new idea or concept lay at the core (as PCR and mitochondrial gene sequencing lie at the core of "African Eve's" invention). As the mania gained currency (figuratively and then quite literally), whatever plausible core had existed at the outset (in the financial cases, promise of a new product or territory, in palaeoanthropology, a new dimension of diversity) became exaggerated by retelling and sustained by spreading belief. No doubt it will be argued that highly educated people, particularly scientists, would not be drawn into error by such common enthusiasms. That argument would be wrong based on known facts. No less a scientist than Sir Isaac Newton was swept up in the South Sea Bubble scam of 1711–1720. *Twice*. The first time he bought shares at a low price and sold them at a higher one, making a profit. But then, as South Sea Company shares continued to rise, he bought in again, very near the peak of the bubble. Prices plunged, including those owned by the discoverer of the Laws of Motion. Newton went bankrupt. Even genius can be overcome by failure to question widespread belief. Against the seeming safety afforded by numbers, Mackay's judgment stands: "Men, it has been well said, think in herds; it will be seen that they go mad in herds, while they only recover their senses slowly, and one by one."

As anyone who has read our Foreword to this point will realize, the introductory evaluative summary that we provide here does not constitute a *pro forma* promotional piece dashed off to aid a colleague sell a few more copies of a book. Both of us have been aware of Professor Bednarik's work for more than a decade; however, we never have met him in person. We both reviewed the prospectus of this book and recommended its publication by Routledge. We have read the entire text – several times – with great professional attention. In the final analysis, we respect the work because it is rooted in critical examination of existing dogmas, is rich in information previously overlooked, and is bristling with new ideas that are deserving of the widest possible critical scrutiny, rather than being ignored. This is not a book that can be allowed to go down the memory hole.

> Who controls the past controls the future. Who controls the present controls the past.
>
> (George Orwell)

Robert B. Eckhardt
Professor of Developmental Genetics and Evolutionary Morphology
Department of Kinesiology and Huck Institute of the Life Sciences
Pennsylvania State University, USA

Maciej Henneberg
Emeritus Wood Jones Professor of Anthropological and Comparative Anatomy
University of Adelaide, Australia
Institute of Evolutionary Medicine
Zurich, Switzerland

1 Introduction
The empirical context

About palaeoanthropology

Every discipline needs to struggle with a number of inherent axiomatic issues determining its credibility, including, for example, the basis of the taxonomies it creates and observes; the level of testability it can manage to bring to bear upon its propositions or hypotheses; or the degree to which it relies on academic authority or consensus. One of the rough-and-ready measures available to assess a discipline's performance would be provided by its history, by reviewing its past performance: how many flawed paradigms have been upheld and defended by it in the past, how obstinate has it been in accepting corrections? Another is to consider the level of support its initial hypotheses tend to enjoy relative to other fields. For instance, in psychology, psychiatry or business studies, the subsequent experimental support of initial propositions is several times that reported in the hard sciences, suggestive of systematic bias (Sterling 1959; Klamer et al. 1989; Fanelli 2010). These disciplines report empirical confirmation of hypotheses five times as often as, for example, space science, while the biological disciplines rank intermediate. Using the same index of confirmation bias, studies applying behavioural and social methodologies on people rank 3.4 times higher than physical and chemical studies on non-biological material (Press and Tanur 2001). It is therefore not surprising that the social sciences are regarded as qualitatively different from the hard sciences (Shipman 1988; Latour 2000; Simonton 2004; Bishop 2007), and much the same needs to be said about the humanities generally (Bednarik 2011a).

Let us look at palaeoanthropology, the branch of anthropology concerned with the study of extinct forms of humans, from the perspective of generic credibility. Its historical performance is far from reassuring. Since the very idea that we may have been preceded by now-extinct ancestral species was raised, the discipline has stumbled from one controversy to another. Best known among those who initially proposed that early humans coexisted with Ice Age fauna was Jacques Boucher de Crèvecœur de Perthes (1788–1868), a French customs official who found himself posted to Abbeville in 1825. There he met Casimir Picard, a local medical doctor, who showed him strangely shaped flints he had collected in the area. Having a lot of spare time on his hands, Boucher de Perthes began to visit the Somme valley's quarries, canal diggings and gravel pits. As his collection of these stones grew, he began to realise that they occurred together with and in the same river deposits as the bones and teeth of now-extinct animals, attributed to the Diluvium period (now named the Pleistocene). He commenced systematic excavations in the region about 1832. Six years later he presented his theory and evidence to the Société d'Émulation in Paris, and in 1839 to the Paris Institute. Each time both his ideas and finds were rejected by the experts, as was a subsequent series of five volumes entitled *On the Creation*. It was easy for his contemporaries to reject de Perthes, because he espoused a few other eccentric ideas, such as pursuing universal peace, raising the living standards of the working class and giving women rights.

Nevertheless, by the 1840s, becoming emboldened by the three-age system (Stone, Bronze and Iron Ages) of Christian Jürgensen Thomsen (1788–1865), Boucher de Perthes (1846) proposed that humans have existed for hundreds of thousands of years. He deduced this from the known geological ages of the deposits yielding the stone tools he had identified. Thomsen, another non-archaeologist, had his *Ledetraad til Nordisk Oldkyndighed* (*Guide to Northern Antiquity*) published in 1836 (English translation in 1848), and the three-age system he introduced remains widely used in principle (Rowley-Conwy 2007). However, Boucher de Perthes continued to be vigorously opposed, and in the year before Darwin's main work appeared, the French archaeologists issued at their congress of 1858 a unanimous declaration that stated that the hand-axes and other stone artefacts from the Abbeville region were "a worthless collection of randomly picked up pebbles". Little did they know that two British geologists, Hugh Falconer and Joseph Prestwich, were quietly doing what good scientists do. Testing Boucher de Perthes' claims they were excavating alongside one of his trenches. Prestwich (1859) confirmed that the persistent Frenchman, who had endured decades of unscholarly treatment from the discipline, had been right all along: humans had coexisted with Pleistocene fauna hundreds of millennia ago in the north of France.

In the same year, as another amateur's book (Darwin 1859) appeared, German schoolteacher Johann Carl Fuhlrott (1803–1877) published a paper introducing human remains from the Kleine Feldhofer Cave in the Neanderthal region in Germany. His view that they represented an early form of human was eschewed even by the journal reluctantly publishing his paper, rejecting his interpretation of the find in a footnote (Fuhlrott 1859). Although the founder of geology, Charles Lyell, and Thomas Henry Huxley, Darwin's most outspoken supporter, accepted Fuhlrott's view, the disciplines of archaeology and human anatomy rejected it outright. The Neanderthal remains were attributed, by various scholars, to a Celt, a Dutchman, a Friesian, a Mongolian Cossack and an idiot. Various bone diseases and a lifetime of riding horses helped explain the unique bone structure. The pre-eminent German anatomist, Professor Rudolf Virchow (1821–1902), determined in 1872 the entire health history of the individual since his childhood from the bones available, and the matter remained closed in Germany for another three decades. It was one of

Virchow's students, Gustav Schwalbe (1844–1916), who published a reassessment of the original Neander valley skeleton, defining it as the species *Homo neanderthalensis* (Schwalbe 1901). In contrast to Boucher de Perthes, Fuhlrott never witnessed his vindication.

One would have thought that these experiences of being monumentally mistaken would have taught the discipline to be more careful in its collective pronouncements, but the precise opposite trend can be observed through history up to the present. The discovery of Pleistocene cave art by Don Marcelino Santiago Tomás Sanz de Sautuola (1831–1888) in Altamira, northern Spain, unleashed yet another fury by archaeology directed against an amateur, and it resulted in his premature death. Beginning 1875, de Sautuola discovered the remains of Pleistocene animals in the cave on his property, and after visiting the 1878 World Exhibition in Paris where he saw stone tools found in France (by now recognised by the discipline), he began excavating in Altamira. He paid no attention to the paintings on the ceiling of the cave for a year, until in November 1879 they were pointed out to him by his 13-year-old daughter, Maria. In 1880, he published a booklet, avoiding the claim of linking the cave art to the Palaeolithic tools. For its illustrations, he employed a mute French painter he had befriended earlier. That was a fatal mistake: the elite of French prehistorians, who had roundly rejected his claims from the start, concluded that the painter had produced the rock art and that de Sautuola was a charlatan. He tried desperately to save his honour, by presenting his case at a conference in Algiers in 1882, then producing another self-funded booklet. It only yielded more disparagement from his numerous adversaries who did not even bother to examine Altamira. In 1888, de Sautuola, the discoverer of Pleistocene rock art, died prematurely, a bitter and broken man whose death continues to weigh heavily on the discipline. Yet as recently as 1996 archaeology has been responsible for the death of a scholar, Dr David Rindos, who died of extreme stress when he sought to recover his reputation after having been sacked unfairly by the University of Western Australia (Bednarik 2013a).

The next significant discovery in archaeology and palaeoanthropology occurred when Eugène Dubois (1858–1941), a Dutch physician who deliberately set out to find the 'missing link' between apes and humans, succeeded in unearthing the first remains of *Homo erectus* in Java. Not only were the odds of succeeding in his quest heavily stacked against him; when he did succeed in 1891, the boffins immediately rejected his find. They scorned his bones from Indonesia for several decades, especially after it became evident that humans could not have first evolved in 'the colonies'. The great discovery of 1912 at Piltdown made it clear that the human lineage originated in England. The report of these fossils demonstrated in no uncertain terms that the braincase of the species evolved to modern human proportions before the rest of the creature from which we descend, which rendered Dubois' chimera from Java merely immaterial. Not only that, when, in 1924, Australian-born anatomist Raymond Arthur Dart (1893–1988) discovered an even earlier human predecessor in the colonies, its publication was ignored. Dart's South African

Australopithecus africanus, the southern ape of Africa, was not accepted for decades, while the fake from Piltdown held sway in palaeoanthropology. Dubois had become isolated and irrelevant, a misanthrope who for decades refused access to the bones he had found. Palaeoanthropology was unable to develop as a discipline until the Piltdown fake (more likely a hoax; Bednarik 2013a: 71–74) was eventually debunked (Weiner et al. 1953).

These examples illustrate some of the problems with both Pleistocene archaeology and palaeoanthropology, but many more could be cited (Bednarik 2013a). Both disciplines rely heavily on the authority of individual scholars, while the practices of refutation, of testing hypotheses, are eschewed and resisted. Both branches of learning have remained unable to shed this state, examples of which still occur up to the present time (a good historical example is provided by the fiasco concerning the Glozel site in France; (ibid.: 74–78). For instance, archaeology Professor Reiner Protsch 'von Zieten' proposed that modern humans evolved exclusively in some unspecified Sub-Saharan region and from there colonised the world (Protsch von Zieten 1973; Protsch 1975). Palaeoanthropologist G. Bräuer (1980; 1984a) developed his idea and in the late 1980s it became virtual dogma (Cann et al. 1987), in the form of the 'African Eve' model. This proposal suggested that somehow, an African mutant population had become unable to breed with any other humans. Owing to its members' mental superiority, it expanded across Africa, outcompeting or extinguishing all other human populations, and then conquered Asia in the same fashion. About 60,000 years ago, upon reaching south-eastern Asia, our super-human ancestors promptly set out to colonise Australia. Having thus exterminated all other humans in Africa and Asia, they finally marched into Europe, succeeding just 30,000 years ago in annihilating those brutish Neanderthals that a century earlier the discipline had decreed never even existed. Therefore, all extant humans descend from that African Eve and her cohorts, who also happened to resemble God physically.

After the disciplines of palaeoanthropology and Pleistocene archaeology had been enthralled by this aberration for many years, quelling all opposition to the African Eve model through the refereeing system of the major journals, the model was found to have been based on fake data. The radiocarbon datings of hominin remains provided by Protsch, its originator, had been concocted by him, and dozens of such finds attributed to the Pleistocene were in fact of the Holocene – in some cases being only a few centuries old (Terberger and Street 2003; Bednarik 2008a; 2011b). Protsch was sacked in disgrace the following year (Schulz 2004). The slow retreat of the African Eve advocates began, but over fifteen years later, it remains incomplete. Instead of accepting that, in their enthusiasm, they had been deceived, the African Eve supporters focused on preserving their careers by admitting as little as possible. Both the popular science and the mass media are too confused by these obfuscations to realise that the African Eve model has been a hoax from the start, perpetrated by a few journals in the Anglo-American sphere of interest. Chinese specialists, for instance, never accepted the deception. Since the African Eve hoax was the only model of the origins of human modernity which asserted that Robusts and Graciles (see Chapter 2) were separate species, unable to interbreed, the multiregional model remained the only viable alternative. This is because any of the several moderate variations of the 'out of Africa' 'modern' origins story (e.g. those of Bräuer 1984b: 158; Pennisi 1999; Eswaran 2002; Templeton 2002; Smith et al. 2005) are merely variations of the multiregional theory (Relethford and Jorde 1999; Relethford 2001), proposing a substantial inflow of African genes. They, therefore, agree with Weidenreich's (1946) original trellis diagram (cf. Wolpoff and Caspari 1996: 200–201).

However, the continuing support for the African Eve hoax (e.g. Stringer 2014) confirms Max Planck's famous quote that "a new scientific truth does not triumph by convincing its opponents and making them see the light, but rather because its opponents eventually die, and a new generation grows up that is familiar with it" (Kuhn 1970). To witness the demise of the replacement hypothesis – that the robust hominins of the world were replaced by the outcome of human speciation originating in Sub-Saharan Africa – we shall have to wait for its supporters to fall off their perches first. That is how all previous major blunders in the discipline were resolved, and not by reasoned discussion. Looking at the historical precedents in Pleistocene archaeology, it took an average of around four decades to abandon key hypotheses, just as long as it took to expose the Piltdown hoax or the Glozel debacle. If the discipline's mainstream had been able to preserve its view, we would not today have Pleistocene archaeology; we would not have hominin ancestors; we would not have a Pleistocene rock art; we would not have human fossils from Africa and Asia. On all major issues in this field, the vast majority of its practitioners were wrong. Yet in all cases, the evidence against them had been defined clearly at the time by the iconoclasts who turned out to be right. In all cases, the corrections came from non-archaeologists and non-anthropologists, and in all of them, these scholars were treated with disdain and disparagement, even driven into a premature death on occasion. This is how one of these amateur dissenters, without whose perseverance we would have no Pleistocene archaeology at all, summed up his experience:

> They employed against me a weapon more potent than objections, than criticism, than satire or even persecution – the weapon of disdain. They did not discuss my facts; they did not even take the trouble to deny them. They disregarded them.
>
> (Boucher de Perthes 1846)

In that respect, nothing has changed since the mid-1800s. Many ideas and models about archaeology – especially Pleistocene archaeology – held by the vast majority of practitioners today are evidently falsities. However, this does not deter their promotion or their inclusion in the curricula of university departments. In the course of exploring the errors of the discipline in this book, we will encounter numerous examples of this. We will examine them and consider their causes. These flawed or invalid hypotheses are not always without far-reaching consequences, because far from being just a hobby of people preoccupied with the past, archaeology is primarily a political discipline (e.g. Silberman 1982; 1989; 1995; Trigger 1984; 1985; 1989). This truism has many aspects. To consider just one of them: for much of the first half of the twentieth century, palaeoanthropology believed that humans evolved first in England, as we have seen. Shortly after proper scientists corrected this mistake, the focus of attention shifted to Africa, and very successfully so. The primary evidence for the misconception was the rather poorly prepared fake objects from Piltdown, but the more general reason for the ready acceptance of this evidence was the colonialist perspective that it was quite natural to expect to find the earliest humans in Europe. However, when it comes to another, equally consequential Eurocentrism, cultural origins, the colonial view has persisted to the present time and seems impossible to eradicate. After strenuously rejecting the Pleistocene cave art found first in Spain, then France, it was finally accepted at the turn of the century (Cartailhac 1902). This led to the development of a cult-like preoccupation with what became regarded as the earliest art in the world, underpinning the notion that advanced culture and symbolism were invented in Europe. All the evidence that Pleistocene palaeoart (art-like productions) occurs in all continents except Antarctica, and that it is significantly older and more numerous in some of them was simply ignored. Australia, for instance, is thought to possess much more Ice Age rock art than Europe, and people with Mode 3 technocomplexes created all of it. These are called Middle Palaeolithic in Eurasia and Middle Stone Age in much of Africa. In Australia, they persist to mid-Holocene times and in Tasmania to the time of the British invasion (Bednarik 2013a). In other words, *all* Tasmanian rock art is of the 'Middle Palaeolithic', whereas, in Europe, known Pleistocene art is Upper Palaeolithic (with a few exceptions). Moreover, in India and South Africa, rock art has been discovered that is several times the age of the earliest reported in Europe.

But while there are dozens of sites of Pleistocene rock art in Europe on the UNESCO World Heritage List, not a single one has ever been listed from the rest of the world. Not only does this stifle research of early palaeoart of the other continents; it reinforces the public's perception that culture was invented in that most deserving of places: Europe. We have many thousands of publications about European Ice Age 'art', yet *all* pancontinental reviews of Pleistocene palaeoart in Africa, Asia, Australia and the Americas have been written by one single author (Bednarik 2013b; 2013c; 2014a; 2014b; 2017a). To make matters even worse, many of the European rock art sites on the World Heritage List for being of the Pleistocene are in fact of the Holocene, pointing to some of the blunders of archaeology mentioned above. These post-Pleistocene sites (the series of sites on the lower Côa river in northern Portugal and Siega Verde nearby but in Spain) help to prop up the neo-colonialist self-deception of Europeans that their continent has given rise to the earliest art traditions.

Maintaining the belief in European cultural precedence is, of course, a political device, and this belief seems immune to any challenges. To re-phrase Boucher de Perthes' words: the scholars do not even bother to deny these facts; they simply disregard them to preserve their fantasy.

The preceding is only one illustration of how archaeology and palaeoanthropology are political pursuits, engaged in 'explaining' the 'Others' – those human groups that share neither the beliefs nor the reality constructs of people conditioned by contemporary Western modes of thinking. Like all human groups, the latter exist in the belief that the world is as they experience it. This self-deception encourages them to be the arbiters of history. They even decree that a specific portion of history is 'prehistory', based on the untestable proposition that written histories are more credible than oral histories. This dominant world-view's righteousness is in many ways a significant barrier to understanding indigenous world-views, and yet it arrogates the role of supreme arbiter of what is true. In the field of archaeology, it presumes this role is justified by a perceived scientific function, but that discipline lacks a scientific basis: its propositions are not internally falsifiable, and their testability is often poor. Illustrating the political role of history, Hobsbawm (1992: 3) noted that "historians are to nationalism what poppy growers in Pakistan are to heroin addicts; we supply the essential raw material for the market". Archaeologists Kohl and Fawcett (2000: 13) elaborated further: "Rather than just the producers of raw materials, historians and archaeologists may occasionally resemble more the pushers of these mind-bending substances on urban streets, if not the mob capos running all stages of the sordid operation." The political uses made of archaeology's "findings have facilitated ethnic clashes and cleansing, bigotry and nationalism far more often than they have promoted social justice" (ibid.). The alignment between nationalism, totalitarian regimes, religion and archaeology has been well documented, and the great archaeologist Bruce Trigger (e.g. 1984; 1989) discerned essentially three forms of archaeology: nationalist, colonialist and imperialist approaches. Silberman (2000) added to these a touristic archaeology and an "archaeology of protest". Kohl, Fawcett, Trigger and Silberman thus present clarity of self-assessment that is rare in archaeology, or any other humanist discipline (Bednarik 2011a).

This political role of archaeology, a field that emerged from the need of new European nations created after the Napoleonic period to acquire national pasts, perhaps helps explain the latent Eurocentrism attached to the belief that art emerged first in Europe. This conviction that the cave art of France and Spain defines the first art in the world is intimately intertwined with the principle that its appearance coincides with the arrival of the African 'super-humans', the African Eve's progeny. As noted, this conviction is as flawed as the idea that art originates in Europe: two defective and factually unfounded hypotheses were linked, confirming each other, instead of being subjected to even the most rudimentary testing. It is self-evident that if art-like productions originated with African Eve's 'new species' (which is not a genetically recognised entity), one would expect to find a trail of ancient art finds from Africa to south-western

Europe. No such find has ever been reported, and even over-simplistic interpretation of the available evidence implies that the cave art is a local 'invention' (a more sophisticated explanation is available, as we will see in this volume).

Moreover, the African Eve hypothesis has recently dissolved into thin air, with the genetic proof that the so-called modern humans and the conspecific robust humans formed a single breeding unit (Krause et al. 2007; Green et al. 2010; Reich et al. 2010). Therefore, the genetic separation into two or more species, so fundamental to this theory, has become unsupportable. No African Eve should be assumed to have ever existed, nor is there any archaeological, genetic or palaeoanthropological evidence that 'anatomically modern humans' from Africa 'colonised' Eurasia (Bednarik 2008a; 2011a; 2011b).

We will examine these issues below; suffice it to note that in addition to the collapse of the replacement hypothesis, it is fundamentally wrong to think that art-like productions of the distant past, collectively named 'palaeoart', first appear on the record in Europe. In other words, both of the components of the false origins myth of 'modern' humans lack any basis in the available evidence. The disciplines of Pleistocene archaeology and palaeoanthropology have been led on a monumental wild goose chase (Bednarik 2017b) in recent decades, continuing these pursuits' long history of creating falsities about the human past. This book is an attempt to show this by analysing the relevant issues and explaining how the most recent major blunder in the field came about. However, more importantly, it presents a wide-ranging alternative explanation of human evolution of the final third of the Late Pleistocene, i.e. its last thirty or forty millennia. This account will not only be much more logically persuasive than the still-dominant model; it will even be supported by all the available empirical information about this general subject. Still more important than that, the highest deficit of the replacement model is that in the final analysis, it does not explain anything; it merely presents a just-so model. The alternative explanation offered in this book will explain literally hundreds of aspects of the "human condition" (Bednarik 2011c) that remain inexplicable otherwise. By the conclusion of this book, it is hoped that the reader will ask, why was such an obvious, all-clarifying explanation not thought of before? Why were instead demographically improbable scenarios of wandering tribes invented, and of worldwide genocide? What should we think of disciplines that cannot see the forest for the trees, and invent unlikely settings that explain none of the questions we have about why we humans are the way we are? After all, the reason why we investigate our origins, our past is to begin understanding our present and future. If the explanations offered in the process are unable to explain anything, what does this say about these efforts?

This book is intended to elucidate the course of recent human evolution and to illuminate the reasons for our present condition. To explain human behaviour (Bednarik 2011c; 2012a; 2013d) only through its present-day manifestations is rather like considering illness through its symptoms instead of its causes: it fails to provide clarification. The disciplines concerned with the human past have flunked in delivering an understanding of the aetiology of the human condition. The human capacity for great evil and good, our 'conscious' experience of both past and future, the feelings and emotions associated with our existence, the circumstances and origins of our capacities and limitations all need to be understood in terms of their aetiologies, if they are to be understood at all. The foundations of the very constructs of the realities we see ourselves exist in and our constant endeavours to construct meanings where, in fact, none exist have all remained profoundly unexplained by the disciplines that should have delivered these aetiologies. Indeed, they have been concerned mostly with trivialities. The present volume will attempt to remedy this state.

Pliocene and Pleistocene archaeology

The story of human evolution unfolded during two geological epochs, the Pliocene and the Pleistocene. The Pliocene lasted from about 5.333 Ma (million years) ago to 2.588 Ma ago (International Chronographic Chart 2014). Although not as yet documented by actual fossil finds, the human ancestor that first split from the apes is assumed to have lived in the last part of the preceding Miocene period. There are numerous fossil hominids from the Pliocene of Africa, and none are currently known from other continents (although there has been a recent controversial claim from Greece and Bulgaria, concerning 7.2 Ma-old *Graecopithecus freybergi*; Fuss et al. 2017); it is, therefore, reasonable to assume that the initial ancestors of the human species may have evolved exclusively in Africa. In breaking with convention, we will review them not in this introductory chapter, but later in this volume.

All hominin fossils found in Eurasia postdate the beginning of the Pleistocene 2.588 Ma ago, the geological period that ended 11.7 ka (thousand years) ago with the end of the last Ice Age. At present, the earliest human finds from Asia are remains found in Georgia, China and Java. A mandibular fragment with two teeth and one maxillary incisor from the Longgupo Cave site in China (Huang and Fang 1991; Huang et al. 1995; Wood and Turner 1995) have been attributed by palaeomagnetic dating to the Olduvai subchron of the Matuyama chron, which means they should be between 1.96 and 1.78 Ma old. The faunal remains from no less than 116 mammalian species support that age, as does a minimum age of one million years secured by electron spin resonance dating from a stratum 3 m above the fossiliferous stratum. However, the hominin status of the mandible has been refuted, and it is now attributed to an ape, perhaps *Lufengpithecus* (Dennis A. Etler and Milford H. Wolpoff, pers. comm., Nov. 1996; Ciochon 2009), but that still raises the attribution of the human incisor and two stone tools also found. The partial skull from the Gongwangling site in Lantian County, 50 km southeast of Xi'an, was earlier estimated to be 1.10–1.15 Ma old, but this was recently revised to 1.63 Ma (Zhou 2014). An age of about 1.7 Ma has been suggested for two human incisors found with stone artefacts near Danawu Village in Yuanmou Country, Yunnan Province (Qian et al. 1991: 17–50).

Similarly, occupation evidence from China's Nihewan basin has been dated palaeomagnetically to 1.66 Ma (Zhu et al. 2004). Finally, stone artefacts found in Renzi Cave, in the Fanchang County, Anhui Province, eastern China, have been assigned a date of 2.25 Ma (Jin et al. 2000). This cave has acted like an animal trap that was accessed by hominins that left behind butchering floors. Their archaic stone knives and scrapers were found among the remains of typical species of the very Early Pleistocene, but the identity of these earliest hominins known in eastern Asia remains unknown. Nevertheless, this discovery has fuelled a debate about the origins of *Homo erectus*, the Asian equivalent of Africa's *Homo ergaster*. In China, *H. erectus* has been reported first from Zhoukoudian 1 Site, later at Gongwangling, Chenjiawo, Donghecun, Qizianshan, Tangshan, Longtandong, Xiaohuashan, Xichuan, Danawu, Quyuan River Mouth, Yunxia, Longgudong-Yunxian, Longgudong-Jianshi and Bailongdong. The first remains found of the species, however, are those from Java, presented, as noted above, by Dubois in the 1890s, and *H. erectus* finds from Southeast Asia are now abundant.

At the opposite end of Asia, the well-preserved remains of several humans from Dmanisi in Georgia (named *Homo georgicus*) with a brain volume ranging from 600 to 780 ml are about 1.7 Ma old (Bosinski 1992; Dean and Delson 1995; Gabunia and Vekua 1995). They seem to resemble *H. habilis* rather than the more developed *H. ergaster* or *H. erectus*. Moreover, the stone tool industry at Dmanisi is similar to the Oldowan cobble tools of eastern Africa but lacks the bifaces associated with the latter human types. Their antiquity seems to roughly coincide with that of the earliest *H. erectus* found in Java (Swisher et al. 1994). Therefore, one possibility is that *H. habilis* was the first human to leave Africa and then developed into *H. erectus* in Asia, with gene-flow continuing between the two continents. The issue is complicated by two factors: the inadequate fossil record from most of Asia, and the putative dwarf species '*Homo floresiensis*', which tends to resist placing it into a realistic cladistics system. However, the latter impediment may well be illusory. Although the popular science writers are infatuated with the 'Hobbit' from Flores, its identification seems likely to be another mistake of palaeoanthropology. What renders this possibility realistic is the cavalier rejection by the mainstream of the more probable alternative interpretation: that the 'Hobbit' was the outcome of pathology (Jacob et al. 2006; Martin et al. 2006; Eckhardt 2007; Eckhardt and Henneberg 2010; Eckhardt et al. 2014; Henneberg and Thorne 2014), caused by genetic impoverishment of an endemic, insular population, as has been observed elsewhere (Bednarik 2011c).

The area intermediate between westernmost Asia and the Far East has yielded almost no early hominin remains so far. The only exceptions are the two specimens found at Hathnora on the Narmada River of central India. The Narmada calvarium and clavicle were both recovered from the Unit I Boulder Conglomerate at Hathnora (H. de Lumley and Sonakia 1985; Sankhyan 1999), about 40 kilometres south of Bhimbetka, where Lower Palaeolithic petroglyphs were first identified (Bednarik 1993a). The partially preserved cranium (Figure 1.1) was initially described as *H. erectus narmadensis* (Sonakia 1984; 1997; M.-A. de Lumley and Sonakia 1985), but is now considered to be of an archaic *H. sapiens* with pronounced erectoid features (Kennedy et al. 1991; Bednarik 1997a; 1997b). Its cranial capacity of 1,200–1,400 ml is conspicuously large, especially considering that this is thought to be a female specimen. The adult clavicle, however, is clearly from a 'pygmy' individual, being under two-thirds of the size of most modern human groups. It is of an individual of a body size similar to '*H. floresiensis*'. Both specimens are among the most remarkable hominin finds ever made, yet both remain widely ignored. They date from the Brunhes normal chron (<730 ka; Agrawal et al. 1988; Agrawal et al. 1989) and have been suggested to be about 200 ka old but their age remains unknown.

Therefore, our current knowledge of the Middle and Early Pleistocene hominin occupation of India remains relatively limited, and even the Late Pleistocene remains inadequately explored. Although the Lower and Middle Palaeolithic stone tool traditions are widespread (Petraglia 1998), represented in massive quantities and typologically well explored in India (Korisettar 2002), their absolute chronology has remained mostly unresolved so far. This

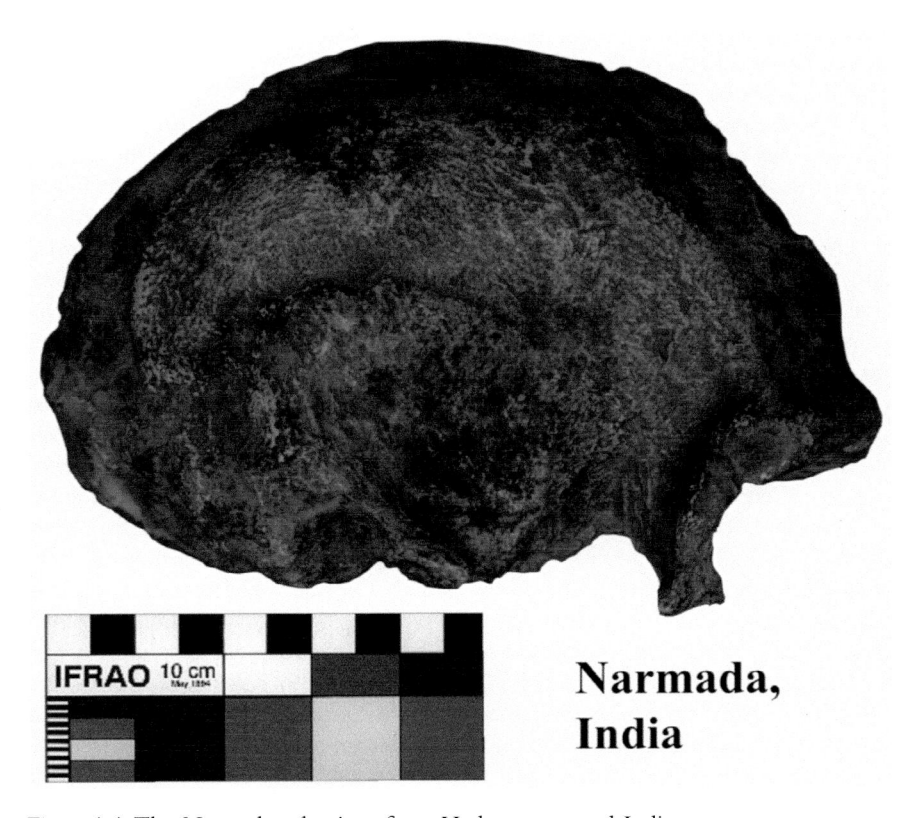

Figure 1.1 The Narmada calvarium from Hathnora, central India

is due both to a paucity of excavated sites (most known sites are surface scatters) and a pronounced lack of well-dated sites. There are some preliminary indications that the Middle Palaeolithic commenced before 160 ka ago. At Didwana (Misra et al. 1982; Misra et al. 1988; Gaillard et al. 1986; Gaillard et al. 1990), thorium-uranium dates for calcrete associated with Middle Palaeolithic industries (Misra 1989) range from 144,000 years upwards. Their validity is reinforced by a thermoluminescence date of 163,000±21,000 years BP from just below the level dated by ^{230}Th/^{234}U to 144,000±12,000 years BP. A single thermoluminescence date for a Middle Palaeolithic deposit in a dune in Rajasthan has been reported to be >100,000 years BP (Misra 1995; Korisettar 2002).

Prospects for a comprehensive temporal framework are at least as bleak for the Lower Palaeolithic period, which is represented primarily by Acheulian industries. However, this dominance of Acheulian forms may well be an artefact of collecting activities that may have favoured the easily recognisable Acheulian types, notably well-made hand-axes. Several attempts to use the uranium-series method, at Didwana, Yedurwadi and Nevasa (Raghvan et al. 1989; Mishra 1992), placed the Acheulian beyond the method's practical range (which ends at about 350–400 ka BP). But one of the molars from Teggihalli did yield such a date (of *Bos*, 287,731+27,169/-18,180 ^{230}Th/^{234}U years BP), as did a molar from Sadab (of *Elaphus*, 290,405+20,999/-18,186 years BP) (Szabo et al. 1990). However, an *Elaphus* molar from the Acheulian of Tegihalli is over 350 ka old. Uranium-series dates from three calcareous conglomerates containing Acheulian artefacts suggest ages in the order of 200 ka (Korisettar 2002). The most recent date so far for an Indian Acheulian deposit is perhaps the U-series result from conglomerate travertine in the Hunsgi valley (Karnataka), which seems to overlie a Late Acheulian deposit (Paddayya 1991). The travertine's age of about 150 ka at Kaldevanahalli appears to confirm that the change from the Lower to the Middle Palaeolithic occurred between 200 and 150 ka ago.

In addition to these very sparse dates from the earliest periods of Indian history, there are several presumed 'relative datings', but these were always subject to a variety of qualifications. Early research emphasised the relation of artefacts to lateritic horizons (but cf. Guzder 1980) and biostratigraphic evidence (de Terra and Paterson 1939; Zeuner 1950; Badam 1973; 1979; Sankalia 1974), which often resulted in doubtful attributions. Sahasrabudhe and Rajaguru (1990), for instance, showed that there were at least two episodes of laterisation evident in Maharashtra and that extensive fluvial reworking occurred. Attempts to overcome these limitations included the use of fluorine/phosphate ratios (Kshirsagar 1993; Kshirsagar and Paddayya 1988–1989; Kshirsagar and Gogte 1990), the utility of which was affected by issues of re-deposition of osteal materials (cf. Kshirsagar and Badam 1990; Badam 1995). Similarly, attempts to use weathering states of stone tools as a measure of the antiquity of lithics (e.g. Rajaguru 1985; Mishra 1982; 1994) have been plagued by the significant taphonomic problems involved in weathering processes.

Thus, there remains wide disagreement about the antiquity of the Early Acheulian (Pappu et al. 2011) and the Mode 1 industries in India. Based on the potassium-argon dating of volcanic ash in the Kukdi valley near Pune to 1.4 Ma ago (Badam and Rajaguru 1994; Mishra and Rajaguru 1994) and the palaeomagnetic measurements and direct ^{26}Al/^{10}Be dates from Attirampakkam to an average of 1.51 Ma (Pappu et al. 2011), some favour that magnitude of age for the earliest phase of the Indian Acheulian. An age of well over 400 ka seems also assured by thorium-uranium dating (Mishra 1992; Mishra and Rajaguru 1994). Others, especially Acharyya and Basu (1993), reject such great antiquity for the Early Acheulian in the subcontinent. Similarly, Chauhan (2009) cautions that the ESR date of c. 1.2 Ma for Early Acheulian finds at Isampur (Paddaya et al. 2002) remains tentative. However, Chauhan and Patnaik (2008) and Patnaik et al. (2009) have shown that lithics at the Narmada site Dhansi, less than 3 km south of the hominin site Hathnora, occur in a major formation of the Matuyama Chron, presumably placing them in the Early Pleistocene.

By the time we arrive at the earliest phase of human presence in India, the available record fades almost into non-existence. There are tantalising glimpses of cobble tool and Oldowan-like traditions from India and Pakistan (i.e. Mode 1 industries; Clark 1977; Foley and Lahr 1997), which so far have not received the attention they deserve. They consist of two occurrences of stratified archaic cobble tools, well below Acheulian evidence and separated from it by sterile sediments at the first site it was described stratigraphically (Wakankar 1975; Bednarik et al. 2005). These quartzite tools from Auditorium Cave at Bhimbetka and Daraki-Chattan relate to petroglyphs at both sites, and they are partially saprolithised. Since it is logical to expect human occupation evidence in India for at least 1.8 Ma and perhaps up to 2.25 Ma ago (because of the early hominin occupation of China and Java), it is to be expected that cobble tools should precede the bifaces of the Acheuloid traditions, and one would have assumed that these have attracted some attention. In reality, they have remained practically ignored.

The Mode 1 assemblages of India, consisting of archaic chopping tools, cores and flake tools, are sometimes referred to as Soanian, which some researchers mistakenly define as post-Acheulian. Most of these occurrences are surface finds (e.g. Salel, Chowke Nullah, Haddi, see Guzder 1980; or Nangwalbibra A, see Sharma and Roy 1985; or Pabbi Hills in Pakistan, see Hurcombe 2004); or come from alluvial or colluvial deposits, including conglomerate horizons (e.g. Durkadi, Armand 1983; or Mahadeo-Piparia, Khatri 1963). These earliest stone tools relate to the finds from Riwat and Pabbi Hills in Pakistan where they date from the Plio-Pleistocene and the Early Pleistocene (Rendell et al. 1989; Dennell 1995; Hurcombe 2004; cf. Chauhan 2007), matching the age of the earliest Chinese finds (Jin et al. 2000; Zhu et al. 2001). The cobble tools at Riwat have been assigned a minimum age of 1.9 Ma, their most likely age being 2.3 to 2.4 Ma. Those of Pabbi Hills have a less secure stratigraphic context, having eroded from a deposit thought to be up to 2.2 Ma old

(Dennell 2004). The finds from Pakistan almost match the antiquity of the flakes and cores from the Yiron gravel in Israel, currently the oldest artefacts known in Asia, at >2.4 Ma (Ronen 1991). In India, there appears to be confusion between 'primary' Mode 1 assemblages (those that precede Mode 2 occurrences stratigraphically) and 'regressive' Mode 1 features (of necessarily much later, perhaps impoverished pockets of technology, which can be found in any part of the world and until well into the Holocene; cf. Guzder 1980; Corvinus 2002; Gaillard 2006). The former are recognised by deep weathering, early geological or stratigraphic context, and by specific features, such as the massive choppers from Daraki-Chattan with their distinctive bi-marginal trimming (Kumar et al. 2016; also reported from other sites of the central region, such as Mahadeo-Piparia, Khatri 1963) and lack of any Levallois features.

The fragmented and precarious nature of the record characterising the presumed initial human colonisation of Asia is repeated in most continents. Universal agreement about that point only applies in Australia, where it is widely believed that first settlement occurred about 50 or 60 ka ago. The timing of the first occupation of the Americas by humans remains unresolved, and the mainstream opinion may well be wrong. In Europe, too, the Asian dilemma is reflected in recent debates in southern Europe.

The earliest credible stone tools reported from Europe are from Chilhac 3 in the Haute-Loire region of central France. They appear to be c. 1.9 Ma old (Guérin et al. 2004), which would place them earlier than those from Dmanisi in Georgia (1.6–1.8 Ma old). The securely minimum-dated Mode 1 cobble tools and flakes from locus 2 at Lézignan-la-Cèbe in southern France (Crochet et al. 2009) are from sediment capped by a basalt flow dated by ^{39}Ar-^{40}Ar to 1.57 Ma. They are therefore older than the basalt. Another site in central France, Nolhac-Briard, has yielded stone tools from a deposit c. 1.5 Ma old (Bonifay 2002). Mammoth remains subsequently recovered from the site are not of the same age as the stone tools (Mol and Lacombat 2009). A site in southern Italy, Pirro Nord in Apricena has produced hominin occupation evidence of between 1.7 and 1.3 Ma age (Arzarello et al. 2007; Arzarello, Peretto and Moncel 2015). Some 340 chert artefacts of Mode 1 technology have been found there in the Pirro 13 fissure. Age estimation by palaeomagnetism places the assemblage in the Matuyama chron, postdating the Olduvai subchron, and this is confirmed by the deposit's biochronology. The earliest human remains so far reported from western Europe are those attributed to *Homo antecessor* from Gran Dolina in the Sierra de Atapuerca, Spain (Carbonell et al. 2008), dated up to about 1.2 to 1.1 million years old.

There are several European cobble tool (Mode 1) sites dating from between 1.5 and 1.0 Ma ago (Bonifay and Vandermeersch 1991): in Spain: the Duero valley, the Orce Basin, Cueva Victoria, Venta Micena, Barranco Leon-5, Fuente Nueva-3 and Sima del Elefante at Atapuerca; in France: Chillac; in Italy: Monte Poggiolo; in the Czech Republic: Stránská skála and Beroun; and in Russia: Korolevo. Other occurrences that could be added to this list include the lithic artefacts found in the Early Pleistocene gravels of various localities across Europe.

Incommensurabilities

What emerges from these details is that the timing of presumed first hominin colonisation events remains unresolved in all continents except Australia (<80 ka) and Antarctica. The presently available evidence suggests overwhelmingly that the genus *Homo* first evolved in Africa, perhaps in the form of *H. habilis* around 2.4 to 2.2 Ma ago, followed by *H. rudolfensis* 1.9 Ma ago. This proposition is supported by the preceding hominin genera – *Australopithecus, Paranthropus* and *Ardipithecus* – so far not having been found in other continents. In other words, it is defined by the absence of evidence. The favoured model then sees the rise of *H. ergaster* in Africa at 1.9–1.8 Ma and its migration to Asia, where it grades into *H. erectus*.

Absence of evidence, of course, is not evidence of absence, and some tantalising recent finds remind us of that. These include the discovery of remains of *Graecopithecus freybergi* at Pyrgos Vassilissis (Greece) and *Graecopithecus* sp. at Azmaka (Bulgaria), dated to the early Messinian at 7.175 Ma and 7.24 Ma (Fuss et al. 2017). This hominid appears to be significantly more human-like than its known African contemporaries, raising the possibility that humans could have evolved first in southern Europe rather than Africa. Then there are the apparently hominin Miocene footprints found on rock at Trachilos, western Crete, soundly dated to 5.7 Ma. They were made at a time when Crete was still attached to the European mainland and are of an upright walking creature with feet resembling those of humans (Gierliński et al. 2017). Once again, established dogma would be stood on its head if this information were correct. Moreover, there is the report from Germany, also in 2017, of two teeth of Hominoidea apparently 9.7 Ma old. Their evolutionary stage resembles that reported from Africa at several million years later (Lutz et al. 2017). Irrespective of the outcome of these investigations, it can no longer be postulated that hominins first appeared in Africa; and the notion that 'anatomically modern humans' first evolved in Africa has never been matched by any evidence, and always was fundamentally flawed.

The preferred paradigm of what happened in the human past suffers from a variety of incommensurabilities. Theories are incommensurable if they are embedded in starkly contrasting conceptual frameworks whose languages do not overlap sufficiently to permit scientists to compare the theories directly or to cite empirical evidence favouring one theory over the other. We now have several dozen 'species' of hominins and hominids, none of which are presented as testable constructs. Ideally, a species is defined as a group of organisms consisting of individuals capable of fertile interbreeding. Unfortunately, even in biology, this canon is often disregarded, and numerous Linnaean species produce fertile offspring by interbreeding with other 'species'. A century ago we had some 300 'species' of grizzly bears, when in fact the grizzly is the same species as the brown bear (*Ursus arctos*) and other ursine sub-species. A similar state is gradually developing in palaeoanthropology, where the 'splitters' who outnumber the 'lumpers' are inventing new species based not on empirical evidence (e.g. genetics), but perhaps on their potential to enhance the academic careers of their discoverers (Henneberg and Schofield 2008). Many of the hominin species proclaimed are much more likely to be sub-species, or even just local variants. A classical expression of the effects of 'splitting' species is the replacement hypothesis, a major incommensurability of archaeology and palaeoanthropology discussed below.

As we have seen, in Europe, we have tantalising glimpses of the earliest human presence in the form of Mode 1 stone artefacts that seem to begin 1.9 Ma ago (Chilhac 3). In Asia, such evidence appears to commence even earlier, >2.4 Ma in the Levant (Yiron) and 2.25 Ma in China (Renzi Cave). Unless the excavators erred significantly or the dating analyses yielded incorrect results, such a scenario seems to exclude the possibility of *H. erectus* being the human that first left Africa. On the contrary, it seems more likely that this group evolved in Asia. Similarly, the habiline characteristics of the putative *H. georgicus* of Georgia could well imply that this population derives from African *H. habilis*. Several factors suggest that the currently favoured framework of hominin evolution in Eurasia and Africa is significantly compromised and that more plausible alternatives do exist.

Another incommensurability concerns the notion fostered about these dozens of purported human species of genetically isolated groups moving through largely unoccupied landscapes of the Old World continents, occasionally making contact with other, genetically incompatible groups. This is very probably an unrealistic scenario because at least by the end of the Middle Pleistocene – and quite possibly very much earlier – every reasonably habitable patch of land in these continents seems to have been occupied by humans. We know this because by 135 ka ago, Neanderthals lived inside the Arctic Circle, where even nowadays temperatures fall below -40°C (Schulz 2002; Schulz et al. 2002; cf. Pavlov et al. 2001). We also know that other robust populations, presumed to be Denisovans, occupied the Tibetan highlands at least 160 ka ago (Fahu et al. 2019). If such highly marginal and extremely inhospitable expanses were inhabited, we could safely assume that, except fully arid or extremely elevated lands, all areas of the Old World were settled by the late Middle Pleistocene. Various other eurytopic animals, besides humans, have successfully colonised climatic zones ranging from the tropics to the tundra; the tiger, bear and wolf come to mind. They did so not by speciation, but by introgression (introgressive hybridisation; Anderson 1949), and it is this process rather than mass migration that best accounts for the mosaic of hominin forms we can observe through time. Demographic genetics, i.e. allele drift based on generational mating site distance, easily accounts for archaeologically observed population changes (Harpending et al. 1998; Relethford 2012; Säll and Bengtsson 2017). Times of environmental stress, punctuating much of the Pleistocene, tend to increase the effects of reticular introgression. That can lead to previously deleterious variants, be they mutation- or introgression-derived, becoming adaptive. Introgression across contiguous populations subjected to demographic and genetic adjustments provides a much better explanation than the simplistic wandering-tribes paradigm.

Another significant incommensurability separating Pleistocene archaeology and palaeoanthropology from proper science is that both disciplines regard evolution as a teleological process, leading to ever-better cultures and humans. The sciences perceive evolution as a dysteleological progression, an inherently stochastic development reacting to random environmental variables: there is no design, no teleology, and no ultimate purpose. In speaking of 'cultural evolution', archaeologists have coined an oxymoron. Also, there is other evidence suggesting that archaeology, as an entire discipline, misunderstands the concept of culture and, therefore, effectively its mandate. The notion of culture as a domain of phenomena hinders the development of a science of human group behaviour (Bednarik 2013d). The lack of unifying causal principles in this domain should remind us, as cognitive scientist Dan Sperber (1996) says, that cultures are "epidemics of mental representations". Culture is defined by the non-genetic transfer of practice (Handwerker 1989), i.e. by learning; but both concepts, culture and learning, are problematic because, in a scientific framework, they explain nothing. Learning is built into the ToM system (Theory of Mind; Baron-Cohen 1991; Heyes 1998; Siegal and Varley 2002; Bednarik 2015a) and caused by unexplained mechanisms (changes in the information state of the brain) in the individual organism's interaction with the environment. Somehow certain information states are reconstructed in other brains as similar states, and the manifestations of this process are perceived as a culture. Many commentators assume that because specific aspects of culture or learning are perceived as taxonomically connected, they must be the same thing. However, the putative domain of culture comprises arrays of very different things, and it is not possible to construct a unitary concept of either culture or learning. Science does not proceed by creating taxonomic entities of all things that are black or of all people younger than Socrates. What most readily separates the sciences from the humanities are the respective taxonomising strategies. For instance, in the sciences, all observations and findings to do with interacting surfaces in relative motion are the subjects of the science of tribology (Jost 1966; Bhushan 2013; Bednarik 2015b; 2019). Practitioners of the humanities may find it strange how to them disassociated phenomena such as the application of lipstick, the lubrication of ball bearings and the movement of tectonic plates could be subsumed as categories under the same heading. This differs markedly from the taxonomies of the humanities, illustrating the incommensurabilities between them and the sciences again.

However, there are many more of these disparities. Consider the basis of the 'cultures' archaeologists have invented for the period they call 'prehistory'. From the Lower Palaeolithic to the Metal Ages, literally hundreds of such 'cultures' have been conceived since the mid-nineteenth century (Petrie 1899) – more in fact than hominin/hominid species have been proposed. Their definition is generally based on variables that are not 'culture-specific', particularly putative tool types and their relative occurrences within excavated assemblages (Sackett 1981; 1988). It should have always been self-evident that tools do not define 'cultures'; they cannot, therefore, be diagnostic variables of culture (Thompson 2012). Some cultures, such as the famous 'potato-washing' culture of some Japanese macaques (*Macaca fuscata*), do not involve tools. More fundamentally, the tool types applied by archaeologists are etic constructs created by them, and they differ from those perceived by the makers or users of lithic artefacts (Bednarik 2013e: 165). In effect, archaeological tool types are merely "observer-relative, institutional facts" (Searle 1995). In their endeavours to create cultural taxonomies for the millions of years of human evolution, archaeologists have entirely ignored the available surviving record of genuinely cultural products of the people concerned, particularly the rich corpus of palaeoart. If orthodox Pleistocene archaeology does show an interest in palaeoart, it trivialises it by assigning it to a category comprehensible in a simplistic reference frame, defining it as 'art'. ('Palaeoart' is no more art than a peanut is either a pea or a nut; it is a collective term describing all art-like phenomena of the deep past.) There is no evidence to justify such a label; the term 'art' always derives from an ethnocentric concept:

> The status of an artifact as a work of art results from the ideas a culture applies to it, rather than its inherent physical or perceptible qualities. Cultural interpretation (an art theory of some kind) is therefore constitutive of an object's arthood.
>
> (Danto 1988)

No modern (e.g. Westernised) human could fathom the ideas Pleistocene hominins applied to palaeoart (Helvenston 2013). One cannot even establish the status of recent ethnographic 'art' works with any objective understanding (Dutton 1993): interpretation is inseparable from the artwork (Danto 1986). To regard palaeoart as art is, therefore, an application of an etic and ethnocentric idea to products of societies about whose emic parameters nothing is known in most cases ('emic' refers to knowledge and interpretation within a culture; 'etic' refers to the interpretation by members of another culture).

Similarly, the term 'prehistoric' has no scientific currency. It derives from the intuitive belief that written traditions provide more reliable records than oral traditions, an untested proposition that may well be wrong. One privileged enclave of humanity regards the period of written records as 'history', and the time before that as 'prehistory'. In this, it ignores several objections: that most human beings were illiterate, and in Europe so until very recent centuries. Moreover, history is written by its winners; the losers of history tend to contribute little to it. Some Andean rulers took this so far that they extinguished the writers of history in their realms as soon as they took power, in order to rewrite 'history'. While the transfer of oral history is often set in rhyme and rhythm and is known to have preserved correct historical information for more than ten thousand years, written texts such as the Bible that are only two millennia old may have become almost incomprehensible because many expressed meanings have changed. In Australia, we have numerous cases of stories that have correctly preserved geological and topographical information for enormous timespans, in the absence of any written communication. The underlying argument, that written history is necessarily more reliable than oral is, therefore, open to challenge. Does a society in which only the clerics could read and write have a written history? Were such groups as the Mayas 'historical' or 'prehistoric' (Bouissac 1997)? Finally, the determination by a small minority of humanity of what constitutes one or the other does not need to be accepted; the majority of all humans that ever existed would be entirely justified to consider that self-appointed minority's view offensive if history is perceived more noble or valuable than prehistory.

There is a convention avoiding the drawbacks associated with this ethnocentric perspective. One can capitalise the word 'History', in the same way as specific periods of human history, such as the Renaissance or the Iron Age are capitalised. The period before this time then becomes 'pre-History', i.e. an arbitrary division of real history. All of human history is history, and there can be no human history before it. Perhaps one way to clarify the misperception is to remember that in some languages, such as German, Spanish and Italian, the words for 'story' and 'history' are the same, *Geschichte, historia* and *storia* respectively. As this convention correctly implies, history is always a story, a narrative about the past, an interpretation edited by many factors; it is not a factual account.

History is, admittedly, much more credible than pre-History, which when created by archaeologists is a field awash with misinformation, misunderstandings, fabrication and spin. Of all the disciplines of the humanities, archaeology is the 'softest' and the most accident-prone. Since its beginnings in the mid-1800s, prehistory has cultivated an ability to get it wrong most of the time that is unparalleled in other fields. As noted above, all of the key discoveries and consequential changes in Pleistocene archaeology have been introduced by non-archaeologists, whose work was in all cases wholly rejected by the 'experts', usually for several decades before it was accepted (Bednarik 2013e). This tradition began perhaps with Boucher de Perthes who correctly proposed that humans coexisted with Pleistocene animals – a proposal that was caustically rejected by the archaeologists for decades. This tradition of scorning new ideas or finds has been continued right up to the present time. The probably most consequential of the most recent blunders in the field is the 'replacement hypothesis', dubbed the 'African Eve' theory by the media. Introduced by an archaeological charlatan, Professor Reiner R. R. Protsch 'von Zieten', its eager embrace by Anglo-American Pleistocene archaeology and palaeoanthropology as well as by the popular science industry has established this meme as virtual dogma.

Not even the discovery that the putative speciation between robust and gracile *Homo sapiens*, which is indispensable to maintaining this doctrine, never took place has put an end to this fad. Instead of conceding that modern humans belong to the same species like Neanderthals and other Robusts, African Eve's supporters now proclaim that the two forms of humans lived side by side and 'interbred'. They simply cannot envision the possibility that at 50 ka ago, the world's humans were all robust types, but at 25 ka ago they were all gracile.

They cannot grasp the notion that gracility did not appear suddenly, but that it developed gradually – albeit in a time interval that to them seems too short for that scale of evolution to occur. Neanderthals interbred with 'anatomically modern humans' no more than great-grandparents interbreed with their great-grandchildren. There is, however, an alternative explanation: evolution is not the only way whole populations can change into very different forms. This book presents such an alternative, and its purpose is to persuade the reader that it is much more credible and plausible than other versions of modern human origins. These other accounts range from those provided by the various mainstream religions at one end, to the African Eve model still believed by most archaeologists. They all have one thing in common: they are not supported by science or by carefully assembled archaeological and palaeoanthropological information, tempered by a dose of healthy scepticism.

Formulating a credible synthesis

Although the archaeological record may be unrepresentative and is certainly inchoate, it does comprise valuable raw data. It would, therefore, be wrong to throw the baby out with the bathwater, and what is needed is a review of all relevant data, ascertaining that archaeological doctrines did not contaminate them. That involves scrutiny of any contingencies in how data were acquired that are dependent upon untested hypotheses. So, for instance, if an assumption in the validity of the African Eve meme is evident in the acquisition of specific evidence, it is likely that such data are tainted by dogma. Since such ambiguities have with time crept into most ideas in archaeology, the backtracking required to create clarity is likely to be very extensive. The endemic lack of falsifiability in archaeological interpretation has encouraged the positing of thousands of frivolous but ardently defended premises based on often very slim evidence. The support cited in archaeology is itself often illusory because the elements archaeologists might perceive in what they call 'material evidence' are merely creations of other archaeologists (Lewis-Williams 1993), i.e. "observer relative or institutional facts" (Searle 1995). As Searle explains, an object may be made partly of wood, partly of metal (the "brute facts"), but its property of being a screwdriver only exists because it is represented as such. There is no testable evidence that the ideas of archaeologists about very ancient stone tools would be shared by their makers or users, nor for that matter are they even diagnostic of cultures. Archaeology invents taxonomies of these stone tools, then taxonomies of cultures represented by these tools or their relative representation within assemblages, and then finally concocts peoples imbued with these invented cultures – yet in all of this, it is ignored that tools do not define cultures. Those variables that do, like palaeoart, were either ignored or forced into the straitjacket of archaeological dogma. Frankel (1993) defines his work and that of his peers as resembling that of the sculptor who "discovers" a statue in a block of marble. An epistemological examination of Pleistocene archaeology, especially, amply confirms this creative aspect of the field (Bednarik 2013a).

This generic example illustrates that the task of distilling from the mainstream model the part that can be regarded as reliable is not as easy as it might seem. Clearing more than one and a half centuries of accumulated deadwood (Bahn 1993) from a non-falsifiable pursuit of one of the humanities is not easily accomplished. It is far beyond the scope of the present volume. Here we are only concerned with forming a credible *general* hypothesis. A foremost need would be to excise the effects of the popular science industry which Michael R. Rose so tellingly refers to as "intellectual lumpenproletariat" (Rose 2016: 70). Its degrading influence shaping the modern worldview through its symbiotic relationship with institutionalised science did perhaps not exist as today's all-pervasive force in Huxley's time when he declared that nothing does more damage to science than cliques and schools of thought. However, today his view is more poignant than ever when we consider how contemporary public science tends to operate: as a contest of countless 'authorities' and vested interests vying for public attention and the funding it engenders, with varying doses of jingoism and nationalism thrown in for good measure. As Nobel Prize recipient, cell biologist Randy Schekman observed, science is being "disfigured by inappropriate incentives" (Schekman 2013). He notes how the brands of "luxury journals" are "marketed with a gimmick called 'impact factor' – a score for each journal, measuring the number of times its papers are cited by subsequent research". Schekman singles out the journals *Nature, Science* and *Cell*, and deservedly so: they have published numerous scientific blunders and tend to be reluctant to admit so. That phenomenon is even more common in the 'leading journals' in such fields as archaeology and human evolution. Schekman likens journals that aggressively market their brands to "fashion designers who create limited-edition handbags or suits, to stoke demand through scarcity".

The performance-related bean-counting of the impact factor or similar indices, such as the h-index (an h-index of 50 means that at least 50 of one's papers have been cited at least 50 times) undermines the integrity of science in various ways. A journal's impact factor score is an average; it has no bearing on the merits of individual contributions, and it is easily manipulated. For instance, a citation may be in an approving fashion, or it may imply rejection, so by itself, it cannot possibly be intended as a measure of credibility. Self-citation is another obvious distorting factor, but more important are the effects of refereeing cliques. These are loosely structured coalitions of like-minded scholars who control what is published in the establishment journals they gather around.

The editors of 'leading journals' favour contributions that 'make waves', because these tend to be cited more often, they sell copy and, most importantly, they satisfy the feeding frenzy of the popular science lobby. The effects on public information are profound. One only needs to consider the frantic pieces appearing in the conduits of Rose's intellectual lumpenproletariats every time one of these 'paradigm-changing' articles is published. The popular science industry churns out hundreds of 'interpretations' replete with misunderstandings, most of which are contradicted by previous and subsequent announcements. Consider, for example, the amplification of genetics hypotheses reporting one day that Eurasians descend from Africans, the next day it is the other way around; beef is healthy to eat, then it is not; or indeed any conceivable topic, from weight loss to climate change, is disseminated in this sensationalising manner. One day we are told that modern humans share up to 20 per cent of their genes with Neanderthals, the previous day we heard that we share 98 per cent with chimpanzees, so the public is perfectly entitled to ask, does this mean that we are closer to chimpanzees than to Robusts? So how do we reconcile this with the statement that we share 60 per cent of genes with bananas? Plainly if geneticists cannot explain their results and claims adequately, or if they are unwilling to take popular science to task, they deserve no attention. Yet the public does not even seem to mind all too much that it is bombarded with all this nonsense dressed up as science.

Perhaps science publishing is too lucrative a business to care about veracity or integrity. One of the most significant players, Elsevier, posted profit margins above 40 per cent in recent years (e.g. 2012 and 2013, amounting to billions of dollars). To put this into perspective, this is an industry that receives its raw material at no cost from its customers (the scientists), then has these same customers carry out the quality control of those materials at no cost (through the refereeing system), only to sell the same materials back to the customers at outrageous prices. So much for the academic principle of unfettered access to scientific knowledge. Not surprisingly, there are now demands for remedial action. For instance, the European Union has a plan for all science publishing to be funded by the states. Another initiative facilitating a degree of democratisation of science publishing is the trend of recent years towards Open Access publishing.

One more aspect of Schekman's disfigurement of science "by inappropriate incentives" concerns the academic reward system as it is emerging. In editing several academic journals, the author of the present book discovered that some of the writers whose work he published received significant financial rewards from their respective institutions as incentives. It is inescapable that publication of their work entails, in any case, a variety of enticements, concerning promotions and grants, but it is a somewhat worrisome development that cash bonuses are being introduced for academic authors. Such prizes are very substantial when publication occurs in a 'leading journal', a trend that can only amplify the distortions Schekman berates.

However, to return to Huxley, he also stated: "The struggle for existence holds as much in the intellectual as in the physical world. A theory is a species of thinking, and its right to exist is coextensive with its power of resisting extinction by its rivals." The extremely competitive nature of the academic juggernaut ensures that such 'species of thinking' defend their existence to the last breath, which explains the phenomenon Max Planck's famous quote addresses: that to see a new insight to triumph, we shall have to wait for its opponents to die. This, as noted, is especially true in fields such as Pleistocene archaeology and human evolution, partisan pursuits as they are at the best of times. If we see this in the context of the power of spin and vested interests, the magnitude of the problem becomes fully apparent. The combined authority of The Academy, the "high priests" of these disciplines (Thompson 2014) and the science publishing industry at all levels are a formidable force by any standards. To consult Huxley again, "Every great advance in natural knowledge has involved the absolute rejection of authority." Challenging the behemoth of these gatekeepers with the power of arbitrating what is to be regarded as being true about the human past would be imprudent for any practitioner of these disciplines. However, an amateur doing so must be out of his mind. Nevertheless, to cite Huxley once again, "The investigation of nature is an infinite pasture-ground where all may graze, and where the more bite, the longer the grass grows, the sweeter is its flavour, and the more it nourishes." In the face of academic reality, this may sound wholesome, idealistic and optimistic, but it is quite unrealistic.

The present book is an attempt by a non-archaeologist and non-palaeoanthropologist to posit that these disciplines have completely misconstrued the explanation of how we, *Homo sapiens sapiens*, became what we are today. Its author is entirely aware that what he attempts here is an unenviable task, but he takes comfort in the long history of these disciplines, which grudgingly accepted every single major innovation a generation or two after first strenuously rejecting it – and that non-professionals proposed all of these revolutions. This author thinks that the high priests might be wrong once again and that a generation from now it will be accepted that once more an amateur had it right and they had it wrong. He also appreciates, however, that this will *not result in a realistic model* becoming the mainstream. Once again, it will lead to a corrupted version that will endeavour to preserve as many of the previous falsities as possible.

So how will one distil a credible synthesis explaining hominin history by culling from the multiple hypotheses about it those that are tainted by dogma? Starting with the very basics, it seems safe to assume that hominins arose from other primates and that this occurred in the Old World (although the jury is still out on where this could have been). The elevation of the many subspecies of the Pliocene and Pleistocene to the status of species, in the absence of genetic evidence, is perhaps premature, as the skeletal evidence they are based on is fortuitously acquired and no doubt unrepresentative. Now that it appears that the interfertility of 'Moderns', 'Neanderthals', 'Denisovans' and *Homo antecessor* (or should that be *Homo sapiens antecessor*?) has been established genetically (Reich et al. 2010; Vernot and Akey 2014; Sankararaman et al. 2014), are we ready to accept that there has only been one human species for *at least* the last half a million years (Henneberg and Thackeray 1995)?

There are far too many fundamental shortcomings in the available record, ranging from misinterpretations to selectivity of the available material, through to taphonomy. The taphonomy of the 'material record' needs to be much better understood, in the sense that it can never be a representative sample (Bednarik 1994a). Another need is the establishment of palaeo-neuroscientific research into the capabilities of hominins (Bednarik 2015a). Many highly relevant specific issues have not even been considered in any consequential fashion. For instance, if it is true that encephalisation, i.e. the enlargement of the brain, was the most prominent feature of humanisation, why is it that at the very time the modern cranial traits began to become evident, this trend reversed dramatically, and the current phase of rapid atrophy of the human brain began? If it is true that encephalisation was necessary to support the growing cognitive demands placed on later humans, why did the presumed acceleration of these demands in the last third of the Late Pleistocene usher in a radical *reduction* of cranial volume? How did exclusive homosexuality arise in humans, and what led to the abandonment of oestrus? More importantly, if the direction of human evolution was dictated by natural selection, what were the evolutionary benefits of acquiring or preserving the thousands of deleterious genes that are responsible for such extant conditions as neurodegenerative ailments, Mendelian disorders, mental illnesses or genetic disorders so numerous that they cry out for explanations? Come to think of it, the physical changes from Robusts to Graciles are hardly improvements. Physical strength waned even more than brain volume, as did robusticity of skeletal architecture. More specifically, the most delicate organ in the body, the brain, suffered a rather significant reduction in the protection offered by the cranium, which was vastly more resistant to trauma in the Robusts, due to both shape and thickness. Why did evolution tolerate this apparent devolution? Also, how did modern humans manage to cope with their increasingly complex cultural and social realities when their brains shrank? If a larger brain was not needed to cope with the supposedly increasing cultural complexities, why did human evolution allow the enormous burden of relentless encephalisation in the first place? This issue alone shows that human evolution has been fundamentally misunderstood. And then there are the questions concerning the cognitive evolution of hominins: what were their true capabilities, ignoring predictions deriving squarely from dogma? On what empirical evidence could such discussions be based constructively, bearing in mind that mainstream archaeology is so inadequately informed about the relevant corpus of evidence that its false predictions were wholly predictable?

These are just some of the questions about human history that archaeology and palaeoanthropology have not only comprehensively failed to answer; most of them are not even being asked, which demonstrates more than anything else the impotence of these disciplines. What do they believe the purpose of their deliberations is? What is the point of having disciplines exploring the human primate that fail so consistently in asking the critical questions in exploring the human condition (Bednarik 2011c) as we know it? This state of the art does, however, resemble a more widespread malaise in the academic project. For instance, there are the axiomatic and endemic uncertainties dominating theoretical physics and cosmology. These disciplines, too, seem unable to credibly address their key issues, such as the origins of the cosmos. However, at a more fundamental level, we can reasonably assume that our individual constructs of reality are generated within our brains, and yet science has no understanding of how these constructs relate to the real world or how they are formed. Indeed, science has not even produced a credible explanation for how the brain constructs reality from the information it receives via the sensory equipment connected to it. It can reasonably well explain the mechanics of perception, for instance, but not how or why perception forms the reality in which we see ourselves exist. Surely such understanding would be utterly fundamental to comprehending the nature of what neuroscientist Todd M. Preuss called the "undiscovered ape".

Admittedly, these epistemological tasks are a great deal more complex and demanding than those that archaeology might be expected to tackle, but the comparison seems justified. In both cases, the most fundamental issues almost seem to be avoided. It is much more likely, however, that they are too overwhelming and most scholars are content with pursuing much more modest quests. In this book, we will try to nudge that paradigm a little. A fundamental epistemological problem relates to the issue of aetiology. This term refers to the cause or causes of a condition, and while it is primarily applied in medicine, its applicability is universal. It defines what is wrong with much of the way medicine operates, such as specifically symptom-driven diagnosis or treatment of symptoms rather than causes. Symptoms help in diagnosis, but by themselves, they tell us little about causes, and science keenly prefers causal reasoning. This issue is particularly apparent in disciplines such as psychology and psychiatry, where the aetiological understanding of conditions ranges from the inadequate to the non-existent and where proper science has made limited headway. If we extend the concept of aetiology to specific human conditions such as consciousness, self-awareness, Theory of Mind, mental illnesses, advanced cognitive abilities or any such states in extant conspecifics, we discover that their aetiologies are almost entirely unknown. Indeed, they have not been investigated in any consequential fashion (but see Bednarik 2012a; 2015a). This is a fair indication of how unknown the human ape indeed is: we understand very little about it in a scientific, causal sense.

This issue is to be broadly addressed in the present book, but to illuminate the underlying difficulties, it is useful first to consider the reasons for this state. We know that, just as in medical research, we are dealing primarily with present manifestations, but in order to be genuinely effective in prompting changes to them, we might need to know how they came about. Contrary to popular belief, medicine is hardly an objective science. Like the second-most funded area of 'research', military research including the space race, its fundamental purpose as a discipline is not the acquisition of an understanding of the world, but one species' self-centred desire of self-preservation and enhancement. This is of no benefit to the rest of this world but often places an enormous burden upon it — as we are beginning to learn. Leaving these philosophical issues aside, the comparison with our state of understanding the human condition could be useful. Very few of the fundamental questions we might have about how or why we became what we ended up being have been answered by Pleistocene archaeology and palaeoanthropology over the past couple of centuries. Surely this should be central to the whole project of these disciplines. However, instead of trying to explain the big issues about human nature, they provided us with an unlikely narrative about some invented African tribe that for unexplained reasons became unable to breed with other humans. It acquired miraculously advanced abilities which enabled it to take over all of Africa, and then it marched on Asia. Upon reaching the south-easternmost part of that continent, it promptly invented seafaring and sailed for Australia. Finally, these mythical *Übermenschen* from Sub-Saharan Africa declared war on the Neanderthal brutes of Europe and wiped them out as well, having already replaced all other human populations (Bednarik 2008a). This science-fiction tale was invented by a German archaeology professor in the 1970s, who has since been sacked in disgrace (Schulz 2004). It is this level of credibility that shows how world archaeology is in no position to tackle the big issues about humans. What is particularly frustrating is that such just-so paradigms serve to explain almost nothing about human nature. Science is meant to explain something with its hypotheses; what does archaeology explain with its feeble and often downright false interpretations?

Therefore, we shall have to begin the task of presenting a much more robust paradigm of 'modern' human origins by examining the merits of the African Eve or replacement hypothesis, which is so much in need of replacement. This book is dedicated to outlining that replacement. In the process it will even explain some of the big issues about humans. Please fasten your seatbelts!

2 The gracilisation of humans

The 'African Eve' hoax

The idea that the technocomplex which archaeologists call the 'Upper Palaeolithic' was first established in Sub-Saharan Africa and then introduced from there to Europe has been debated since the early twentieth century. For instance, Dorothy Garrod, noted for her infamous role in the Glozel affair in 1927 (Bednarik 2013a: 76), subscribed to this notion. However, the actual 'African Eve' hoax, as we shall tactfully call it, derives from the work of Professor Reiner Protsch 'von Zieten'. His aristocratic title was as bogus as his second doctorate, and a German court heavily fined him for claiming to have the latter. Presenting a series of false datings of fossil human remains over following years, he proposed that modern humans arose in Sub-Saharan Africa (Protsch von Zieten 1973; Protsch 1975; Protsch and Glowatzki 1974; Protsch and Semmel 1978; Henke and Protsch 1978). Thirty years later, it was shown that all of his datings had been concocted and he was dismissed by the University of Frankfurt (Terberger and Street 2003; Schulz 2004). However, during the intervening decades an entire academic cottage industry had been established around the original idea, spawning numerous variations on the general theme: the 'Afro-European *sapiens*' model (Bräuer 1984b), the 'African Eve' complete replacement scenario (Cann et al. 1987; Stringer and Andrews 1988; Mellars and Stringer 1989; Vigilant et al. 1991; Tattersall 1995; Krings et al. 1997); the Pennisi (1999) model; the 'wave theory' (Eswaran 2002); the Templeton (2002) model; and the 'assimilation theory' (Smith et al. 2005), among others.

Of these, the mitochondrial African Eve model is the most extreme, contending that the purported African invaders of Asia, then Australia and finally Europe, were a new species, unable to interbreed with the rest of humanity. They replaced all other humans, either by exterminating or out-competing them (be it economically or epidemiologically). There were significant methodological problems with this 'African Eve theory', as the media dubbed it, right from the start. The computer modelling was botched by Cann et al. (1987) and its haplotype trees were irrelevant. Based on 136 extant mitochondrial DNA samples, it arbitrarily selected one of 10^{267} alternative and equally credible haplotype trees (which are very much more than the number of elementary particles of the entire universe, about 10^{70}!). Maddison (1991) demonstrated that a re-analysis of the Cann et al. model could produce 10,000 haplotype trees that are more parsimonious than the one chosen by these authors.

Moreover, there is no reason to assume that the most parsimonious tree should define what occurred, bearing in mind the entirely dysteleological nature of evolution (Hartl and Clark 1997: 372). Several of Cann et al.'s assumptions were false, such as exclusive maternal transference of mitochondria; the constancy of mutation rates of mtDNA (Rodriguez-Trelles et al. 2001; 2002); and the purported models of human demography or the timing or number of colonisation events are entirely unknown. Another flaw of the replacement model was that Cann et al. had misestimated the diversity per nucleotide (single locus on a string of DNA), incorrectly using the method developed by Ewens (1983) and thereby falsely claiming greater genetic diversity of Africans, compared to Asians and Europeans. This oft-repeated claim (e.g. Hellenthal et al. 2008; Campbell and Tishkoff 2010) is inherently false. At 0.0046 for both Africans and Asians, and 0.0044 for Europeans, the genetic diversity coefficients are very similar. Even the premise that genetic diversity of extant humans decreases with increasing geographical distance from Africa (e.g. Atkinson 2011) is just plain false (see Chapter 4).

Gibbons (1998) has noted that by using the modified putative genetic clock, Eve would not have lived 200,000 years ago, as Cann et al. had contended, but only 6,000 years ago (which would match the prediction of the Old Testament precisely, but is universally rejected by the boffins). Indeed, the hypothetical split between Eve's tribe and other humans has been placed at times ranging from 17,000 to 889,000 years ago by various writers, all without credible justification (e.g. Vigilant et al. 1991; Barinaga 1992; Ayala 1996; Brookfield 1997). This applies to the contentions concerning mitochondrial DNA (African Eve) as much as to those citing Y-chromosomes ('African Adam'; Hammer 1995). The divergence times projected from the diversity found in nuclear DNA, mtDNA and DNA on the non-recombining part of the Y-chromosome differ so much that a time regression of any type is most problematic. Contamination of mtDNA with paternal DNA has been demonstrated in extant species (Gyllensten et al. 1991; Awadalla et al. 1999; Morris and Lightowlers 2000; Williams 2002), in one recorded case amounting to 90 per cent (Schwartz and Vissing 2002). The issues of base substitution (Lindhal and Nyberg 1972) and fragmentation of DNA (Golenberg et al. 1996) have long been known (see Gutierrez and Marin 1998). Other problems with interpreting or conducting analyses of palaeogenetic materials are alterations or distortions through the adsorption of DNA by a mineral matrix, its chemical rearrangement, microbial or lysosomal enzymes degradation, and lesions by free radicals and oxidation (Geigl 2002; Carlier et al. 2007).

In its original forms, the replacement hypothesis derived not from genetics, but from palaeoanthropology in the form of fossil skeletal evidence. Genetics was only enlisted in the late 1980s, and its application was bungled from the beginning (Baringa 1992; Hedges et al. 1992; Maddison et al. 1992; Templeton 1992; 1993; 1996; 2002; 2005; Brookfield 1997; Klyosov and Rozhanskii 2012a; 2012b; Klyosov et al. 2012; Klyosov and Tomezzoli 2013). The essential claim of the replacement theory that the 'Neanderthals' were genetically so different from the 'Moderns' that the two were separate species has been under strain since Gutierrez et al. (2002) demonstrated that the pair-wise genetic distance distributions of the two human groups overlap more than claimed, if the high substitution rate variation observed in the mitochondrial D-loop region (see Walberg and Clayton 1981; Zischler et al. 1995) and lack of an estimation of the parameters of the nucleotide substitution model are taken into account. The more reliable genetic studies of living humans have shown that both Europeans and Africans have retained significant alleles from multiple populations of Robusts (Garrigan et al. 2005; Hardy et al. 2005; cf. Templeton 2005). After the Neanderthal genome yielded results that seemed to include an excess of gracile single nucleotide polymorphisms (Green et al. 2006), more recent analyses confirmed that 'Neanderthal' genes persist in extant Europeans, Asians and Papuans (Green et al. 2010). 'Neanderthals' (Robusts) are said to have interbred with the ancestors of Europeans and Asians, but not with those of Africans (Gibbons 2010; cf. Krings et al. 1997).

Sankararaman et al. (2012) report that comparisons of DNA sequences between 'Neanderthals' and present-day humans have shown that the former share more genetic variants with non-Africans than with Africans. Further evidence of 'interbreeding', from a 40-ka-old *Homo sapiens sapiens* fossil, implies a 10 per cent contribution of 'Neanderthal' genes, suggesting 'interbreeding' occurred just four generations previous (Viegas 2015). Sankararaman et al. (2014) report finding 'Neanderthal' haplotypes in the genomes of 1,004 present-day humans. Prüfer et al. (2014) demonstrated several gene flow events among Neanderthals, Denisovans and early modern humans, possibly including gene flow into Denisovans from an unknown archaic group. Kuhlwilm et al. (2016) analysed the genomes of a Neanderthal and a Denisovan from the Altai Mountains in Siberia together with the sequences of chromosome 21 of two 'Neanderthals' from Spain and Croatia. They believe that a population that diverged early from other Moderns in Africa contributed genetically to the ancestors of Neanderthals from the Altai Mountains roughly 100 ka ago. They did not detect such a genetic contribution in the Denisovan or the two European Neanderthals. Finally, Vernot et al. (2016) report the occurrence of 'Neanderthal' and 'Denisovan' DNA from present Melanesian genomes. All of this could suggest that gracile Europeans and Asians evolved mostly from robust local populations, which had long been evident from previously available evidence (e.g. the close link in tooth enamel cellular traits between Neanderthaloids and present Europeans; Weiss and Mann 1978).

So why was Protsch's African Hoax recycled for several decades? Cavalli-Sforza et al. (1988) had already considered that the phylogenetic tree separates Africans from non-Africans, a view reinforced by Klyosov et al. (2012). However, whereas the first authors interpreted this as placing the origin of 'modern humans' in Africa, Klyosov et al. showed that this separation continued for 160 ±12 ka since the split of the haplogroups A from haplogroups BT (Cruciani et al. 2002); therefore, Africans and non-Africans evolved mainly separately. As Klyosov et al. (2012) most pertinently observe, "a boy is not a descendant of his older brother". Therefore, contrary to Chiaroni et al. (2009), haplogroup B is neither restricted to Africa nor is it at 64 ka remotely as old as the haplogroups A are (some of these may be older than 160 ka).

It is interesting to note that the 'genetic clock' archaeologists so ardently subscribe to is rejected by them when it is applied to the dog, implying its split from the wolf occurred 135 ka ago. In the second case, they disallow it because there is no paleontological evidence for dogs before about 15,000 years ago (e. g. Napierala and Uerpmann 2010). However, in this, they seem to be mistaken (see e.g. Germonpré et al. 2009). This raises the question of why the same restraint in relying on genetic indices has not been exercised with hominins, and why an implausible catastrophist scenario has been so eagerly embraced concerning the human species. After all, we are only one of many species that have managed to colonise a great variety of environments, from the Arctic to the tropics. In all other cases, genetic diversity is thought to be the result of introgression and gene flow between neighbouring populations, and not of conquest by 'superior' species. To see whether this discrepancy in approach could be attributable to humanistic fervour (Bednarik 2011a) rather than to science, we could review the palaeoanthropological and 'cultural' contexts of the African Eve hoax.

Initially, as noted, the African Eve hoax derived from false age determinations of numerous hominin remains, especially in Europe. These included the four Stetten specimens from Vogelherd, Germany, widely claimed to be about 32 ka old (e.g. Churchill and Smith 2000a; 2000b), when in fact their Neolithic provenience had long been noted (Gieseler 1974; Czarnetzki 1983: 231) and their ages range credibly from 3,980±35 to 4,995±35 carbon-years (Conard et al. 2004). The Hahnöfersand calvarium, the "northernmost Neanderthal specimen found" and dated to 36,300±600 BP or 35,000±2,000 BP (Bräuer 1980) by Protsch, is at 7,470±100 BP or 7,500±55 BP a Mesolithic 'Neanderthal' (Terberger and Street 2003). The Paderborn-Sande skull fragment, purportedly 27,400±600 years old (Henke and Protsch 1978) is, in reality, only 238±39 carbon-years old (Terberger and Street 2003). It is so fresh that when it was drilled for resampling, it emitted a putrid smell. The Kelsterbach skull, dated to 31,200±1,600 years BP by Protsch (Protsch and Semmel 1978; Henke and Rothe 1994), is probably of the Metal Ages (Terberger and Street 2003) but has mysteriously disappeared from its safe in Protsch's former institute. The cranial fragment from Binshof, claimed by Protsch to be 21,300 ±20 years old, turned out to be only 3,090±45 years.

Further afield, the 'modern' Robust specimen from Velika Pećina, Croatia, is now known to be only 5,045±40 radiocarbon years old (Smith et al. 1999). Those from Roche-Courbon (Geay 1957) and Combe-Capelle are now thought to be Holocene burials (Asmus 1964; Perpère 1971), as probably also applies to the partial skeleton from Les Cottés (Perpère 1973). The Crô-Magnon specimens, widely regarded as the 'type fossils' of anatomically modern humans, are neither of the Aurignacian technocomplex nor are they anatomically modern; especially cranium 3 is quite Neanderthaloid. Being about 27,760 carbon-years old (Henry-Gambier 2002), they are more likely to be of the Gravettian than of the Aurignacian. A similar pattern of intermediate forms between Robusts and Graciles pertains to the numerous relevant Czech specimens, most of which were lost in the Mikulov Castle fire of 1945. The surviving sample includes the Mladeč specimens, now dated to between 26,330 and 31,500 BP (Wild et al. 2005); the very robust specimens from Pavlov and Předmostí (both between 26 and 27 ka); Podbaba (undated); and the slightly more gracile and more recent population from Dolní Věstonice. The same pattern of 'intermediate' forms continues in the specimens from Cioclovina (Romania), Bacho Kiro levels 6/7 (Bulgaria) and Miesslingtal (Austria).

Therefore, there are no anatomically modern finds from Europe that are more than 28,000 carbon-years old. The earliest *liminal* 'post-Neanderthal' finds currently available in Europe, still very robust, are the Peştera cu Oase mandible from Romania (Trinkaus et al. 2003), thought to be in the order of 35 ka old but re-dated to c. 40 ka (Fu et al. 2015), and the partial cranium subsequently found in another part of the same extensive cave (Rougier et al. 2007). Both lack an archaeological context and are clearly not 'anatomically modern'. Also intermediate between robust and gracile types are the six human bones from another Romanian cave, Peştera Muierii (c. 30 ^{14}C ka BP; Soficaru et al. 2006). Looking at Eurasia as a whole, there are hundreds of specimens from the last third of the Late Pleistocene that are intermediate between robust *Homo sapiens* and *H. sapiens sapiens*. Collectively, they show that a simplistic division between 'Moderns' and 'Neanderthals' is obstructive to a comprehension of these finds. They include those from Lagar Velho, Crete, Starosel'e, Rozhok, Akhshtyr', Romankovo, Samara, Sungir', Podkumok, Khvalynsk, Skhodnya and Narmada, as well as several Chinese remains such as those from the Jinniushan and Tianyuan Caves, Maludong (Red Deer Cave) and Longlin Cave. These range in age from the entire Late Pleistocene, right up to the Bølling-Allerød interval, indicating that the 'replacement' is a complete myth. Incredibly, these numerous intermediate or liminal forms contradicting the belief that robust and gracile populations were separate species were ignored by those wishing to preserve this dogma. More importantly, they failed to appreciate that not a single fully gracile specimen in Eurasia can credibly be linked to any of the Early Upper Palaeolithic tool tradition, be it the Aurignacian, Châtelperronian, Uluzzian, Proto-Aurignacian, Olschewian, Bachokirian, Bohunician, Streletsian, Gorodtsovian, Brynzenian, Spitzinian, Telmanian, Szeletian, Eastern Szeletian, Kostenkian, Jankovichian, Altmühlian, Lincombian or Jerzmanovician (Bednarik 2011b). The proposition that these industries were introduced from Sub-Saharan Africa is therefore without support, a notion squarely contradicted by the lack of any geographically intermediate Later Stone Age finds from right across northern Africa until more than 20,000 years after the Upper Palaeolithic had been established all over Eurasia. Similarly, the African Eve advocates ignored that at least seven Early Upper Palaeolithic sites have yielded human skeletal remains attributed to Neanderthals: the Châtelperronian layers of Saint Césaire (c. 36 ka) and Arcy-sur-Cure (c. 34 ka) in France; the Aurignacian of Trou de l'Abîme in Belgium; the Hungarian Jankovichian of Máriaremete Upper Cave (c. 38 ka; Gábori-Csánk 1993); the Streletsian of Sungir' in Russia (which yielded a Neanderthaloid tibia from a triple grave of 'Moderns'; Bader 1978); and the Olschewian of Vindija in Croatia (Smith et al. 1999; Ahern et al. 2004; Smith et al. 2005) and Cotencher in Switzerland. Like other late specimens, those from Vindija are more gracile than most earlier finds – so much so, that many consider them transitional (e.g. Smith and Raynard 1980; Wolpoff et al. 1981; Frayer et al. 1993; Wolpoff 1999; Smith et al. 2005).

The replacement paradigm is not even supported by the paleoanthropological finds from Africa, which generally mirror the gradual changes in Eurasia through time. It is often claimed that anatomically modern humans (AMHs) date from up to 200 ka ago, yet no such specimens have been presented. The skulls from Omo Kibish offer some relatively modern features as well as substantially archaic ones; in particular, Omo 2 (an undated surface find) is very robust indeed (McDougall et al. 2005). The more complete and better dated Herto skull, BOU-VP-16/1, is outside the range of all recent humans in several cranial indices (White et al. 2003). The lack of AMHs from Sub-Saharan Africa prior to the exodus of Eve's scions is conspicuous: the Border Cave specimens have no stratigraphic context; Dar es Soltan is undated; the mandibles of Klasies River Mouth lack cranial and post-cranial remains; and the Hofmeyr skull features the same intermediate morphology evident at the same time, 36 ka ago, in Eurasia (Grine et al. 2007; 2010).

As the Levant is on the route the exodus would presumably have taken, the lack of African fossils of the African Eve 'species' prompted the replacement advocates to turn to that region for support. The Mount Carmel finds from Qafzeh Cave and Skhul Shelter were recruited as 'Moderns', yet all of these skulls present prominent tori and receding chins, even Qafzeh 9, claimed to be of the most modern appearance. The distinct prognathism of Skhul 9 matches that of 'classic Neanderthals', and the series of teeth from that cave has consistently larger dimensions than Neanderthaloid teeth, an archaic trait. Supposedly much later 'Neanderthal' burials in nearby Tabun Cave as well as the Qafzeh and Skhul material are all associated with the same Mousterian tools, and the TL datings of all Mount Carmel sites are far from soundly established, with their many discrepancies and inversions. The claims of 90-ka-old 'modern' humans from Mount Carmel, a cornerstone in the Eve model, are in every respect unsound. This population is best seen as transitional between robust and gracile forms, from a time when gracilisation had commenced elsewhere as well.

Another nail in the African Eve's coffin is presented by the extant AMH Australians, with their average cranial capacity well within the range of *Homo erectus*, who possess molars and other indices of robusticity matching those of Europeans several hundred millennia ago. Their tool traditions were of Mode 3 types ('Middle Palaeolithic') until mid-Holocene times and remained so in Tasmania until British colonisation two centuries ago. The guiding principle of the replacement advocates that Mode 4 technologies were introduced together with 'modern' anatomy is a falsity – in Europe, as well as elsewhere. This brings us to the next major contradiction of the African Eve model.

One of the chief obstacles the African Eve theory has always faced is that the Early Upper Palaeolithic (EUP) tool traditions of Eurasia, claimed to indicate the arrival of AMHs, all seem to have evolved locally. They first appear relatively simultaneously between 45 ka and 40 ka BP, even earlier, at widely dispersed locations from Spain to Siberia (e.g. Makarovo 4/6, Kara Bom, Denisova Cave, Ust'-Karakol, Tolbaga, Kamenka, Khotyk, Podzvonkaya, Tolbor Dorolge; Bednarik 1994b). The earliest radiocarbon date was provided by the EUP deposit of Senftenberg, Austria, at >54 ka BP (Felgenhauer 1959). In Spain, the Aurignacian commenced at least 43 ka ago (Bischoff et al. 1994; Cabrera Valdés and Bischoff 1989). EUP variants such as the Uluzzian (Palma di Cesnola 1976; 1989), the Uluzzo-Aurignacian, and the Proto-Aurignacian (43–33 ka BP) have been reported from southern Italy (Kuhn and Bietti 2000; Kuhn and Stiner 2001). The montane Aurignacoid tradition of central Europe, the Olschewian (42–35 ka BP), clearly developed from the region's final Mousterian (Bayer 1929; Kyrle 1931; Bächler 1940; Zotz 1951; Brodar 1957; Malez 1959, Vértes 1959; Bednarik 1993b). The Bachokirian of the Pontic region (>43 ka BP), the Bohunician of east-central Europe (44–38 ka BP; Svoboda 1990; 1993), and various traditions of the Russian Plains complete the picture to the east. Some of the latter industries, such as the Streletsian, Gorodtsovian and Brynzenian derived unambiguously from Mousteroid technologies, whereas the Spitzinian or Telmanian are free of Mode 3 bifaces (Anikovich 2005). The gradual development of Mode 3 industries into Mode 4 traditions can be observed at various sites along the Don River, in the Crimea and northern Caucasus, with no less than seven accepted tool assemblages coexisting there between 36 and 28 ka ago: Mousterian, Micoquian, Spitzinian, Streletsian, Gorodtsovian, Eastern Szeletian and Aurignacian (Krems-Dufour variant). A mosaic of early Mode 4 industries began before 40 ka BP on the Russian Plain and ended only 24–23 ka ago. In the Crimea, moreover, the Middle Palaeolithic is thought to have ended only between 20–18 ka BP, which is about the same time the Middle Stone Age ended across northern Africa. Elsewhere in the Russian Plain, the first fully developed Upper Palaeolithic tradition, the Kostenkian, appears only about 24 ka ago. All of this shows that there cannot possibly be a correlation between UP technologies and 'Moderns'.

The Russian succession of traditions connecting Mode 3 and 4 technocomplexes is repeated in the Szeletian of eastern Europe (43–35 ka BP; Allsworth-Jones 1986), the Jankovician of Hungary; and the Altmühlian (c. 38 ka BP), Lincombian (c. 38 ka BP) and Jerzmanovician (38–36 ka BP) further north-west. Similarly, the gradual development from the Middle Palaeolithic at 48 ka BP (with 'Neanderthal' footprints of small children) to the Upper Palaeolithic is documented in Theopetra Cave, Greece (Kyparissi-Apostolika 2000; Facorellis et al. 2001). Thus there is a complete absence of evidence of an intrusive technology in the postulated eastern or south-eastern entry region of Europe, through which an invading tradition arriving from the Middle East would have had to pass. Nor should this even be expected, considering that in the Levant both Mode 3 and Mode 4 industries were used by robust as well as gracile populations: all credible archaeological evidence refutes the replacement advocates' notion that their 'Moderns' introduced Mode 4 in Europe. The Mousteroid traditions of the Levant developed gradually into blade industries, e.g. at El Wad, Emireh, Ksar Akil, Abu Halka and Bileni Caves, and that region's Ahmarian is a transitional technology. This can be observed elsewhere in south-western Asia, for instance, the Aurignacoid Baradostian tradition of Iran develops *in situ* from Middle Palaeolithic antecedents. The late Mousterian of Europe is universally marked by regionalisation (Kozłowski 1990; Stiner 1994; Kuhn 1995; Riel-Salvatore and Clark 2001), miniaturisation and increasing use of blades, as well as by improved hafting technique. This includes the use of backed or blunted-back retouch on microliths that were set in birch resin in Germany, almost as early as the first use of microlithic implements in the Howieson's Poort tradition of far southern Africa. Therefore, the notion that a genetically and palaeoanthropologically unproven people with a Mode 4 toolset travelled from Sub-Saharan Africa across northern Africa is entirely unsupported, while there is unanimous proof that these traditions developed *in situ* in many Eurasian regions long before they reached northern Africa or the Levant.

Finally, the African Eve advocates failed to consider that the record of palaeoart provides precisely the same finding as derived from genetics, physical palaeoanthropology and technological development. Having fallen victim to Protsch's 'hoax', they relied on the unassailability of their belief that the EUP traditions, especially the Aurignacian, were by AMHs (Graciles). Among the EUP traditions previously attributed to Moderns, the Châtelperronian was in 1979 discovered to be the work of Neanderthals. The Châtelperronian of Arcy-sur-Cure in France had produced numerous portable palaeoart objects, including beads and pendants (Figure 2.1). Instead of conceding that they had been wrong, the Eve supporters then argued that the Neanderthals, who were incapable of symbolling according to their dogma, must have 'scavenged' these artefacts (White 1993; Hublin et al. 1996). They failed to explain, however, what such supposedly primitive creatures would do with symbolic objects.

This is one of the numerous examples of the accommodative reasoning of the replacement advocates. Others are the claims that beads of the Lower Palaeolithic are unrealistic before Eve (d'Errico 1995; d'Errico and Villa 1997; Rigaud et al. 2009), or the assertion that Early Pleistocene seafaring colonisers (Bednarik 1997a; 1999; 2003a) might have drifted on vegetation (Davidson 2003). After we first observed the lack of evidence linking early Aurignacian finds to the purported Moderns (Bednarik 1995a), we proposed that no such link exists to any EUP industry (Bednarik 2007; 2008a). The contention that

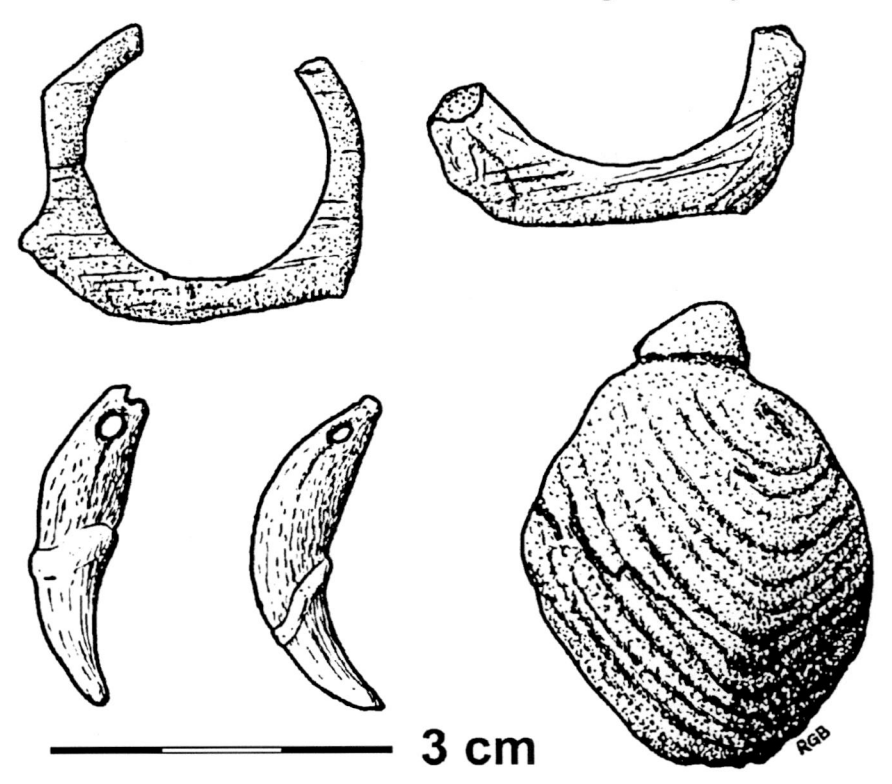

Figure 2.1 Ivory ring fragments, perforated animal canines and fossil shell pendant made by Neanderthals, Grotte du Renne, Arcy-sur-Cure, France

the Aurignacian rock art (e.g. in Chauvet Cave, El Castillo and several other sites) and portable palaeoart (e.g. in Hohlenstein), arguably the most complex and sophisticated of the entire Upper Palaeolithic, is the work either of 'Neanderthals' or their direct descendants (Bednarik 2007; 2008a; 2011b; 2011c; 2017a) has removed the last vestiges of support the African Eve hypothesis has ever claimed. The notion that palaeoart was introduced in Eurasia by AMHs now stands refuted, and one can only wonder how such an unlikely and thoroughly incongruous construct could ever have captivated Pleistocene archaeology and palaeoanthropology. The archaeological record shows unambiguously that the Upper Palaeolithic tool traditions of Eurasia developed *in situ*, that the hominins in question evolved *in situ*, that the genetic evidence is ambiguous at best, and introgression and gene flow can account fully for the genetic observations. 'Modern' humans derive from archaic *H. sapiens* in four continents, and they interbred with them no more than grandchildren breed with their grandparents.

Klyosov et al. (2012), who demonstrate genetically that recent human evolution in Eurasia must have occurred *in situ*, list no less than 24 papers asserting that AMHs entered Europe between 27 and 112 ka ago. Most of these nominate 40–70 ka as the time of the 'African invasion'. Ancient DNA recovery and analyses are subject to significant difficulties and what is 'reconstructed' from aDNA analyses reflects (perhaps subconscious) suggestions of earlier literature (M. Henneberg, pers. comm. 2019). The aDNA sequences are heavily 'reconstructed' using modern information. It would seem that propositions about timing merely reflect archaeological estimates of a phenomenon that never actually occurred. Hundreds, if not thousands, of authors merely fell victim to a hoax that exceeds, in the magnitude of its effects, the Piltdown hoax of 1912 (Bednarik 2013a).

Palaeoart and seafaring

A comprehensive survey of the world's known palaeoart preceding Mode 4 technocomplexes has been attempted (Bednarik 2017a). Presenting thousands of earlier specimens, it well and truly retires the notion that art-like creations began at the same time as Upper Palaeolithic modes of production or the appearance of AMHs in Europe. The latter issue is particularly instructive. The simplistic consensus view in archaeology is that the world's first palaeoart appears in south-western Europe at the same time as AMHs enter that stage, i.e. at some time between 40 and 30 ka ago. This entirely false belief is the basis of a subliminal, almost subconscious, conviction shared by the public and archaeology that art and modern culture first appeared in France and Spain. How did this idea arise?

For half a century, beginning in 1912, the discipline believed that humans evolved in England, of all places (see Chapter 1). As a result of the Piltdown hoax, palaeoanthropology was led up the garden path by a whimsical notion that served political aspirations – a Eurocentric idea that flew in the face of common sense: that humans originate from a little island off that small appendage of Eurasia called Europe. Apart from retarding the discipline for half a century, for instance, by ignoring the discovery of australopithecines (Dart 1925), this had the effect of preserving the illusion that England was still the centre of the world, even as its empire was beginning to crumble. By the time the Piltdown matter, the second-biggest hoax in the history of palaeoanthropology (the African Eve hoax takes first place), was finally put to rest, the British Empire lay in ruins. Furthermore, the thought that humans emerged in 'the colonies' had become acceptable.

Precisely the same kind of falsities constitute the notion that culture and art began in Europe. It, too, is Eurocentric and replete with political connotations. These include the subliminal message that Europe's societies are superior and have been so for a long time. One of this falsity's most important manifestations can be found in the World Heritage List by UNESCO, which features dozens of south-western European rock art sites claimed to be of the Pleistocene.

However, not a single site of Pleistocene rock art from the rest of the world is listed, despite such sites being far more numerous in other continents (Bednarik 2013b; 2013c; 2014a; 2014b; 2014c). Thus, the illusion was created, in the minds of the public, popular science writers and even among most archaeologists, of 'art' first having appeared in such places as the French Dordogne or Spanish Cantabria. Not only is this merely false, in reinforcing the point, archaeologists have even nominated many rock art sites in the Iberian Peninsula as having Palaeolithic content, when in reality their petroglyphs can only be relatively recent and have in some cases been demonstrated to be of the twentieth century (Bednarik 2009). In nominating such sites scholars have served the political objective to glorify the European past at the expense of other pasts – and by using false evidence. We are again reminded of the discipline's political roles (for excellent commentaries, see the work of archaeologists Trigger (1984; 1989); Silbermann (2000); and Kohl and Fawcett (2000)). A discipline whose chosen task it is to "molest the past" (Campbell 2006) has no choice but to "dig up someone else's past, which means nothing but trouble" (ibid.). It usurps and writes the histories of the societies that were supplanted or extinguished by the victorious nation-states. There are very few of these states in the world whose sovereignty was not acquired by conquest or the violence of colonialism. Therefore, from the perspective of the indigenes concerned, the re-interpretations of their histories by archaeologists and their cultures by anthropologists are unwelcome. There have been many heated battles between these protagonists (Bednarik 2013a).

To return to the African Eve hoax. One of the most devastating blows its model has suffered is our proposal that the most sophisticated 'artistic' productions of the Upper Palaeolithic are not by AMHs, but by more robust humans, either late Neanderthals or their still Neanderthaloid descendants (Bednarik 2007; 2008a). Not only does this destroy the idea of continuous teleological development of the complexity of cultures; it pulls the rug right out from under the African Eve's feet. There is universal agreement that the most 'advanced', the most sophisticated Palaeolithic 'art', is not from the end of that 'period', but from the *Early* Upper Palaeolithic. It includes the portable masterworks from Hohlenstein-Stadel, Geißenklösterle and Vogelherd in Swabia, Germany, and the stunning rock art creations of Chauvet and Cussard Caves in France. This turns one of the guiding principles of archaeology on its head, and also exposes one of the most severe epistemological flaws in Pleistocene archaeology: the concept of 'cultural evolution'. Many archaeologists see human history as a continuous ascent, in both the cultural and the somatic sense. They see in it an upward development of ever-better culture and ever more perfect body. In this, they are subconsciously guided by religious dogma: God, after all, created us in his image, and physical perfection is not represented by some lowly robust hominins, but by the perfectly flawless AMH. We evolved towards a predetermined ideal, a likeness of God.

Similarly, the primitive tools evolved towards ever more sophisticated kinds, and since simplistic archaeology bases its invented cultures on tools and their

38 *The gracilisation of humans*

relative numbers within excavated assemblages, this also provides a measure of 'cultural evolution'. With these two teleological constructs of development, archaeologists demonstrate, unwittingly, the massive incommensurability between their humanistic pursuit and science (cf. Claidière and André 2012; Guillo 2012). In science, evolution is a purely dysteleological process; there is no definable end product, and the process is entirely random, dependent upon the vagaries of many environmental and other factors. Therefore, the formulation 'cultural evolution', found in all textbooks of archaeology, is an oxymoron, as noted in Chapter 1: it cannot exist.

This kind of thinking in terms of an ascending development, of seeing AMHs as some perfect outcome, is not only religiously based; it is infantile. Modern humans are anything but perfect; in fact, if there could be such a thing as devolution, they should be seen as devolved humans. We will frequently return to this point in later chapters. For the moment, let us focus on the palaeoart of the Pleistocene, i.e. the period also known as the Ice Age. There is no evidence of the presence of AMHs before 30 ka ago in Europe, and even most hominin specimens between 30 and 25 ka old still feature distinctly robust traits. The most recent presumed Neanderthals have been reported from Vindija Cave, Croatia, first dated at 28–29 ka, later at 32.5 ka, and finally at ~44 ka (Devièse et al. 2017). However, the sole bone tool also dated was 29,500±400 years old, roughly matching the original dating, and the actual age of the cave's Neanderthal occupation remains thus unresolved. There is a trend among replacement advocates currently to define human remains from the interval 30–40 ka ago as modern, although all of them are quite robust and intermediate. The issue is complicated by the occurrence of Neanderthaloid specimens as recently as in the Mesolithic and even later. Therefore, the differentiation between robust and gracile specimens is somewhat subjective and depends on the disposition of the adjudicator.

Similarly, the antiquity of the European rock art of that same interval is not adequately resolved. The best-dated site is Chauvet Cave, which has yielded over 80 radiocarbon results from its floor deposit, from charcoal rock art and torch marks on the cave wall. Moreover, the cave's entrance was sealed more than 21.5 ka ago (Sadier et al. 2012), and yet 'Palaeolithic art experts' try hard to dispute these datings (Zuechner 1996; Pettitt and Bahn 2003; Pettitt et al. 2009; Combier and Jouve 2012). The results for the older phase of the Chauvet cave art cluster at 32–33 ka (Clottes et al. 1995; Clottes 2001), but the distortion by the Campanian Ignimbrite event at 40 ka BP (Fedele et al. 2008) suggests that these estimates are too low. An age of between 35 and 40 ka might be more realistic (Bednarik 2007). Be that as it may, the Chauvet rock art is of the Aurignacian tool assemblage and thus not the work of fully AMHs. The human tracks seen on the existing floor of the cave were made by robust adolescents (Bednarik 2007).

Numerous other examples of cave art have been attributed to the Aurignacian or a contemporary Early Upper Palaeolithic (EUP) tool tradition. For instance, the cave art of l'Aldène at Cesseras (Hérault, France) has to be more than 30,300 years old because the cave became closed at that time (Ambert et al. 2005; Ambert and Guendon 2005). The single decorated chamber of Baume Latrone (Gard, southern France), 240 m from the cave's entrance and of challenging access, comprises widely separated traditions, the oldest being finger flutings, most probably of the Middle Palaeolithic. They are about four times as old as the most recent engravings, still of the Palaeolithic (Bednarik 1986). The nearby clay pictograms of 'mammoths' are thought to be Aurignacian (Clottes 2008: 60). Several engraved blocks of limestone, bearing apparent depictions of vulvae, cupules and crudely made zoomorphs from La Ferrassie and the Abris Cellier, Belcayre, Poisson and Blanchard have also been attributed to the Aurignacian, but there is no proof that they could not be Middle Palaeolithic. This applies especially to La Ferrassie, because of the Neanderthal cemetery there, and the limestone block over the Neanderthal child grave No. 6 that bears 18, mostly paired cupules (Capitan and Peyrony 1921; Peyrony 1934). A block bearing red and black pigment traces exfoliated from the ceiling of Abri Castanet, also in France, has been reported from a 37-ka-old deposit. The most recently dated Aurignacian cave art motif is a zoomorph executed in charcoal in Coliboaia Cave, Bihor, Romania (Clottes et al. 2011).

In recent years several new relevant claims have appeared. For instance, a bovid figure drawn with a finger in Cueva de la Clotilde in Cantabria has been attributed to the Aurignacian but this attribution is only based on style, which is unreliable. There are unconfirmed reports of Neanderthal petroglyphs in Zarzamora Cave (Segovia, Spain). In Gorham's Cave, Gibraltar, a deeply engraved design of eight lines was excavated below a layer dated to 39 ka BP, at which time the site was occupied by Neanderthals (Rodríguez-Vidal et al. 2014). Sanchidrián has reported apparent torch soot next to ichthyform paintings that dates from between 43,500 and 42,300 BP (Collado Giraldo 2015). Other Spanish pictogram sites that have been suggested to include EUP rock art motifs attributable to Neanderthaloids comprise Pondra Cave in Cantabria (González Sainz and San Miguel 2001: 116–118), and in Asturias La Peña (Fortea Pérez 2007), Abrigo de la Viña (Fortea Pérez 1999), El Conde Cave (Fernández Rey et al. 2005), yellow bovid figures and charcoal dots in Peña de Candama, and possibly the cave of El Sidrón (Fortea Pérez 2007). Finally, Maltravieso Cave in Extremadura contains on panel 3 in its 'hall of paintings' some painted motifs that have been suggested to be more than 37,000 years old (Collado Giraldo 2015: 200).

Other pictograms, at Tito Bustillo in Asturias, have also been attributed to the Aurignacian (Balbín Behrmann et al. 2003) and have been claimed to be as early as pre-37,700 BP (Pike et al. 2012: 1412). A triangular motif of red pigment in Altamira, Cantabria, has been reported to be more than 36,160±610 years old (Pike et al. 2012: 1410). Another Cantabrian cave, El Castillo, has yielded minimum uranium-series dates of up to 41,400±570 BP from its rock art (ibid.). This proposal would place the sampled red 'disc' motifs before the earliest dated Aurignacian occupation evidence at the site (Hedges et al. 1994).

More recently, claims of much higher ages have emerged from three Spanish cave art sites, again based on U-Th (uranium-thorium) analysis results. Minimum ages ranging from 51–79 ka were presented from La Pasiega; 41–70 ka from Maltravieso; and 40–68 ka from Ardales (Hoffmann et al. 2018). In all cases the speleothem sampled was superimposed over red paint residues, which means that the paint must have been applied earlier, perhaps even much earlier. Surprisingly high ages, secured by the same method, have also been reported from Indonesia. Aubert et al. (2014) have attributed minimum ages ranging up to 44 ka (at Leang Barugayya 2) and 40 ka (at Leang Timpuseng) to a series of painted motifs in seven caves in the Maros karst of Sulawesi.

All these results are not cited approvingly here, despite supporting our case for very early rock art traditions. The use of uranium-thorium datings from very thin carbonate speleothem needs to be reviewed (Bednarik 2012b; Clottes 2012). Where such dates were checked against radiocarbon ages, they were shown to be much higher than those produced by an alternative and more reliable method (Bednarik 1984a; 1997c; Plagnes et al. 2003; Balter 2012). Nevertheless, even if these excessive claims from Spain are disregarded, the growing number of proposals of rock art in France and Spain that seems to have been created by robust *Homo sapiens*, including Neanderthals, cannot be ignored, as it is also reflected in portable and usually more securely dated palaeoart of western Europe. The first ever reported cave art from central Europe is also much more safely attributed. The finger flutings in Drachenhöhle in eastern Austria are of the Olschewian (Bednarik in press), which, as we have seen above, has yielded only Neanderthal remains elsewhere (at Vindija and Cotencher). Therefore, even if the dating of some of the listed examples remains controversial, there are numerous rock art and portable palaeoart instances that are safely attributable to Robusts.

Perhaps most importantly, in the present context, numerous sophisticated portable palaeoart objects derive from the Châtelperronian deposits in Arcy-sur-Cure, central France (see Figure 2.1). This tool industry is universally agreed to be Upper rather than Middle Palaeolithic, yet it is the work of Neanderthals. However, it is another cultural tradition of robust and intermediate humans, the Aurignacian, which has yielded the most complex palaeoart objects of the entire Palaeolithic. Outstanding among them are the therianthropes from Hohlenstein-Stadel (Figure 2.2) and Hohle Fels in the German Swabian Alb (Hahn 1988a; 1988b; 1991; Conard et al. 2003). Their depictions of creatures that combined human with zoomorphic characteristics evidence a complex cosmovision or ideology, far removed from the traditional conceptualisations of their primitive makers. Again, it is conspicuous that the most 'developed' tradition is not at the end of the Upper Palaeolithic, but in its early phases. This is perhaps most conspicuous in the cave art of Chauvet, for example, in the rhinoceros with many horns (Figure 2.3). This composition is perhaps best understood as an attempt to make a moving picture rather than a still picture: it seems to depict a moving, perhaps attacking animal. If so, this would be the first known endeavour to depict movement. Such sophistication was not achieved again for the next

The gracilisation of humans 41

30,000 years. Another example is the tiny green serpentine female figurine from Galgenberg, Lower Austria (Bednarik 1989), also dynamic rather than the usual static sculptures found until historical times.

The above finds provided an inkling of the creativity of humans when they were intermediate between the robust Neanderthals and the almost 'modern' humans of the Holocene period. Nevertheless, the production of palaeoart does not begin with the Châtelperronian or any other EUP tradition – not even in Europe. The Middle Palaeolithic traditions provide a good number of exogrammic finds, such as 111 perforated phalanges of the saiga antelope and three engraved items from Prolom 2, Crimean Peninsula (Stepanchuk 1993); several other perforated bones or teeth from German, French, Bulgarian and Spanish sites (Bednarik 2014c); a Mousterian bone flute from Divje babe 1, Slovenia

Figure 2.2 Aurignacian lion-headed human ivory figurine, possibly female, from Hohlenstein-Stadel, Germany

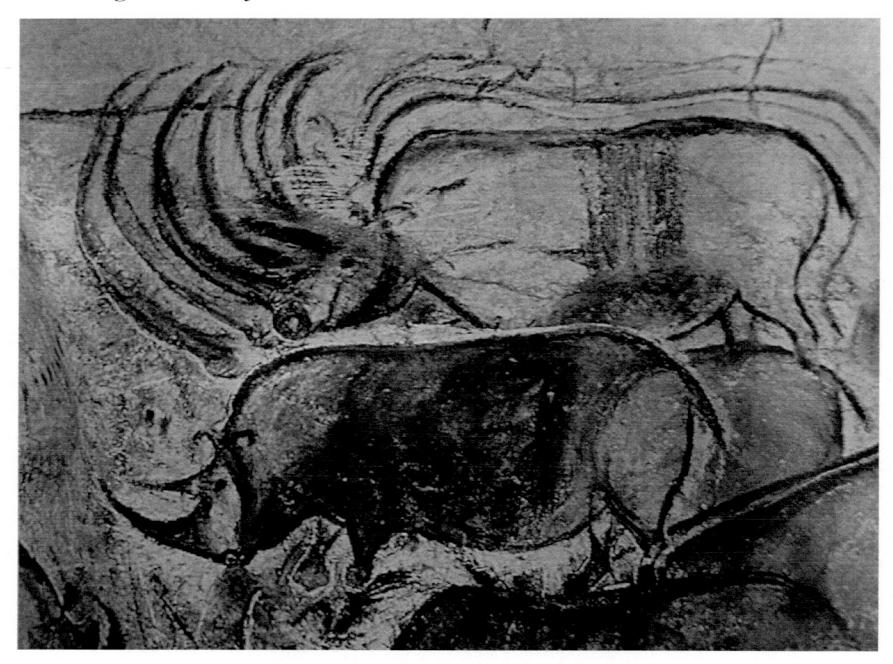

Figure 2.3 Aurignacian pictograms apparently depicting the movement of a charging woolly rhinoceros, Chauvet Cave, France

(Turk et al. 1995; Turk 1997; Turk and Dimkaroski 2011) made from a cave bear bone; and a collection of 13 beads or pendants from Kostenki 17, Russia (Praslov and Rogachev 1982). Conceptually more complicated may be a circular, engraved translucent nummulite from Tata, Hungary (Vértes 1964; Bednarik 2003b: Figure 28). According to U-series dating, this tiny palaeoart object should be between 68,000 and 114,400 years old (Moncel 2003). A finely polished and bevelled, elongated plaque made from a lamella of a mammoth molar and rubbed with red ochre comes from the same site (Vértes 1964). The Mousterian deposit of Axlor, Spain, has provided a circular sandstone pebble with a cross-shaped petroglyph, >47,400 years old (Barandiarín 1980; García-Diez et al. 2013). This object is reminiscent of the Tata nummulite, in its combination of a circular and cross shapes (Figure 2.4). Then there are numerous engraved portable objects: three bone fragments from the Micoquian of Oldisleben 1, Thuringia, Germany, one of which appears to feature a stickman (Bednarik 2006a); a bone artefact from Taubach near Weimar and an ivory fragment from Whylen near Lörrach, both in Germany (Moog 1939); and engraved bone fragments from the French sites of Le Moustier, Abri Suard, Abri Lartet, Marillac, Petit Puymoyen, Montgaudier, Peyrere 1 Cave, La Quina, Cotencher, Grotte Vaufrey, La Ferrassie, L'Hermitage and Abri Blanchard. Bones with Middle Palaeolithic engravings have also been reported from Tagliente Shelter in Italy (Leonardi 1988); Cueva Morín (Freeman and Gonzalez Echegaray 1983) and El Castillo (Cabrera Valdés et al. 2006) in Spain; Schulen in Belgium (Huyge 1990); and Bacho Kiro in Bulgaria. Leonardi (1988) has reported engravings on limestone and flint from several sites in Italy, and Vértes (1965) has described similar finds from the Mousterian of Hungary. A quartzite pebble recovered from El Castillo, Spain, bears five very small cupules (Cabrera Valdés et al. 2006). Much more complex are the engravings on a schist plaque from Temnata Cave near Karlukovo, Bulgaria (Crémades et al. 1995).

These portable art-like objects from the Middle Palaeolithic are not, however, the earliest finds of this nature from Europe. The oldest discoveries are from the preceding Lower Palaeolithic, and chronologically they begin with two marked bone fragments from Kozarnika Cave, north-western Bulgaria, reported to be between 1.1 and 1.4 million years old (Guadelli and Guadelli 2004). One is a bovid bone 8 cm long with about ten grooves, the other a cervid bone fragment bearing 27 notches along an edge. The marks are anthropogenic, but the finds need to be independently assessed to establish their precise status. Both the age and the significance of the several engraved pieces from the Steinrinne near Bilzingsleben in Germany (Mania and Mania 1988; Bednarik 1995a) are much better authenticated. They are of the Holstein interglacial (isotope stage 11, from 424–374 ka ago; Lisiecki 2005) and were excavated with thousands of stone implements and numerous hominin remains so robust that many have described them as *Homo erectus* specimens (Figure 2.5). Five large bone fragments, some of them of the forest elephant (*Palaeoloxodon antiquus*), bear a variety of arranged line engravings made with stone points (Mania and Mania 1988; Brühl 2018). Their deliberate nature has been demonstrated by laser-microscopic study (Steguweit 1999). A broken polished ivory point and a quartzite slab both bear arcuate markings (Bednarik 1995a). A similarly engraved forest elephant vertebra has been found at Stránská skála, the Czech Republic (Valoch 1987; Bednarik 1995a). More recent is another bone fragment, probably of a horse, from the Late Acheulian of Sainte Anne I, at Polignac, Haute-Loire, France (Raynal and Séguy 1986; cf. Crémades 1996). It bears ten regularly spaced cuts along one edge.

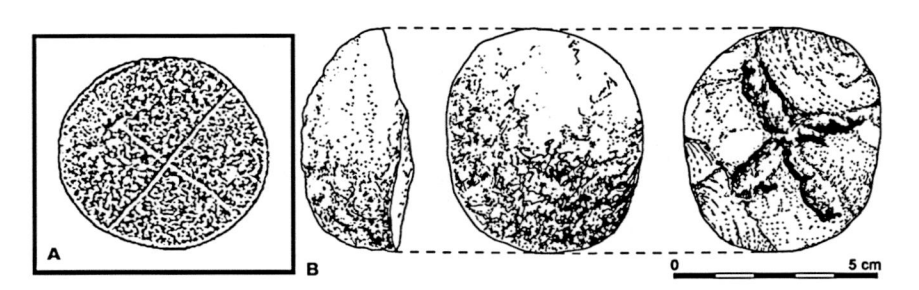

Figure 2.4 Middle Palaeolithic combinations of circular and cross forms from (A) Tata (Hungary) and (B) Axlor (Spain)

44 *The gracilisation of humans*

Figure 2.5 Engraved bone fragment of the forest elephant and hominin skull fragment of the Lower Palaeolithic of Bilzingsleben, Germany

Other palaeoart finds of the Lower Palaeolithic of Europe are the numerous stone beads of the Acheulian from a series of sites, including Les Boves near Amiens; from St. Acheul; the Loire River; Soissons near Aisne, Picardie; and from near Paris (all in France); as well as from the Biddenham quarry at Bedford and other sites in England (Bednarik 2005). They are all of perforated *Porosphaera globularis* fossils and have been reported for well over a century and a half (Boucher de Perthes 1846; Prestwich 1859: 52; Smith 1894: 272–276), yet some authors continued to reject them until recently (Rigaud 2006–2007; Rigaud et al. 2009) (Figure 2.6). Similarly, the same French team rejects the significance of the expertly perforated wolf incisor from Repolust Cave in Styria, Austria (Mottl 1951). It is from a hand-axe-free Lower Palaeolithic tool industry and thought to be roughly 300 ka old (Bednarik 2001). It was found together with a triangularly shaped, pointed bone fragment, also perforated at one end. D'Errico and Villa (1997) attribute the drilled holes to animal gnawing, although they omit to explain why they perceive identical Upper Palaeolithic beads as man-made; why any carnivore would risk its canines by biting through the root of a tooth; or why Lower Palaeolithic people elsewhere used perforated objects that cannot be the result of animal gnawing.

D'Errico had similarly explained away the above-mentioned Mousterian bone flute from Divje babe I, as having been caused by animal chewing (d'Errico and Villa 1997; d'Errico et al. 1998; 2003; see also Chase and Nowell 1998). This is despite the holes in the flute clearly not made by compression and ignores the flute's two and a half-octave compass that extends to over three octaves by over-blowing. Denial is a common reaction to finds that challenge the replacement hypothesis, for instance, Davidson (1990) explains the partially perforated fox tooth from the Mousterian of La Quina as the outcome of animal chewing, though he admits that "why an animal would chew a tooth is less obvious!" A major factor causing such irrational reactions is that most commentators are unaware of the massive corpus of evidence of palaeoart production in other continents that predates the 'Upper Palaeolithic revolution' of Europe. If they were to take note that all of the Pleistocene palaeoart production of just one country, Australia, numbering hundreds of thousands of specimens, is 'pre-Upper Palaeolithic' (i.e. of Mode 3 technology), they might

Figure 2.6 Lower Palaeolithic stone beads made from modified *Porosphaera globularis* fossils, Biddenham, England

appreciate that the evidence from Europe is in any case not all that important when viewed in the bigger picture. None of the oldest known exogrammic manuport, engraving, pigment use, petroglyph or proto-figurine is from Europe; all these earliest manifestations of the human ability to master the use of exograms have been provided by the much less comprehensively researched 'colonies'. Certainly, there is no reason to assume that 'culture', 'art', 'symbolling' or language originated in Europe. Nor should cognitive or technological evolution be expected to have been pioneered there. One of the most significant technological developments of the Pleistocene was acquiring the ability to colonise new lands by crossing the open sea. It demands the use of advanced inter-individual communication and the capability to ferry an adequate number of individuals, especially fertile females, to the new shores. This capability emerged in the first half of the Pleistocene, most probably in the Indonesian Archipelago, although the island of Socotra at the Horn of Africa is perhaps a contender in this.

Similarly, the earliest apparent use of exograms comes not from Europe, but Asia and Africa (Bednarik 2013b; 2013c; 2017a). For instance, the oldest known portable engraving was found on a shell from the *Homo erectus* deposit at Trinil, Java (Joordens et al. 2014). Dating by ^{40}Ar/^{39}Ar and luminescence suggests that it is roughly between 540 and 430 ka old. Petroglyphs seem to have arisen first in India, at sites such as Auditorium Cave, Bhimbetka (Bednarik 1993a) and Daraki-Chattan (Bednarik et al. 2005). They are of people with Mode 1 tool industries but remain essentially undated, and they are followed by the late Mode 2 traditions that created identical petroglyphs in the southern Kalahari Desert (Figure 2.7). Found at two sites in the Korannaberg hills and north of them, at Potholes Hoek and Nchwaneng, these petroglyphs are attributed to the climatic incursion of 410–400 ka ago (Beaumont and Bednarik 2015). The earliest currently known proto-figurines have been reported from Morocco (Bednarik 2003c) and Israel (Goren-Inbar 1986; Goren-Inbar and Peltz 1995). The quartzite object from Tan-Tan, coated in haematite pigment, is of the Moroccan Middle Acheulean (Figure 2.8); the Berekhat Ram proto-figurine is of the Late Acheulian. A proto-figurine is necessarily evidence of a pareidolic reaction to a natural feature resembling another object in shape, as expressed in action emphasising those features. The oldest indication of pigment use comes from southern Africa, from such sites as Kathu Pan 1 in South Africa (0.8–1.3 Ma old; McBrearty and Brooks 2000); Kabwe at Broken Hill, Zambia (Clark et al. 1947; McBrearty and Brooks 2000); and Wonderwerk Cave in South Africa (~1.1 Ma ago; Beaumont 1990; 2004; 2011). However, the pareidolic perception of natural visual properties as resembling ('seeing as') specific objects seems to extend furthest, suggesting the use of exograms as early as three million years ago. The most relevant find is the jaspilite cobble from the dolomite cave Makapansgat in northern South Africa (Eitzman 1958; Bednarik 1998). This manuport had been transported a considerable distance, no doubt due to its rare material, bright reddish colour and the strong resemblance to a primate's head. Excavated from Member 4, between 2.4 and 2.9 Ma old, it co-occurred with two stone artefacts (Maguire 1980) of Oldowan typology. Its most important role here is that the primate that handled it (be it an *Australopithecus* or *Homo*) detected its pareidolic properties, which presupposes apperceptive capability. It also implies self-awareness and developed Theory of Mind (Bednarik 2015c) (Figure 2.9).

Obviously, the currently earliest examples of evidence implying the use of exograms by hominins are not necessarily proof that similarly early finds are not possible in Europe, but the significant empirical bias in favour of Asia and Africa renders that somewhat unlikely. First, the massive body of more recent palaeoart specimens from these continents (Bednarik 2013b; 2013c; 2017a) that pre-dates the appearance of AMHs in Europe underlines the disparity. Second, other indicators also imply that significant advances in human abilities first began in the two main centres (notably southern Asia and Africa) and most likely spread from there. As mentioned, a significant development was the proficiency of harnessing the energies of currents, waves and wind with the property of buoyancy to invent assisted locomotion, and of then applying that skill to deliberate maritime migration. This was a significantly greater leap for humanity than Neil Armstrong's "small step for [a] man" because it set humankind on a trajectory of harnessing nature deliberately (Bednarik 2003a), of which more recent exploits are merely inevitable outcomes.

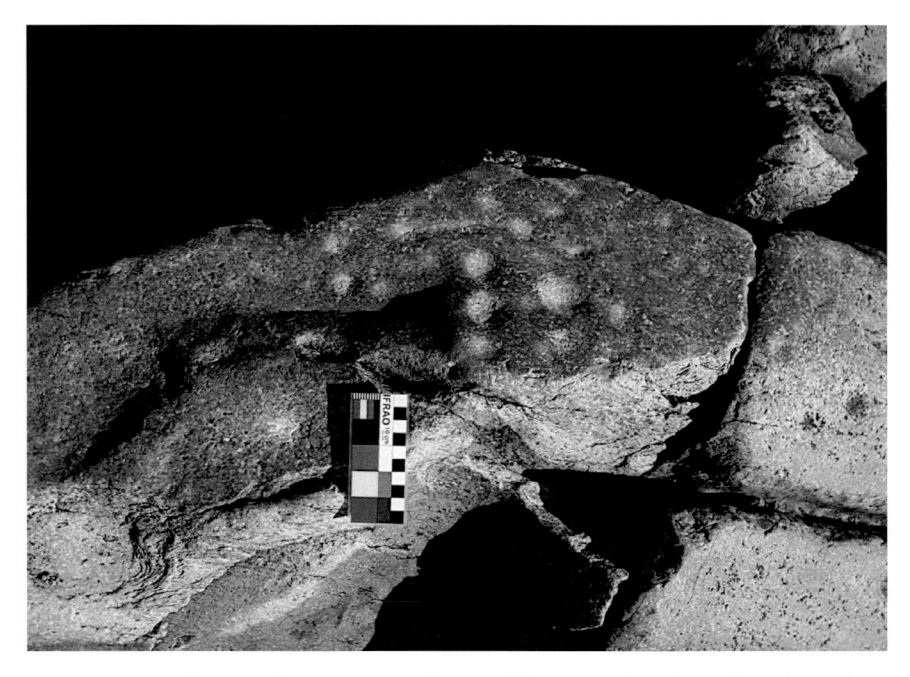

Figure 2.7 Cupules at Nchwaneng, southern Kalahari, South Africa, thought to be c. 400,000 years old

48 *The gracilisation of humans*

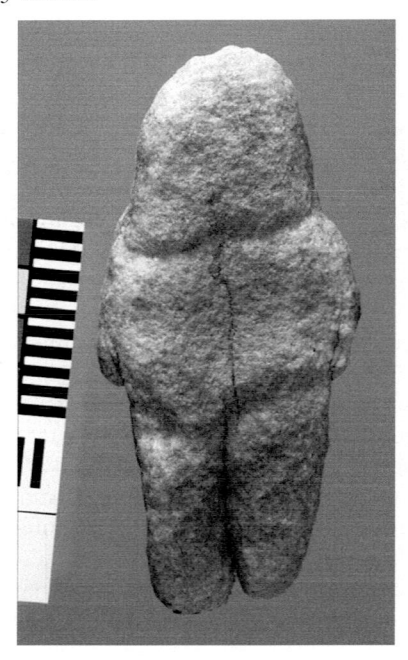

Figure 2.8 Modified manuport of the Middle Acheulian, Tan-Tan, Morocco

For any pelagic colonisation attempt to be archaeologically visible today, it had to first result in a well-established population that managed to persist for many generations. Small groups that barely succeeded in the sea crossing; that found it hard to establish themselves; or that were unable to produce a genetically viable and large enough, lasting populace are unlikely to emerge in archaeological sampling. Groups that lacked genetic variability were prone to developing endemic dwarfism (Berger et al. 2008) or more severe deleterious conditions, including the so-called 'Hobbit' (mistakenly regarded as a distinct species, *Homo floresiensis*; Morwood et al. 2004; see Eckhardt and Henneberg 2010; Eckhardt et al. 2014; Henneberg et al. 2014) and other small-bodied humans on islands. There is no consensus on the minimum number of individuals needed to constitute a viable breeding unit, but the scenarios of the lone pregnant female drifting on a log, or the hapless small group drifting on vegetation matter in the wake of a tsunami are not adequate to found a sufficient base for long-term propagation.

One of the most important aspects of maritime colonisation is that the world's sea straits are all dominated by unpredictable and robust cross-currents, the outcomes of relative tides in the seas connected by these narrows, the rotation of the Earth and the relative position of the Moon. Frequently these currents run in opposite directions at each side of such straits. They change direction unpredictably and cannot be forecast by the local fishermen possessing generationally accumulated knowledge. This means, first of all, that it is

Figure 2.9 The Makapangat jaspilite cobble, a manuport deposited in a cave between 2.4 and 2.9 Ma ago

impossible to cross any such sea strait without using an effective means of propelling a watercraft. The energy for such propulsion can come from wind power, rowing or paddling, and it also allows a vessel to be steered. The inability of predicting the directions of strong currents means that the travellers cannot compensate for the transverse cross-currents by selecting an upstream starting location. They could not predict the current's direction along the destination shore; therefore, the success of any such crossing was truly a matter of luck. Indeed, the survival rate among Stone Age seafarers was appallingly low. Tindale (1962) recorded a staggering death rate of 50 per cent on sea voyages of Kaiadilt Aborigines covering the minor distance of just 13 km from the Australian mainland to Bentinck Island.

50 The gracilisation of humans

Another example is provided by an incident in 2001 when a flotilla of an unknown number of bamboo rafts carrying refugees fleeing the religious and ethnic violence then occurring in Sulawesi drifted into the Pacific. Thirteen of the rafts were washed up on a beach at Woleai, one of the 600 islands of Micronesia, still bearing skeletal remains of some of their many sailors, together with food remains (Bednarik 2003a). Not a single person survived the journey of 1,600 km, despite the use of modern aids such as water containers. Indeed, we can be sure that countless Pleistocene aspirants of crossing sea barriers lost their lives in the process. This is confirmed by the 'death rate' of this author's experiments with sailing stone-tool built rafts: half of his eight expeditions failed, and if he had travelled without escort vessels he and his fellow mariners would have inevitably perished at sea. Indeed, when he did succeed in reaching the target shore, this was accomplished only barely (Bednarik 2014d). Such successes involved complex manoeuvres to avoid getting swept out into the open ocean and incredible sacrifices by the crews (one paddler fell into a coma for two days through extreme exertion). On one occasion, one of these 'Palaeolithic' rafts got caught in a maelstrom at the target shore (Figure 2.10).

If armchair archaeologists who have suggested that people might have crossed sea straits by drifting on vegetation flotsam were scientists, they would be required to demonstrate by replication that this is possible to do. It is of course entirely impossible, and even if it were feasible, it would not clarify the issue. The problem is not just how to succeed in the crossing, but how to ferry across a group of people large enough to found a viable population. More importantly, the armchair archaeologists would need to explain why, in the case of Wallacea, it is that of the thousands of terrestrial animal species on the Asian plate, only those up to rat or lizard size have ever been able to colonise any of

Figure 2.10 Bamboo raft built with Lower Palaeolithic stone tools off the coast of Flores, April 2008

the over ten thousand islands, in the course of millions of years. Wallacea comprises the islands between the most important biogeographical divide in the world, Wallace's Line, and the limit of the Australian-New Guinean mainland fauna, Lydekker's Line. If humans can colonise Wallacea on drifting vegetation matter, then so can other Asian primates as well as countless further land animals. Small animals, such as rodents or snakes, can survive on such drifts for many years. Many of the large species of the Sunda or Asian plate are excellent swimmers, among them the rhinoceros, pig, deer, hippo and tapir, to name just a few. None of them has ever colonised Wallacea. Birds, insects or bats managed to fly across the sea barriers, and humans introduced some species in relatively recent times (e.g. dog, pig, macaque, deer). The only exceptions are proboscideans who managed to cross unaided to many of the Wallacean Islands and also colonised the Philippines. They evolved into several species of elephants and Stegodontidae, including endemic dwarf forms. Elephants can travel 48 km on the open sea (Johnson 1980), in which they are assisted by their trunks used as snorkels and by digestive gases in their intestines that help in buoyancy. The habit of elephants of travelling in herd formation is of great benefit in colonising new lands, as is their inclination to tow others to allow them to rest when needed.

Hominins lacked all these characteristics, and they were, by nature, relatively weak swimmers. As the only large terrestrial species, besides proboscideans, to succeed in colonising islands more than 30 km distant, hominins most certainly could not cross sea barriers without propelled watercraft (unless they rode on elephants, as suggested by one archaeologist). As the first sea crossings happened many hundreds of millennia ago, the available archaeological record of watercraft is of no help in speculating about the nature of these very early vessels. The earliest logboats have been discovered at Pesse (Holland), 8,265±275 years old (Van Zeist 1957); Noyen-sur-Seine (France), 7,960±100 BP (Bednarik 1997b); Kuahuqiao (China), 7,600–7,700 years old (Tang et al. 2016); Dufuna (Chad), between 7,670±110 and 7,264±55 years old (Breunig 1996); Lough Neagh (Northern Ireland), attributed to the eighth millennium BP (Callaghan and Scarre 2009); La Marmotta on Lake Bracciano (Italy), late seventh millennium BP (Fuggazzola Delpino and Mineo 1995); Lystrup 1 (Denmark), 6,110 ±100 BP (Arnold 1996); and from Torihama (Japan), 5,500–6,000 years old (Morikawa and Hashimoto 1994).

A worked reindeer antler from Husum (Germany), thought to be a boat rib of a skin-boat of the Ahrensburgian (Ellmers 1980), is even believed to be in the order of 10,500 years old. But both skin-boats and logboats (or dugouts) can be assumed to be relatively late developments. More critical is the taphonomic (related to preservation and selective survival; Bednarik 1994a) dictum that the rising sea-level of the early Holocene, beginning around 11,000 years ago, has destroyed most archaeological evidence along all coasts, and the probability of finding the remains of older watercraft is close to zero. Therefore, the nature of any vessels of the Pleistocene remains unknown. Nevertheless, it stands to reason that they would have been relatively primitive and simple, and they were almost certainly rafts of some type rather than boats. At this point, it becomes most relevant that the earliest known seafaring evidence happens to be in the Indonesian Archipelago, the region most abundant in large bamboo species, where stalks can exceed 20 cm in diameter. It is hardly a coincidence that bamboo is better suited for the construction of simple rafts than any other material, for several reasons. For instance, it has been demonstrated that it takes as little as three minutes to fell a 16 m long bamboo culm with a Lower Palaeolithic hand-axe (Bednarik 2014d). The air chambers it comprises render bamboo, which is a grass rather than a tree, more buoyant than any timber.

The presence of humans on the Indonesian island of Flores in the Middle or even Early Pleistocene was first recognised by Dutchman Theodor Verhoeven who in March 1957 found stone tools in the fossiliferous deposit at Ola Bula (Verhoeven 1968: 400). Two months earlier, he had recovered remains of Stegodontidae at the same site on Flores (Hooijer 1957; Verhoeven 1958). Henri Breuil recognised among these initial finds some typical Lower Palaeolithic stone implements (ibid.: 265), while van Heekeren (1957) placed the fossiliferous and artefact-bearing stratum between 830 ka and 200 ka. Koenigswald initially assigned Verhoeven's finds to the Middle Pleistocene and then pronounced them to be between 830–500 ka old (von Koenigswald and Ghosh 1973). In 1968, Verhoeven teamed up with Maringer, and they excavated at the sites Verhoeven had discovered in the vicinity (Boa Leza and Mata Menge; Maringer and Verhoeven 1972; 1975). They noted a correspondence of their lithic industry from the Flores Soa basin with the Javan Pacitan tradition, which is universally attributed to *Homo erectus*, thus confirming the age estimates attempted at the time. These were decisively confirmed by the palaeomagnetic analysis of Sondaar et al. (1994) which placed the Mata Menge occupation close to the end of the Early Pleistocene, 780 ka ago.

Aware since the late 1970s of the critical evidence from Flores, we then realised that because nearly all of the relevant publications on the topic had appeared in German, Australian archaeologists commenting on the human colonisation of Wallacea and Australia had remained blissfully unaware of it for several decades. A couple of rather brusque papers from us attracted the ire of one scholar we had chastised (Groves 1995; cf. response in Bednarik 1995b). However, they also had the positive effect of a quantum leap in the understanding of Australian archaeologists, one of whom we were able to convince to abandon his work in the Kimberley in favour of excavating at Verhoeven's sites. His work resulted in further confirmation that hominins were very well established on many sites in Flores by up to 850 ka ago, as determined by fission track dating of sedimentary zircons (Morwood et al. 1998). That evidence came from the sites Boa Leza, Mata Menge, Koba Tuwa and Ngamapa and was subsequently extended to an age of about one million years at Wolo Sege (Brumm et al. 2010).

Whenever the initial colonisation of Flores occurred, it cannot mark the first crossing of the sea by a group of hominins, because to arrive there – by whatever route and at whatever sea-level – other stretches of the sea had to be crossed first.

By far the most likely alternative is to go first from Bali to Lombok, then to Sumbawa, and from there to Komodo which was connected to Flores in the past. Bali was part of Java during lower sea-levels, which in turn was joined to the Asian mainland, which explains the rich eutherian fauna up to Bali. Lombok is clearly visible from Bali, and both Sumbawa and Komodo can be seen from the preceding islands. Even Timor, separated from Flores and Alor by Ombai Strait, a deep graben of 3,000 m, is visible from the shore of departure. Timor, too, was colonised by both proboscideans and hominins in the Middle Pleistocene (Bednarik 1999; 2014d) (Figure 2.11). From there, the great southern land would have beckoned, but there was never any visual contact, and that apparently delayed the settlement of Australia for hundreds of thousands of years. It is assumed on this basis – the target coast having to be visible to inspire confidence – that the Bali-to-Timor route is the most likely taken. Thus, the first successful maritime colonisation most likely occurred by crossing Lombok Strait, where the Wallace Line separates Bali and Lombok, and it took place more than one million years ago.

Figure 2.11 The first stone tool of the Middle Pleistocene found in Timor, at Motaoan

There is, however, one other contender. The island of Socotra lies about 80 km east of the Horn of Africa (Cape Guardafui) and c. 380 km south of Arabia (Ras Fartak), separating the Gulf of Aden from the Indian Ocean. In 2008, Russian archaeologists discovered two major surface occupation sites comprising thousands of Oldowan-type stone tools near Wadi Tharditror. They are located in the island's central part, in the vicinity of Raquf village. Another concentration of the lithics was found at Wadi Aihaft near Nojhar village. The implements lack any bifaces, and the industry is dominated by unifacial and bifacial choppers similar to the earliest assemblages on the African mainland (Amirkhanov et al. 2009). If their typology is a reliable indicator of antiquity, this occupation might precede the first settlement of Wallacea. However, typology alone is not a reliable indicator of antiquity, because more recent traditions could be technologically impoverished versions. Nevertheless, Socotra has never been connected to any mainland for as long as hominins have existed. The island can safely be assumed to have been colonised by them in the Pleistocene.

Similarly, the Mediterranean islands of Crete and nearby Gavdos have recently yielded extensive evidence of Lower Palaeolithic settlement (Mortensen 2008). Like Socotra, Crete has not been connected to any landmass for millions of years. It has been sundered from the Greek mainland since the Messinian (Miocene, 7.2–5.3 Ma ago). However, a primate, known only from its footprints, existed on Crete about 5.7 Ma ago. Its track suggests that it may have been a hominin, preceding similarly evolved species in Africa by several million years (Gierliński et al. 2017). Be that as it may, the hominin presence on the island during the Middle Pleistocene has been documented by two separate teams of researchers (Strasser et al. 2010; 2011; Strasser 2012; Simmons 2014). It consists of hand-axes and other Acheulian tools excavated at coastal sites. Crete is visible from both Greece and Turkey but has never been visible from the northern coast of Africa. It was most likely reached from Greece via the small island of Andikíthira. To the south of Crete, 39 km away, lies the small island of Gavdos in the Libyan Sea. It, too, has provided evidence of early human colonisation, in the form of Lower and Middle Palaeolithic type lithics (Kopaka et al. 1994–1995; Kopaka and Matzanas 2009).

These two Greek islands are not the only Mediterranean isles apparently colonised before the Middle Palaeolithic. A bone from Nurighe Cave near Logudoro on Sardinia has been suggested to be of *Homo erectus* (Ginesu et al. 2003). It coincides with reports of Lower Palaeolithic stone tools from six sites near Perfugas in northern Sardinia (Arca et al. 1982; Martini 1992; Bini et al. 1993; cf. Cherry 1992). At lower sea-levels, Sardinia was periodically connected to Corsica ('Corsardinia') which could have been reached from Elba, then part of the Italian mainland. The distance, c. 35 km, was similar to that managed by the first human colonisers of Lombok more than one million years ago. Then there is the single unifacial chopper excavated near Korission Lake on the Greek island of Corfu, dated to the early Brunhes palaeomagnetic chron (Kourtessi-Philippaki 1999). Although not strong evidence by itself, it derives support from the presumably Lower Palaeolithic pebble tool industry previously reported from this island (Cubuk 1976). The stone tool traditions reported from another Greek island, Euboea, feature Lower and Middle Palaeolithic typologies (Sampson 2006). There is also sound evidence of occupations of these periods at Rodafnidia on Lesbos, which includes many Acheuloid hand-axes (Galanidou 2013), but that island was connected to the Turkish mainland at times of lower sea level.

Middle Palaeolithic occupation evidence occurs on many Mediterranean islands, including Kefallinía (Kavvadias 1984; Warner and Bednarik 1996), Corsica (Coscia Cave; Bonifay et al. 1998), the Sporades (Panagopolou et al. 2001), the Ionian Islands (Ferentinos et al. 2012) and the Cyclades (Seferiadis 1983), besides Gavdos and Euboea, as already mentioned. Therefore, the collective evidence from numerous islands in the Mediterranean indicates that many were settled by hominins of the Middle and even Lower Palaeolithic periods, confirming their ability to colonise islands successfully. Indeed, in the general region of Sahul (Greater Australia, including New Guinea, probably first occupied around 60 ka ago), the abilities of seafarers of essentially Middle Palaeolithic or Mode 3 technologies to cross the open sea are indicative of incredible maritime accomplishments (in the Sahulian region, Mode 3 industries continued to mid-Holocene times). Some of these islands could only be reached by crossing vast stretches of ocean. For instance, the small island of Buka is 180 km from the nearest land, New Ireland. Kilu Cave on Buka yielded human occupation evidence of a technology similar to that used by Neanderthals in Eurasia, and was 29 ka old (Allen et al. 1988; Wickler and Spriggs 1988; Wickler 2001: 68). Similarly, Noala Cave on tiny Campbell Island, 120 km from Australia in the Indian Ocean, was occupied 27 ka ago (Lourandos 1997).

In short, the maritime exploits of hominin colonisers using Lower or Middle Palaeolithic technologies refute the notion of most archaeologists that these people were primitive brutes, devoid of culture and not deserving of human status. Indeed, it needs to be asked how these sentiments arose in the first place. After all, the scientific definition of 'culture' as a transfer of practice by non-genetic means (Handwerker 1989) postulates that the simian habit of washing potatoes is deemed evidence of culture. Archaeologists deal with culture frequently, but they seem to have a completely different notion of what it is. Once again, the incommensurability of the humanities and the sciences is starkly apparent. Not only have they wholly different and incompatible concepts of the nature of culture, archaeologists even believe that culture evolves. It is therefore as difficult to communicate across the epistemological divide created by these and various other incommensurabilities as it is to communicate between belief systems (religions) and the sciences.

Precisely the same applies to the tendency of archaeologists to deny human status to all pre-AMHs, in the face of the scientific evidence that *Homo sapiens neanderthalensis*, *Homo sapiens heidelbergensis*, *Homo sapiens Denisova*, *Homo sapiens antecessor* and *Homo sapiens rhodesiensis* are all subspecies of the same species,

Homo sapiens. That species not only colonised the world; it produced and used exograms, the surviving traces of which we call palaeoart. Moreover, these same archaeologists reject palaeoart evidence predating the appearance of *Homo sapiens sapiens* because it contradicts their mantra that only the enlightened AMHs could have produced or used cultural materials. Not only are their views not backed up by any relevant empirical evidence; they were formed based on fake information (e.g. the claims of Reiner Protsch), and they were shaped from preconceived sentiments. How did all of this come about?

Robusticity and gracility

In Chapter 1, we noted the inconsistencies in biology concerning the definition of the term 'species'. In an ideal world, a species would be defined as a group of organisms consisting of individuals capable of fertile interbreeding, i.e. their offspring would be able to reproduce (Mayr 2001). This definition is clear-cut and precise, but, unfortunately, its application in biology is often vague and rarely tested genetically. In the academic system we have inherited, there is an incentive for an individual researcher to discover a new species, and nowhere is that more pronounced than in the exploration of human ancestry. Palaeoanthropologists can be divided into two groups: the 'splitters' strive to separate genuine species or promote subspecies to species level, and many of them have secured fame and success through discovering such 'new species'. They are counteracted by the 'lumpers', who endeavour to consolidate the numbers of putative species by grouping those that seem to be conspecifics. It comes as no surprise that in this wrangle, the lumpers are greatly outnumbered by the splitters, and also outperformed by them in terms of popularity. The media, understandably, is not interested in stories about a lack of new species, and the research funding agencies are guided by public interest generated by popular science magazines like *Nature, Science* and *Cell* (Schekman 2013).

Moreover, the splitters tend to define the lumpers as 'multiregionalists' in terms that imply that this is intended as a derogatory description. Perhaps they still misunderstand multiregional theory, perceiving it as the cantilever model invented by Howells (1959) when he misinterpreted Weidenreich's (1946) trellis model. Very simply, Howells ignored Weidenreich's numerous diagonal lines indicating genetic exchange between the populations of adjacent regions (Figure 2.12). Palaeoanthropology has suffered from Howells' mistake ever since, and most scholars still today fail to understand that every theory of the origins of 'modern' humans that does not regard them as a unique species is in reality multiregional. All alternatives proposed are simply variations of the multiregional theory with a substantial inflow of genes from one region (Relethford and Jorde 1999; Relethford 2001), entirely in agreement with Weidenreich's diagram. The only exception is the replacement or African Eve hypothesis, which now stands refuted: all humans of the past half-million years seem to belong to one species. Its promoters are now obliged to retreat, which they do ever so reluctantly.

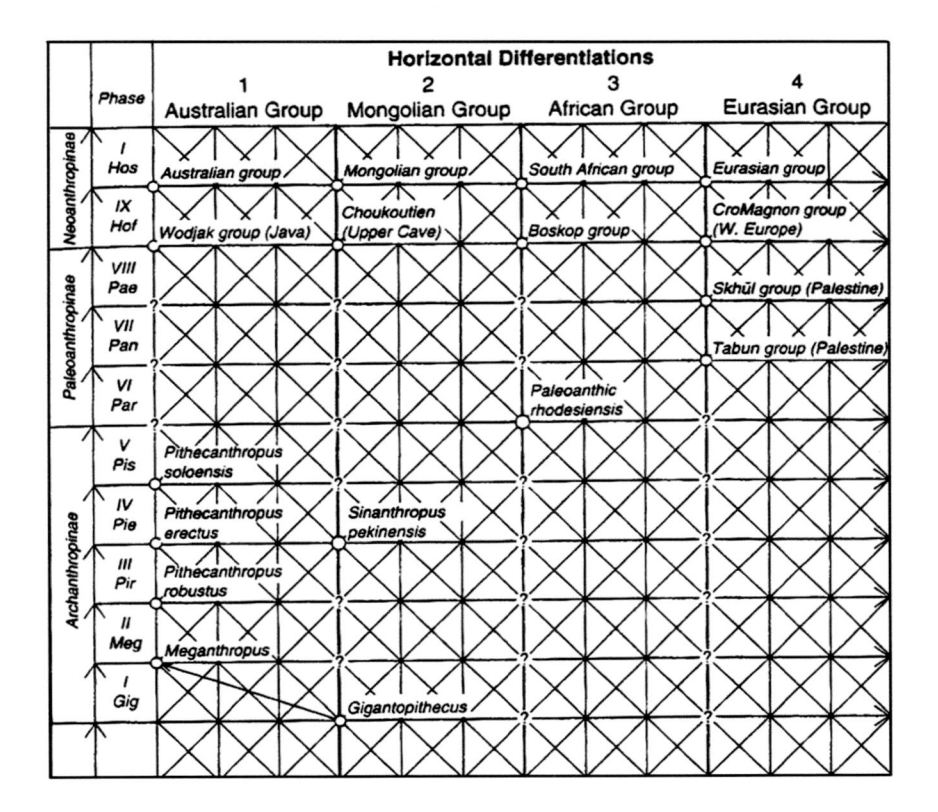

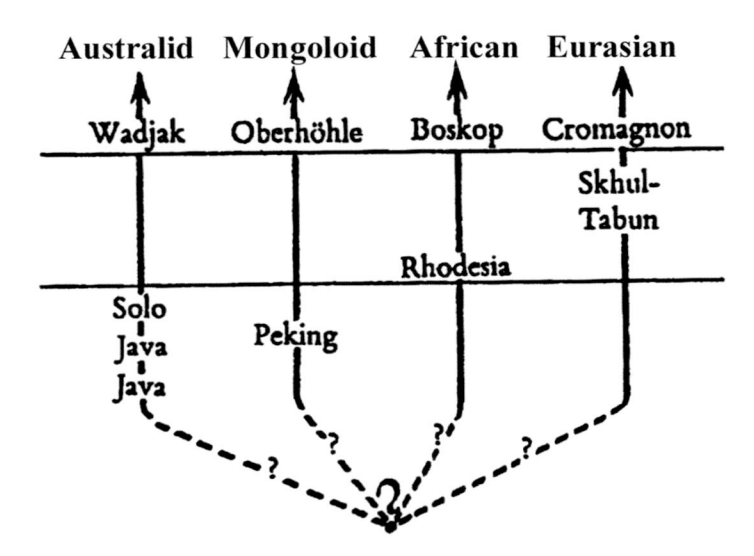

Figure 2.12 Weidenreich's (1946) original trellis model of hominin evolution (above) and Howell's false interpretation of it (below)

Degrees of robusticity or gracility are not reliable delineators of species. What is referred to in the context of human evolution is, of course, robusticity, or the lack of it, in skeletal remains, and the related issue of rugosity. The latter refers to the surface features of bone, in particular, fibrous and fibrocartilaginous enthuses (Imber 2003). We cannot know most details of *non-skeletal* robusticity/gracility. Skeletal robusticity defines the strength of skeletal elements relative to some mechanically relevant measure of body size. It is generally considered to reflect the magnitude of the mechanical loads that are habitually incurred by an element as the organism interacts with its environment. In the case of palaeoanthropology, most attention is given to the quantification of the degree of development of cranial superstructures. Cranial size is biased towards maximum cranial length, various measures of upper facial breadth, basion-nasion length and palate-dental size. The cranial superstructures are not only strongly influenced by specific cranial dimensions but are inter-correlated in their expression (Lahr and Wright 1996). Robusticity is variable across the skeleton within a single individual, between individuals in a population, between populations and between species/subspecies of hominins, all of which reflects great variation in the magnitudes of biomechanical loads and the behaviours that produce them. Some skeletal features, such as the cross-sectional dimensions of the long-bone diaphysis, are developmentally plastic and respond to applied mechanical loadings throughout life, i.e. to behavioural and environmental variables.

Therefore, indices of robusticity or gracility can derive from two sources: the individual's phylogenic history as encoded in its DNA, or its ontogenic history as an individual. This poses the first tangible difficulty in interpreting individual remains of the Pleistocene. We have absolutely no information about roughly one-half of the human populations of that geological period. This is due to the frequent sea-level fluctuations in the course of the Pleistocene, which erased all traces of societies that lived near the coast, along the lower reaches of major rivers or in deltas, i.e. the richest ecological zones. Pleistocene archaeology consists purely of propositions based on data about inland people, the tribes that moved with the animal herds. Absolutely nothing is known about the presumably more sedentary near-coastal tribes. They are likely to have been genetically different but were no doubt subjected to introgression and gene flow. What all of this shows is that the real genetic history of humans remains unknown, being a good deal more complicated than commentators tend to perceive. In this scenario, it is not readily apparent from a specific individual's degree of robusticity where in this artificial spectrum these particular remains belong. Countless modern humans are of a skeletal architecture that would identify them as Robusts. Some have indeed been identified as Neanderthals, for instance, the "northernmost Neanderthal ever found", from Hahnöfersand in Germany (Bräuer 1980), turned out to be a robust calvarium of the Mesolithic (Terberger and Street 2003). Another German 'Neanderthal' is the robust cranial fragment from Drigge, but only 6,250 years old (ibid.; Terberger 1998).

If we add to these complications the understanding that skeletal robusticity as perceived by the palaeoanthropologists is not a reliable measure, that it is quite variable within a population, and that it may not be reflected in the robusticity of the soft tissue, we realise how potentially threadbare their pronouncements may well be. However, the most debilitating aspect is the disconnect between these arbitrary assertions and the verdicts of geneticists, based mostly on present-day genetic data, about the genetics of robusticity (Groucutt et al. 2015). A better understanding of these issues is provided by a review of recent hypotheses concerning the purported places of origin of 'modern', gracile humans. It shows the wide divergences among authors presenting the replacement model, who seem to have exhausted all possibilities. Stringer (2011; 2014) is a vital advocate of the most extreme form, the 'recent' African origin model, complete with speciation event, subsequent dispersal from a central site and conveniently flexible timing. A more moderate version, in which the 'Moderns' are not necessarily a separate species, is suggested by Bräuer (1992), Green et al. (2010) and Reich et al. (2011), among others. The more realistic 'assimilation model' views the spread of 'modern' genes as a result, not of replacement, but gene flow or introgression (Smith 1992; Trinkaus 2005). Lahr and Foley (1998) perceive multiple dispersals by various routes from Africa, occurring between 70 and 50 ka ago. Klein (2009) and Shea (2011) believe that the exodus of the African superhumans only occurred after 50 ka ago, whereas Oppenheimer (2012) advocates a single migration between 75 and 60 ka. Mellars et al. (2013) adhere to a strict diffusionist model, in which microliths represent peoples, so they see a Hansel and Gretel kind of trail of microliths into Eurasia. In sharp contrast, Petraglia et al. (2010) and Reyes-Centeno et al. (2014) propose multiple waves of migration of 'modern' people with Middle Palaeolithic tool kits that occurred between 130 and 45 ka ago. Their model lacks the elements of technological and cognitive superiority of Eve's descendants. Similarly, Armitage et al. (2011) believe the Graciles left Africa via Arabia about 130 ka ago. In a variation of this variation, Usik et al. (2013) claim that the exodus via Arabia occurred 25 ka later, but still by people with Mode 3 technologies.

What all of this shows is that there are many versions of the African Eve model, some more radical than others, and there is much disagreement among the various protagonists. However, all of them emphasise the arbitrary dichotomy between robust and gracile forms and therefore gloss conveniently over intermediate ones. None of these alternatives is any more convincing than, for instance, Field's (1932) much earlier, superseded claim that the cradle of *Homo sapiens* was in the Middle East or northern Africa. In contrast to this Babylonian confusion, some authors have long maintained that humans evolved simultaneously in Eurasia, Africa and eventually Australia, with recurrent gene flow between many regions (e.g. Weidenreich 1946; Thorne and Wolpoff 2003). This emerges as the more compelling but significantly less favoured alternative. It is only with the present volume that the actual processes leading to fully Graciles are coherently explained by taking a completely different approach, one that is not predicated on imagined cultures, tool types or somatic variables.

60 The gracilisation of humans

The dichotomy of robust and gracile recent humans only has relevance as a statistical expression, taking into account the longer-term trend. In the early Holocene, around 10 ka ago, humans were on average some 10 per cent more robust than they are today. By 20 ka ago, their robusticity was in the order of 20 per cent greater, and this clinal trend continues back to the first half of the Late Pleistocene. This is of fundamental importance to understanding the emergence of what most call AMHs. The term itself is devoid of real meaning (Tobias 1995) and as Latour (1993) tellingly notes, "we have never been modern". The concept relies on a sophism: the unfounded belief of most archaeologists that they somehow share the consciousness of those they call 'modern'. In reality, present-day humans do not even share the cosmovision of people of the Middle Ages. Neuroplasticity is such that, for instance, literate and illiterate conspecifics have significantly different brain structures and chemistry (Helvenston 2013), and the neural systems of contemporary academic sophisticates and people who lived 30 ka ago can safely be assumed to be quite different. Those differences would likely increase as one goes back in time, just as robusticity increases gradually.

Therein lies the problem for the replacement advocates, whose catastrophist claim that all robust humans were replaced by gracile ones who could not breed with them is clearly false. These commentators are unable to name the time when this replacement is thought to have occurred; they are unable to name the African region where the putative speciation event occurred that generated their new species, and they cannot produce unambiguously fully modern skeletal remains that are more than 28 ka old. On the contrary, very robust populations continue to the end of the Pleistocene, e.g. the Red Deer Cave people in China's Yunnan Province (Curnoe et al. 2012). Moreover, the African Eve advocates ignore that all skeletal remains related to Early Upper Palaeolithic occupation sites are either robust or intermediate between robust and gracile. As we have seen in the first section of this chapter, the evidence from Europe offers indications of increasing gracility of 'Neanderthal' fossils from the time before about 40 ka ago (and perhaps more recently, such as the very gracile Neanderthals from Vindija Cave). Then follow roughly 10 or 15 millennia of specimens that are genuinely mid-way between robust and gracile, such as the mandible and separate partial cranium from Peștera cu Oase, Romania, or the six bones from Peștera Muierii in the same country. These are followed by numerous finds dated between approximately 30 ka and 25 ka ago that become increasingly gracile with time, including the Crô-Magnon specimens and scores of finds from central Europe, especially the Czech Republic.

What is particularly striking about this development is that the rate of gracilisation is much higher in females. The males retain significant robusticity well past 30 ka and in many cases are distinctively Neanderthaloid: consider, for instance, Crô-Magnon cranium 3, Mladeč 5 or Dolní Věstonice 16. Male crania remain characterised by thick projecting supraorbital tori, posterior flattening, low brain cases and very thick cranial vaults (Trinkaus and Le May

1982; Smith 1982; 1985; Frayer 1986; Jelínek 1987; Jelínek et al. 2005). As in Neanderthals, the cranial capacities of these male specimens exceed those of AMHs. This sexual dimorphism, specially marked between c. 32 ka and 25 ka ago, is quite pronounced (Figure 2.13).

Nevertheless, the females at 25 ka ago are still more robust than the average male European is today. The overall reduction in robusticity is distinctively lower in some recent groups, such as the Aborigines of Australia or the Fuegians of South America. One of the many empirical observations that discredit the replacement hypothesis is that Robusts survived in the Old World continents long after supposedly modern humans reached Australia. Clearly, the relationship between robust and gracile populations is not remotely as simple as that hypothesis would predict. The clinal and very gradual change evident for perhaps the last 50,000 years is as apparent as the clinal development of culture and technology. Nowhere in Eurasia is there a sudden appearance of a new invasive population, nowhere do we see a sudden appearance of a technology introduced from Africa. On the contrary, traces of Upper Palaeolithic technological skills appear in places such as central Europe and southern Siberia as early as 30 millennia before the Middle Palaeolithic/Middle Stone Age ends right across northern Africa 20 ka ago.

Bearing in mind that by 130 ka ago, robust hominins had managed to occupy the Arctic Circle (Pavlov et al. 2001; Schulz 2002; Schulz et al. 2002) and Tibetan highlands 160 ka ago, it is entirely unrealistic to expect that there were any unoccupied desirable lands in Eurasia during the entire Late Pleistocene. If robust humans accepted the incredible hardships of living in regions of temperatures below 40°C, we could safely assume that all other land was occupied, with the possible exception of very high altitudes and fully arid zones. In other words, the notion of unclaimed land that new invaders could effortlessly colonise lacks a logical basis: the human occupation of Eurasia was substantially complete, and the stories of very thin human populations are

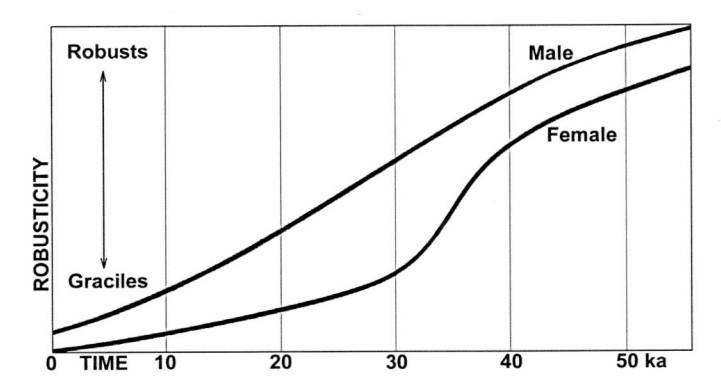

Figure 2.13 Male and female relative cranial robusticity/gracility in Europe during the final Pleistocene

simply inventions to justify the African Eve narrative. We can also be sure that the residents of Eurasia 40 ka ago were very well established and genetically adapted, having lived in these lands for hundreds of thousands of years. The notion that they were overwhelmed and entirely wiped out by trickles of naked Africans who lacked genetic and technological adaptations to cold climates is not only demographically absurd; there is no evidence of technology in Africa, or in the Levant, or especially in northern Africa, of Upper Palaeolithic traditions preceding those of Eurasia. Nor is there any evidence of Upper Palaeolithic rock art in Africa, either before or after the purported invasion.

The complete lack of any African precursors of the Eurasian Upper Palaeolithic and the extensive evidence of local developments from Middle to Upper Palaeolithic technologies in many centres across Eurasia already render the African Eve model without archaeological support. An example is the tradition of cave bear hunters in central Europe, especially in the Alps. Usually defined as Alpine Palaeolithic or Olschewian (Bayer 1929; Bächler 1940; 1957), the pursuit of cave bears in their hibernation dens, especially high up in the mountains, accompanied by a distinctive set of behaviour patterns (e.g. deliberate deposition of bear skulls) was begun by 'Neanderthals' in the Last Interglacial, especially in a series of high-elevation caves in Switzerland. The Olschewian of Vindija in Croatia has provided several fragmentary human remains interpreted as 'Neanderthals', while that of the roughly contemporary occupation of nearby Velika Pećina yielded a frontal fragment of a supposedly 'modern human' (Karavanic 1998; Devièse et al. 2017). The continuation of the Olschewian tradition in the Würm I/II Interstadial (Göttweig) is best represented in the eastern Alps and surrounding regions (Valoch 1968; Rabeder 1985; Markó 2015) and continues almost to 30 ka ago (Fladerer 1997; Pacher and Stuart 2009; Rabeder and Kavcik 2013). At that time, its makers would be expected to have been gracile humans, although no gracile remains have so far been reported with this tradition. Nevertheless, the Alpine Palaeolithic or Olschewian tradition spanned not only the transition from the Middle to the Upper Palaeolithic, but it also witnessed the gracilisation of the Neanderthaloid peoples of central Europe.

Numerous other examples can be cited that illustrate the complexities of these anatomic and cultural transitions. The evidence from Iberia, the Levant, China, Australia and elsewhere, that robust peoples coexisted with more gracile populations and even produced the same tools and forms of palaeoart supports the view that the physical differences were of little, if any, consequence to the development of culture. What the evidence suggests is that there was a broadly based clinal change from robust to gracile that is continuing today. It took place in all parts of the world then occupied by humans, including Australia, and it did so during much the same time. A mass movement of people does not account for it, nor does technological or cognitive 'superiority'.

The one factor that persuaded the African Eve's advocates that they were on the right track was that the change from robust to gracile took so little time, relatively, that it seemed impossible to attribute to Darwinian natural selection. Therefore, they figured, they were on safe ground because the only alternative they perceived was an intrusive population that replaced the resident one. Everything else about the African Eve fable is made up *ex post facto*. Here we will consider that an alternative to the replacement idea does exist to explain the change from robust to gracile – and all the available empirical evidence will even support it. More than that: it will provide realistic explanations and answers for numerous questions and apparent paradoxes, as we will soon see.

Human neoteny

Here is the first issue for which the replacement hypothesis has not found an explanation: why did the metamorphosis from robust to gracile humans occur? Why were features such as globular and thinner skulls, less muscular bodies, flatter faces, more fragile skeletal structure, or smaller teeth and brains selected? To say, because a new species evolved is not an answer; it is an evasive response. Science is concerned with causes and effects, and the African Eve advocates have sidestepped the question of cause. They have, therefore, explained nothing; they have simply proffered a just-so explanation: for unspecified reasons, a new, rather less well-equipped species arose, and it outcompeted or eradicated all other hominins. What were the selective pressures that favoured what looks like a collection of unfavourable traits – and that is before we consider the even more devastating genetic impairments that humanity was saddled with as a result. The only answer we have seen to this is that the new species possessed language and was more intelligent. Leaving aside the linking of language origins to the advent of the Upper Palaeolithic, which practically nobody still defends today, what precisely is the evidence for this enhanced intelligence, and what caused it? Again, there is no clear answer, only accommodative and even circular reasoning. Nothing in the archaeological record proves that there was a sudden jump in intellectual capacity. We know, for instance, that Robusts, including Neanderthals, mastered complex technologies, such as the intricate preparation of resins, of composite tools, of making and using microlithic tools. They used talons and feathers of raptors, presumably for decoration, produced rock art and ornaments and sailed the sea, in some cases, over enormous distances to colonise new lands, among other evidence of a well-developed intellect. Unless we can detect evidence for a quantum leap in the expressions of intellect, we once again fail to pinpoint the timing of the change-over. On the contrary, if we compare the complexity of the palaeoart at the beginning of the Upper Palaeolithic with that of later times of that period, we find that it was highest at the beginning and that this palaeoart is apparently the work of relatively robust people (Bednarik 2007).

In effect, the replacement hypothesis has not explained why the conversion from robust to gracile human in an unspecified part of Africa ever occurred –

what favoured the natural selection of Graciles, why did the purported speciation happen in the first place? It has not even been attempted to clarify this in Darwinian terms, yet the change is attributed to natural selection and genetic drift (Bednarik 2011d). Another detail the replacement advocates have never explained is their presumption of an inevitability of one species outcompeting another completely and utterly. Why then did the earliest hominins not outcompete the apes? Why did one grass-eater not outcompete all others? While it is certainly possible that one species succumbs to the competition of another, this is not at all inevitable. In this volume, we shall attempt to peel away the obfuscating elements and instead explore the causal basis of recent human gracilisation.

Many years ago Desmond Morris created a famous meme with his term 'naked ape' (Morris 1967), the major effect of which it was that the broader public began seeing the human species for what it is biologically: a primate. More recently, that perspective was sharpened by defining the modern, gracile human as 'the neotenous ape' (Bednarik 2008b; 2011c). Neoteny, i.e. pedomorphism, foetalisation or juvenilisation, defines the retention into adulthood of juvenile or foetal physiology (Gould 1979; Ashley Montagu 1989; Ashley Montagu 1997). Humans resemble chimpanzees anatomically most closely in the latter's foetal stage (Haldane 1932; De Beer 1940; Ashley Montagu 1960). The skull of an unborn ape lacks the prominent tori of the adult ape, is thin-walled, globular and thus resembles the cranium of a modern human. Upon birth, its robust features, including bone thickness, develop rapidly. The slow closing of the cranial sutures characteristic of humans after birth is another neotenous feature (controlled by genes RUNX2 and CBRA1). The face of the ape embryo forms an almost vertical plane, with prognathism developing after birth and fully becoming expressed in mature apes. The face of the modern human remains flat throughout adulthood. The brains of foetal apes and adult humans are also much more similar to each other, in terms of proportion and morphology, than they are to those of adult apes. Both the foetal chimpanzee and the adult human have hair on the top of the head and the chin but are otherwise mostly naked. All male apes have a penis bone (*os priapi*), but it is absent in both foetal chimpanzees and most humans. The penis bone of the ape is one of the last parts to form in the ape foetus, yet in the human, it is typically absent and seems to have been compensated for by the significantly increased length and thickness of the penis, relative to apes (Badcock 1980: 47).

Similar preservation of foetal features is evident in the female human genitalia. The hymen is a feature found only in the neonate ape, whereas in the human female it is retained for life in the absence of penetration. The *labia majora* are an infantile feature in the female chimpanzees, whereas in humans they are retained for life. The organs of the lower abdomen, such as rectum, urethra and vagina, are typically aligned with the spine in most adult mammals, including all apes. It is only in foetal apes and humans that they point forward relative to the spine. The human ovary reaches its full size at the age of 5, which approaches the age of sexual maturity of the apes (De Beer 1940: 75). An important indicator of human neoteny relates to the extremities. The legs of unborn apes are relatively short, while the arms are about as long in relation to the body as they are in humans. In the apes, the arms become much longer after birth, but this change does not occur in the human. Similarly, human hands and feet resemble those of embryonic apes closely, but differ very significantly from the hands and feet of mature apes. Of particular interest is the observation that the human foot retains throughout life the general structure found in foetal apes. The significance of this is that it contradicts a key hypothesis in hominin evolution, the hypothesis that the human foot is an adaptation to upright walking. If humans are neotenous apes, as the empirical evidence tends to suggest, it would be more likely that upright walk is an adaptation to neoteny of the hominin foot. This alternative has never been considered by paleoanthropologists – one of many examples featured in this book when lateral thinking can find more credible alternatives to established dogma.

Various other features in humans are neotenous; even the shape of the cartilage of our ears is a neotenous feature. Numerous other species have been subjected to neoteny, which has played an essential role in general evolution. The perhaps most obvious cases are the domesticated vertebrates, whose somatic and behavioural differences vis-à-vis their progenitor species are virtually all attributable to neoteny. As a phylogenic development in which foetal characteristics remain into adulthood, specific processes of anatomical maturation are retarded in neoteny (de Beer 1940). However, it also leads to behavioural effects. In particular, neoteny also favours the retention of plasticity or "morphological evolvability" (de Beer 1930: 93). It is therefore assumed that adaptively useful novelties become available as maturation genes are freed by pedomorphosis (Bednarik 2011c).

Relative to the neonate ape, the newborn human is not remotely as far developed. In contrast to the infant ape, it could not possibly cling to the fur of a mother for many months, if not years. Following the first year after birth, the human brain more than doubles in both volume and weight. This enormous increase, unparalleled in other animals, is due to the foetus having to be expelled relatively early in its development. The time of birth in humans is a finely tuned compromise between encephalisation and the obstetric limitations of the birth canal. After the first year, the brain continues to grow, approaching adult size by the age of 3, but then expands still slightly more up to adolescence and even beyond. This extraordinary development is unheard of in the rest of the animal kingdom. For comparison, rhesus monkey and gibbons achieve 70 per cent of adult brain size at birth and the remaining 30 per cent in the subsequent six months. In the larger apes, the size of the brain approaches adult size after the first year of life. Humans mature sexually about five years after chimpanzees do, and our teeth also erupt much later. Nearly half the genes affecting neoteny are expressed later in life in humans than in apes or monkeys, with some of this genetic activity delayed well into adolescence. The brain transcriptome (the total of all messenger RNA molecules) is dramatically remodelled during postnatal development, and developmental changes in the human brain are delayed relative to other primates (Somel et al. 2009).

To summarise the neotenous traits that characterise modern humans: they include the thinness of skull bones, flattened and broadened face, lack of tori, relatively large eyes, smallish nose, small teeth and jaws, larger brains, and limbs that are proportionally short relative to the torso, especially the arms (Gould 1977; Ashley Montagu 1989; de Panafieu and Gries 2007). The acquisition of some of these characteristics, most especially encephalisation, involved enormous evolutionary costs (prolonged infant dependency, reduced fertility, obstetric fatalities, and so forth; Joffe 1997; O'Connell et al. 1999; Bednarik 2011c). These deleterious developments need to be explained because many of them utterly contradict the principles of evolution.

However, the increase of cranial volume, in particular, is not a feature of recent human evolution. It is, in fact, one of the most frequently cited indices of human evolution, characterising the entire history of the genus (De Miguel and Henneberg 2001). It is almost without parallel in the natural world, the horse is the only animal that experienced greater encephalisation. This relentless increase in brain size had led to a volume of about 750 ml by the beginning of the Pleistocene (Leigh 1992; Hawks and Wolpoff 2001; Rightmire 2004), only to again double subsequently (Ruff et al. 1997; Lee and Wolpoff 2003). The last 800 ka account for much of the increase, which during that time averaged about 7 ml in brain capacity per 10 ka (although the increase was not linear). This significant enlargement of an energetically expensive organ is of great significance for two reasons: it has been credited with underwriting the sophisticated cognitive and intellectual development of hominins, but it also came with high biological, metabolic, social and neurological costs (Bednarik 2011c). Nevertheless, the continuing existence and flourishing of our genus seem to imply that these costs were outweighed by the considerable benefits encephalisation also engendered.

This refrain has been so central to all considerations of human evolution for over a century that it has become the teleological chorus of the discipline, the obligatory hymn to the grandiose self-delusion that dominates Pleistocene archaeology. We are told that it is this great enlargement of our brain that has made anatomically modern humans the masters of the world. However, *nothing could be further from the truth*. Since the time we became 'modern', by whatever process, our brain has continuously shrunk in size (Bednarik 2014e). During the terminal Pleistocene and the following Holocene, our endocranial volume decreased by about 13 per cent, i.e. at a rate *37 times greater* than the previous rate of increase (Beals, Smith and Dodd 1984; Henneberg 1988; 1990; Henneberg and Steyn 1993).

So, in effect, we have been led to believe that our modern lineage outcompeted the robust humans because our ancestors, thanks to their bigger brains, outsmarted those primitive Neanderthals. Not only did they have significantly smaller brains than the Neanderthals; the atrophy of their brains then continued and even accelerated right up to the present time. Instead of attributing our triumph over the benighted Neanderthals to our larger brain, should we perhaps ascribe it to our *smaller* brain? We could say that there are evolutionary advantages in a smaller, more efficient brain, although that would prompt the awkward question, why did hominins then first have to suffer the dire consequences of encephalisation? This is the sort of reasoning the discipline usually engages in to wriggle out of its flawed models. Not only have we been misled about the role of brain size in human evolution; the discipline will have to explain its monumental self-contradiction. It also needs to clarify how the present, apparently accelerating brain atrophy can be squared with our supposed cognitive superiority. Is higher intelligence perhaps a function of smaller brains? And, just to confuse the issue a little more: when the ratio of average brain diameter (calculated as a cube root of average cranial volume) to average body height is compared for various hominin species, there is no significant difference from *Australopithecus afarensis* to modern humans (Henneberg and Saniotis 2009: Figure 2).

This is just one example of how archaeologists and palaeoanthropologists completely misinterpret the empirical evidence and base far-reaching ideas on an inadequate understanding of scientific information; we will encounter many more in the course of the next chapters of this volume.

3 Evolution and pathologies

Narratives of evolution

Paleoanthropologists used to distinguish between the subfamilies of Hominoids (humans and their ancestors) and Anthropoids. The latter comprised the chimpanzee, the gorilla and the orang-utan, according to the old Linnean taxonomy. Since molecular DNA studies have shown that humans, chimpanzees and gorillas are genetically closer to each other than each of them is to orang-utans, the subfamilies Pongidae (orang-utans) and Homininae (humans, their ancestors, chimpanzees and gorillas) have replaced the old system. The latter are divided into Hominini (humans and their ancestors), Panini (chimpanzees and bonobos) and Gorillini (gorillas). A hominin, therefore, is a creature agreed to be either human or a potential human ancestor; it is of the genera *Homo, Australopithecus, Paranthropus* or *Ardipithecus*.

Human evolution took place principally during the Pliocene and Pleistocene periods. The first geological epoch began roughly 5.3 Ma (million years) ago and ended 2.5 Ma ago; the second, also known as the 'Ice Age', ended only 11.7 ka (thousand years) ago and gave way to the current epoch, the Holocene. Two potential ape precursors of the hominin line are *Graecopithecus freybergi* (7.2 Ma) and *Sahelanthropus tchadensis* (7.0 Ma, but uncertain). Since the first was found recently in Greece and Bulgaria (Fuss et al. 2017), the African origin of the hominins is no longer assured. Most palaeoanthropologists have been Africanists for more than half a century, and they resist any alternative model. When apparently hominin and very well-dated tracks (see Chapter 2) were found at Trachilos on the Greek island of Crete several years ago (Gierliński et al. 2017), the 'Africanists' managed to prevent their publication for more than six years through the peer review system. These tracks are 5.7 Ma old, pre-dating the Pliocene, and if they are indeed hominin, they would also question the out-of-Africa hypothesis.

During the Pliocene, the australopithecines and *Ardipithecus* flourished in Africa, and by about 3.5 Ma ago, *Kenyanthropus platyops*, currently the earliest species thought to be directly ancestral to the human line, appeared in east Africa. The position of the current two *Ardipithecus* species, *Ardipithecus ramidus* and *Ardipithecus kadabba*, remains controversial. They lived about 4.4 Ma ago, like the preceding species in forests rather than savannas, and since they also walked upright, this seems to contradict the hypothesis that erect bipedalism is a response to a reduction of woodlands. The gracile australopithecines commenced at about 4.2 Ma. With a brain little more than a third the size of a modern human, they were undoubtedly erect bipeds, as especially the Laetoli tracks amply demonstrate. Several species are being distinguished, the most recent still living 2.5–2.0 Ma ago and found with stone tools. Apparently, the gracile australopithecines evolved into the later robust forms, now subsumed under the genus *Paranthropus* (Henneberg and Thackeray 1995). They were more muscular, their skeletal remains are somewhat larger and more robust, and they developed alongside human species.

The fully human species that existed alongside australopithecines and then later coincided with *Paranthropus* were *Homo habilis, H. rudolfensis* and *H. ergaster*, dating from about 2.5–1.6 Ma. Numerous stone tools of the Oldowan tradition have been found with some of these remains. Some researchers consider them a single species, but the 'splitters' prevail (see Chapter 2). Sexual dimorphism, the great physical differences of the sexes, reduced gradually with the australopithecines and had largely disappeared by the time of *H. ergaster*, who may have continued to 1.2 Ma. This species developed typologically distinctive stone tools, most importantly bifaces (the famous hand-axes) and appeared as *Homo erectus* in Asia. Before the 1950s, Asia shared with Africa the status of a potential theatre of initial human evolution. Finds from two Chinese sites, in Renzi Cave and the Nihewan basin, imply that stone tools there may be as old as 2.2–1.8 Ma. Human remains found with stone tools in the Longgupo Cave site, also in China, are also thought to be between 1.96 and 1.78 Ma old (Huang et al. 1995). If these dates were correct, the origins of the earliest credible human ancestor, *H. erectus*, would remain unresolved, as noted in Chapter 1: he may have originated in either or both continents, Africa and Asia. The remains of several hominins from Dmanisi, Georgia, referred to as *H. georgicus* and 1.7 Ma old, occur on the very doorstep of Europe.

When *H. erectus* appeared around 1.8 Ma ago, its brain capacity was about 850 ml, but its late representatives, in the order of 500 ka old or perhaps even more recent, ranged from 1,100–1,250 ml, i.e. well within the range of modern humans. The increase occurs without significant change in body size (although see Henneberg 1998; Henneberg and Saniotis 2009), *H. erectus* being roughly of modern height. The subsequent history of the grading of *H. erectus* into archaic *H. sapiens* remains unclear in Asia, where several finds contradict the replacement model. Among them are the Hathnora calotte from the Narmada valley in India (de Lumley and Sonakia 1985) and the tiny clavicle (Sankhyan 1999) from the same site and stratum. The former resembles an *H. erectus* with a 'modern' braincase (Bednarik 1997a); the latter is from a dwarf that was little more than 1 metre tall (Sankhyan 1997). It brings to mind the human remains from Liang Bua in Flores (Morwood et al. 2004), probably not of a species at all but a genetically challenged gracile human (Henneberg and Schofield 2008; Henneberg et al. 2014).

In Europe, *H. antecessor* from Gran Dolina, Spain, is dated up to about 1.2–1.1 Ma and has a brain size of just over 1,000 ml (Carbonell et al. 2008). It shares some features, especially triple-rooted molars, with the earlier *H. ergaster*, its double-arched brow-ridges with *H. erectus* and the much later Neanderthals. Other current contenders for the title of first European come from Orce in Spain (only the stone tools are accepted, a bone fragment supposedly matching the Dmanisi finds in age is controversial) and Ceprano in southern Italy. Their status and ages are fiercely debated. Nevertheless, the appearance of humans in south-western Europe with a corresponding lack of early occupation evidence in eastern Europe, the similarity of the tools and technological trajectories both north and south of the western Mediterranean (including, importantly, the use of beads), and the evidence of navigation of an inland lake then existing in the Sahara (Fezzan Lake in Libya; Werry and Kazenwadel 1999) render it likely that Europe was settled via Gibraltar. Around 600 ka ago, *H. heidelbergensis* appears in Europe with a brain capacity ranging from 1,100–1,400 ml, which is within the range of modern humans. This is probably already a subspecies of *H. sapiens*, as the subsequent *H. sapiens neanderthalensis* certainly is. Both *H. heidelbergensis* and *H. antecessor* may have practised disposal of their dead if only to repel scavengers, and they certainly built shelters. Many Neanderthals were deliberately buried, sometimes in cemeteries (Peyrony 1934). At Budrinna, in the Libyan Sahara, a 400-ka-old graveyard of more than 120 interments has been excavated next to a village of stone-walled dwellings of the Acheulian. Conservative archaeologists reject both the cemetery and the stone huts: conventional wisdom is that stone structures were invented in the Neolithic, just a few thousand years ago. However, evidence of stone-walled or other structures does occur at Lower Palaeolithic sites in several countries (Bednarik 2011c), weakening archaeological consensus.

There is also considerable ongoing discussion about the course that human evolution took. Some researchers believe that *H. ergaster* gave rise to *H. heidelbergensis* who went on to evolve into *H. sapiens sapiens* via *H. sapiens neanderthalensis*. They regard *H. erectus* as an off-shoot of the 'main stem' that eventually became extinct. Others consider both *H. ergaster* and *H. heidelbergensis* to be *H. erectus*; while others again see *H. heidelbergensis* as a subspecies of *H. sapiens*. The difference is between the 'splitters' and the 'lumpers'; those who tend to invent separate species where there are none, and those who take a more realistic view, namely, that there can be considerable variation within a species. The most extreme form of 'lumping' is to include *H. erectus* with *H. sapiens* (Wolpoff 1999) and is quite likely correct. However, this is not a popular view among paleoanthropologists, who derive their reputations from presenting and defending their claims, and whose individual prestige can be enhanced by having identified 'a new species'. It may be judicious to lean towards the 'lumpers' because the identification of a species is that its members can interbreed to produce fertile offspring. Since genetically widely separated species, such as the wolf and the coyote (and many others), can do so very successfully, it becomes apparent that the 'splitters' are probably wrong in the case of human evolution. They have made a career out of emphasising minor dissimilarities between fossils, whereas the 'lumpers' perceive geographical variations as they are found within countless non-human species. Moreover, the human fossil record remains woefully inadequate, and there is a great deal still to be discovered.

This is a brief overview of human evolution (for more detail, see Bednarik 1997a; 2014d; Wolpoff 1999; Eckhardt 2000), of the empirical data and their direct significance. It shows that despite the progress made since 1856, when certain bones were found in a limestone quarry in the Neander valley, we are not very much wiser about the general topic. This is not because of a shortage of finds; we have a good supply of skeletal remains from much of the world. It is not because evolutionary theory is inadequately understood; the principles are well understood by modern science. Nor is it because there is a lack of interest. Much enthusiasm and dedication have been invested in the evolution of our genus over the past one and a half centuries.

So why is it then that we have so many competing, wildly different explanations that an observer could be excused for comparing the field to a Babylonian confusion? Let us compare this discipline with others that are significantly younger, not much more than a third its age. Genetics, although founded by Gregor Mendel and Imre Festetics well over a century ago, only became a discipline in the early twentieth century, especially in the 1940s and 1950s. Today it is light years ahead of human evolution studies. Ethology became a scientific discipline only in the mid-twentieth century, after the seminal work of Konrad Lorenz and Nikolaas Tinbergen. Today it is a celebrated main pillar of biology. Alternatively, consider plate tectonics, and the sophisticated work it has produced since Alfred Wegener's pioneering work on continental drift was accepted in the mid-twentieth century, about two decades after his death. Modern geology would be unthinkable without plate tectonics.

These are three examples of disciplines that have massively outperformed palaeoanthropology in a much shorter time, and we need to ask why that is so. The answer might lie in Eckhardt's (2007) request for a theoretical framework for palaeoanthropology. He elaborates on the example of the putative *Homo floresiensis*, the 'Hobbit' of Flores claimed to be a Pleistocene species that survived as recently as 13 ka ago. With a cranial volume of only 430 ml and stone tools such as those found elsewhere in the world from that time, it is an absurdity, and yet nearly all palaeoanthropologists accepted this creature as a new species that had undergone eons of separate evolution, perhaps even since the time of the australopithecines. The pathological nature of the specimen was ignored. Just as in the similarly frivolous African Eve hoax (Bednarik 2013f), the mass media and the popular science outlets have exercised significant influence over how the story unfolded. There is this vicious circle of the media influencing the scholars concerned, the direction and priorities of their research

72 Evolution and pathologies

and the sales pitch of the projects and institutes involved. The spin doctors of research seem to determine its direction and priorities.

All of this accounts for the considerable gap between the available credible data concerning human evolution and the narratives created about this information. However, besides the tales provided by the humanities of archaeology and palaeoanthropology, there are also accounts by the sciences. The humanities tend to see evolution in purely Darwinian terms, whereas some sciences have developed more finely tuned frameworks. Before exploring these, it is expedient to examine the origins of evolutionary theory. It is almost universally attributed to Charles Darwin and Alfred Russel Wallace, which ignores many previous works. Even in the years after Darwin's seminal 1859 volume, evolutionary theory was defined as the "Mohammadan theory of evolution" (Draper 1875: 118). The general idea had been expressed since the heyday of Greek philosophy: consider the 'ladder of nature' (*scala naturae*) of Aristotle. Natural selection, environmental determinism, the struggle for existence, the transformation of species into others were all defined in Arabic literature, initially by Abu 'Uthman 'Amr Ibn Bahr Al Qinanih Al Fuqaymih Al Basrih, known as Al-Jahiz (776–869 CE). His *Kitab al-hayawan* ('Book of animals') is a seven-volume encyclopaedia of at least 350 varieties of animals. Al-Jahiz introduced the concept of food chains and also proposed a scheme of animal evolution that entailed natural selection, struggle for existence and possibly the inheritance of acquired characteristics. For instance, he noticed how the environment was responsible for the different human skin colours, he recognised adaptations, and he reported ethological observations. *One thousand years before Darwin*, Al-Jahiz believed that plants evolved from inanimate matter, animals from plants, and humans from animals. He stated:

> Animals engage in a struggle for existing, and for resources, to avoid being eaten, and to breed ... Environmental factors influence organisms to develop new characteristics to ensure survival, thus transforming them into new species. Animals that survive to breed can pass on their successful characteristics to their offspring.

Many Muslim scholars adopted Al-Jahiz's views subsequently, including Al-Farabi (870–950), Abu Al-Hasan Al-Mas'udi (d. 957), Muhammad al-Nakhshabi (tenth century), Ibn Miskawaih (d. 1032), Ibn Sina (980–1037), Al-Biruni (973–1048), Raghib Al-Isfahani (d. 1108), Ibn Tufayl (1100–1186), Ibn Rushd (1126–1198), Maulana Jalal Al-Din Rumi (1207–1273), Al-Damiri (1344–1405) and Ibn Khaldun (1332–1406). Sections of Al-Jahiz's book on evolution were also translated into Latin in 1617 (Paris) and 1671 (Oxford). Darwin's grandfather, Erasmus Darwin (1731–1802) encountered the idea, and his grandson proceeded to collect a great deal of evidence by observations to support this idea. It needs to be appreciated that Charles Darwin read Arabic and had direct access to Arabic literature. He was introduced to Islamic culture in the faculty of religion at the University of Cambridge and learnt Arabic in order to understand Islam. He was then a student of Samuel Lee, who was well versed in oriental sources (Zirkle 1941).

The idea of evolution had long been in the air, but religion, especially the Christian religion, prevented its fair consideration until Darwin and Wallace presented their overwhelming evidence in its favour. Indeed, the most admirable aspect of Darwin's work is that he overcame the limitations that society, culture and especially religion had imposed on him. Evolution refers to the changes in heritable traits of organisms occurring over generations that are due to natural selection, mutation, gene flow and genetic drift. Although this aspect of evolutionary theory has been honed and refined over the years, in recent decades it has been supplemented by further concepts addressing specific facets ignored by a hard-core Darwinian approach. Darwin himself would have approved of at least some of these considerations. Among them are in particular developmental systems theory (DST), niche construction theory and the gene-culture co-evolutionary model. DST embraces a broad range of positions that expand biological explanations of organismal development, recognising that development is a product of multiple interacting sources; that an organism inherits resources from the environment in addition to genes; or that it contributes to the construction of its environment. It thus replaces the overly restrictive focus on the genes with a model of interacting systems (Oyama 2000; Oyama et al. 2001). DST, therefore, raised some pertinent points, especially concerning the non-genetic inheritance of traits and the cybernetic feedback from organism–environment systems changing over time.

One of these factors raised by DST is considered in greater detail in niche construction, which has been presented as another major force of evolution (Odling-Smee et al. 2003), operating similar to natural selection. Niche construction is the process whereby organisms modify their own (and coincidentally that of other organisms) selectively active environment to such a degree that it changes the selection pressures acting on present and future generations of such organisms. This is of particular interest in the present context, because humans are arguably the most effective niche constructors, having created niches for themselves that have transformed the planet. Laland et al. (2000) see much of niche construction as guided by socially learned knowledge and cultural inheritance (cf. Silk 2007). The perhaps most effective niche created by hominins is their language. Bickerton (2010: 231) rightly emphasises that niche construction and language production are both autocatalytic processes: once started, they drive themselves, creating and fulfilling their demands.

Another view holds that, in addition to genetics, dimensions termed epigenetic (non-DNA cellular transmission of traits), behavioural and symbolic inheritance systems also moderate evolution (Jablonka and Lamb 2005). Until recent decades the idea that the medium of heritable variation may be something other than DNA, such as life experience, was considered Lamarckian and thus unacceptable. However, the trans-generational phenotypic effects that account for evolution have multiple origins and channels of transmission, and even those elements of the phenotype that are conveyed by the genes do not necessarily start that way. All organisms are said to be subject to epigenetic inheritance, which refers to the physiological/biological process beyond the level of DNA. Behavioural inheritance is found in most species and defines the transference of information or behaviour through learning rather than genetically. Research needs to focus on gene expression and associated genomic processes (like methylation, chromosomal condensation, alternative splicing and environmentally modulated mutation rates), converging with research on the development and the phenotype, including the behavioural phenotype. Symbolic inheritance is Jablonka and Lamb's third dimension and of particular importance to human evolution. Cultural development can have properties similar to evolution: cultural innovation (cf. genetic variation), cultural transmission (cf. heredity) and differential multiplication and survival of culture (cf. natural selection). For instance, some chromosomal regions are more sensitive to mutations than others due to epigenetic marks on DNA (methylation marks). Not only the rate and the location but also the timing of mutations can be non-random. The underlying contention of these ways of thinking is that evolution is not a purely genetic process relying on the appearance of mutations.

The idea that human evolution cannot be assumed to have been just a genetic process has been gaining momentum since the 1960s (Dobzhansky 1962: 18; 1972), but it has recently received a new impetus from increasingly sophisticated work. The notion of a progressive moderation of human evolution by culture is the central plank of the gene-culture co-evolutionary model or dual inheritance theory (DIT) (Boyd and Richerson 2005; Richerson and Boyd 2005). DIT is based on the principle that, in human evolution, genes and culture interact continually in a feedback loop. However, the emphasis on 'cultural evolution' is an encumbrance, because that concept is a misapplication of the term 'evolution'. 'Culture' is the non-genetic transmission of practice or socially learned behaviour, and it is distinctively teleological. That means its direction is at least in part the result of deliberate choices and that it strives for improvements (Gabora 2011). In sharp contrast, evolution is entirely dysteleological; it has no purpose, no ultimate goal; its operation is subject to entirely random variables. This difference is of fundamental importance to the sentiments of the present volume, the underlying purpose of which it is to explore the distinction between stochastic and systematic influences on human development.

While the concept of culture is widely applied to non-human animals, cultural adoptions of agriculture and dairying that caused genetic selection for the traits favouring digestion of starch and lactose, often cited in DIT, are limited to humans. DIT defines five significant mechanisms of 'cultural evolution': (1) natural selection; (2) random variation; (3) cultural drift; (4) guided variation; and (5) biased transmission. DIT advocates perceive similarities between these variables and processes of evolution. However, these purported parallels are coincidental, and it is precisely the tension between teleology and dysteleology that distances the two systems. For instance, random variation arising from errors in learning is very different from a mutation in genetic evolution.

Genetic drift is quite different from the vaguely defined process of 'cultural drift': the examples of changes in the popularity of American baby names (Hahn and Bentley 2003), archaeological pottery or the songs of songbirds illustrate this. These misconceptions derive from mistaken beliefs about the nature of culture. Culture is merely an epidemic of mental representations (Sperber 1996). More to the point, 'culture' is simply a short-hand way of expressing that certain information states in one person's brain cause, by mechanisms as yet unexplained, similar information states (or mental representations or 'memes') to be reconstructed in another brain. Humans are not passive receptacles filled by culture; these self-extracting systems cause them to be creating their own realities. Thus culture is self-generated content supplemented with low-cost useful information from others. The purported similarities with evolution are coincidental, and the comparison is misleading and technically wrong.

Not surprisingly, DIT has encountered considerable opposition (cf. e.g. Henrich and Boyd 2002; Gabora 2005; Mesoudi et al. 2006; Henrich et al. 2008). The underlying notion might have its origins in Dawkins' (1976) suggestion that genes propagate themselves in the gene pool by leaping from body to body via sperm or eggs, so memes propagate themselves in the meme pool by leaping from brain to brain. However, 'memes' are not discrete cellular entities, as genes certainly are. In contrast to genes, memes are continuous and change over time, and they are distributed and content-addressable (Gabora 2011). In evolution, what is transmitted from parent to offspring are sets of almost unchangeable self-assembly instructions. A concept of 'generation' cannot be applied to 'culture', because an idea or technology or other memes can disappear and reappear through time, which a gene cannot. Other differences between the mechanisms of evolution and culture concern replication and fitness. Not only is there no mechanisms for a cultural inheritance, the creative processes that effect change are clearly teleological, and far from random. Fuentes (2009) has sought to reconcile the pronounced duality of evolutionary biology and socio-cultural anthropology, pointing out that symbolic and other cultural processes influence behaviour and potentially physiological and even genetic factors. His demand that behavioural plasticity has a specific role in human behaviour again runs counter to neo-Darwinism.

The importance of these debates lies in reminding us that genetics cannot 'explain everything' in hominin evolution, apart from helping us to test competing hypotheses. Irrespective of one's position on any of the listed models, at least some of them cannot be ignored. For instance, there can be no doubt that niche construction plays a significant role in modifying the course of evolution. However, one more factor in evolution has received relatively scant attention, and that is the role of sexual selection (Darwin 1871; Fisher 1930; Miller 2000), in contrast to natural selection. In biological terms, the fitness of an organism is expressed in a straightforward effect: the number of fertile offspring it has in the next generation. The difference between the two forms of selection boils down to the differentiation: natural selection creates better survivors, sexual selection favours those that are more successful in achieving fertilisation. From the perspective of sexual selection, physical 'ornaments' and displays are forms of 'costly signalling' (Zahavi and Zahavi 1997; MacAndrew 2002; Bird and Smith 2005). Such signalling tends to be costly in terms of survival value; it demands metabolic energy and resources and can be quite detrimental to the individual concerned, and yet it is so widespread in the animal kingdom. In many species, it determines behaviour, just as combat between males for access to females is a dominant feature in the lives of the members of thousands of species. If we consider the many species in which such contests lead to the total exhaustion and often the death of the males concerned, it becomes apparent that natural and sexual selection are two alternative and, in some ways, competing avenues to succeeding in biological fitness, i.e. breeding success. The sexual selection strategy can clash severely with the survival strategy, and yet in many species, the former has been refined to great sophistication and elaboration. It tends to emphasise expressions of sexual dimorphisms, such as in plumage, size, aggression or robusticity, and sexual selection can even be found in plants and fungi. Sexual reproduction is not a 'rational' aspect of the natural world, because asexual reproduction is a much more efficient way of propagation (Hartfield and Keightley 2012), and the extremes of sexual selection seem to be inefficient uses of resources. Asexual reproduction may be more efficient in the short term, but it does not provide the necessary mechanism for compensation of numerous genetic defects occurring as a result of mutations being a result of the weak chemical structure of the DNA (M. Henneberg, pers. comm. 2019).

Fisher (1930), a statistical biologist, has explained some of the salient issues of sexual selection. He was one of the fathers of quantitative genetics and established the basic principle of natural selection, Fisher's Theorem. Fisher explained why the sex ratio between males and females is approximately 1:1 in most species that produce their offspring by sexual reproduction. If there were a scarcity of one sex, he showed, individuals of that sex would have better statistical prospects of having offspring than individuals of the other sex. Parents genetically predisposed to having more offspring of the scarcer sex would then have more grandchildren on average, increasing the relative number of that sex until a ratio of 1:1 would prevail in the population.

As a consequence of this effect, populations will strive for a 1:1 ratio of sexually mature males/females. Another of Fisher's insights, the 'sexy son hypothesis', explains that a female's best choice to succeed is a male partner whose genes will produce a male issue with the best prospects of reproductive success, i.e. the 'sexy sons' likely to generate larger numbers of progeny carrying the female's genes. Male caregiving, fidelity, gifts or territory, by contrast, do not benefit that expression of fitness. In his 'runaway' model, Fisher took this line of reasoning further, explaining why the costly display was not only tolerated by evolution but flourished. Darwin had been quite irritated by the example of the male peacock, confiding to a friend that the feathers in the bird's tail "made him sick". Obviously, the excesses of extreme costly display run counter to natural selection and often involve great disadvantages. Fisher explained this conundrum by positing that the peahen chooses the male with the most elaborate tail because its genetically predisposed male offspring will, in turn, attract more female attention, thus causing the 'runaway' effect by positive feedback.

These examples show the importance of sexual selection in considering evolution. The peacock's costly display strategy is repeated in thousands of other species across the animal kingdom, often with absurdly extreme outcomes. This and other factors show that the natural selection and sexual selection strands can be intricately intertwined. Just as it is essential to acknowledge that natural selection determines the survival prospects of the individual, it is also inevitable that sexual selection ultimately determines whose genes are carried forward. The latter's contribution to evolutionary mechanisms is therefore decisive: to win a race, one first has to be in it, but what counts in the race of evolution is still the winning. We will return to this subject in Chapter 4 when the issue of sexual selection is explicitly applied to humans. So far, we have considered these topics from the perspective of the natural sciences. As we are approaching the main topic of this volume, the direction humans developed through time, we are leaving most of the animal kingdom behind and focus on the primates. First of all, we need to consider the differences between humans and other primates from a perspective that is fundamentally important in understanding the 'human condition' (Bednarik 2011c) but has been severely neglected.

The brains of apes and humans

One of the principal preoccupations of archaeology has been to chronicle the ascent of humans from lowly apes to the very apex of evolution, God's masterpiece. In this, the discipline has been dominated by the creation of deep division between modern hominins and their benighted predecessors, robust hominins. This strategy of separating 'us' from the 'primitive other' is evident throughout the history of the field, from the utter rejection of the Altamira cave art almost one and a half centuries ago to the present rejection of any proposals of cognitive, technological or intellectual sophistication prior to the only hominin that can gain access to Heaven. As we have seen at the beginning of Chapter 1, the Altamira evidence was comprehensively denied because it was considered incompatible with the lowly cultural level of a Stone Age technology. The discoverer of the cave art, Marcelino de Sautuola, was accused of faking the rock art and died prematurely, a broken man. He was vindicated 24 years after his death (Bednarik 2013e).

We face a similar dilemma at present: a massive corpus of evidence has been assembled suggesting that much earlier hominins, including *Homo erectus* and very probably earlier forms, possessed far greater cognitive, intellectual and technological competence than archaeology is prepared to countenance (Bednarik 1992a; 1993a; 1995a; 1997a; 1999; 2003a; 2003b; 2005; 2011b; 2017a). Collectively, this evidence attributes the creation and use of ornaments (beads and pendants) and art-like productions as well as the ability to colonise new lands by maritime travel to hominins of the Lower Palaeolithic. It is widely rejected by Pleistocene archaeologists, not based on any proof or refutation, but because they take the position that there was a massive difference between the progeny of their African Eve and any preceding humans. In their view, this difference must be expressed in significant mental and technical disparities. That view is not based on evidence, but on the perceived absence of evidence, and when relevant proof is presented, it is explained away as a "running ahead of time" (Vishnyatsky 1994), or is rationalised as misidentifications, the dating of the material evidence is questioned, or the presenter's competence is queried. The only possibility not explored is that the dogma of excessive primitiveness might be wrong.

The "running ahead of time" argument is particularly illuminating, because of its absurdity. It effectively proposes that instances of symbol use occurred well before symbolling was introduced, essentially because some 'genius' individual invented this or some other innovation. To appreciate the shallowness of this argument, one might imagine one *Homo erectus* man 'inventing' the use of a bead on a string around his neck to indicate that he was the chief. If nobody else in his group shared his genius, how would he have conveyed this meaning? What is apparent from such trivial notions is that their proposers lack the understanding that meaning can only exist if it is shared, and "running ahead of time" is no credible explanation of anything in archaeology.

The differences between humans and the great apes have fascinated both the scholars and the public since the nineteenth century. Countless purported distinctions have been cited to differentiate between humans and apes. Today we regard practically all of them as mirages that dissolved as soon as they were examined more closely. Perhaps the most favoured discriminator variables were tool use, upright walk, forward planning, recursion, Theory of Mind, self-awareness, consciousness, symbolling ability and one specific application of this, language (Bednarik 2011c). Ethological research and especially primatology have broken down many, if not all, of these barriers between our ape cousins and us. Apart from some minor differences in the respective genomes, remarkably few real differences have remained, and they do not even attract much interest from researchers. One apparently fundamental contrast is that we seem to lack clear evidence that the great apes engaged in the production and use of exograms. Most archaeologists and other scholars concerned with hominin evolution seem to be unaware even of the existence of exograms. Exograms determine all of what is simplistically called culture – the entity we have noted above that needs to be expelled from a scientific lexicon because it is so poorly defined. If we replaced the ideas of 'culture' or 'meme' with an understanding of how exograms interact with the human brain, we might improve our comprehension of what this thing 'culture' is, and we might even retire the related idea of 'learning' (which is merely a built-in concept in the Theory of Mind system).

Exograms are memory traces stored outside the brain, and we will consider them in some detail later. For the moment we are only concerned with their role in defining the difference between humans and the great apes. Of course, we have no conclusive, evidence that the apes are incapable of creating or using exograms; perhaps such indications will become available some day. All we can say is that, on present indications, this does not seem very likely. So, what are the actual structural differences between the brains of pongids and recent humans?

Preuss (2001), who defines humans as the "undiscovered primate", considers that the many differences in the cortical organisation found in primates have been ignored in past years. He sees the prefrontal cortex and portions of the posterior association areas in humans enlarged beyond what would be expected in comparison to primary sensorimotor structures. The primary visual areas (Brodmann's area 17) in humans differ from both apes and monkeys in the way they segregate information from the magnocellular and parvocellular layers of the lateral geniculate nucleus (Preuss and Kaas 1999). The human visual striate cortex is 121 per cent less than expected for an ape of human size and the lateral geniculate about 140 per cent less than expected (Holloway 1995; 1996). The motor systems in humans are considerably more highly developed compared to apes, notably the phylogenetically newest area of the motor cortex, the pyramidal motor system. Although more efficient and advanced than that of apes, it occupies less cortical area, leaving more room for prefrontal and parietal association cortex in *Homo sapiens sapiens*. The caudal primary motor area is mediated by corticospinal efferents from 'old' cortical neurons in the extrapyramidal system. The striatum is densely interconnected with the globus pallidus and other nuclei and is known collectively as the basal nuclei. In generating motor neuron activity and motor output, it uses the integrative mechanisms of the spinal cord (Rathelot and Strick 2009).

According to Semendeferi (1997), the frontal lobes or divisions of them may not be larger than would be expected, and the same is suggested for the cortex in the frontal lobes. While anterior gross brain morphology is thought to have been similar for the past 300,000 years (Bookstein et al. 1999), the structures and connections of the frontal lobes, which may be relatively recent in evolutionary time, may be responsible for brain illnesses in humans. Their smaller structures in the frontal lobes such as Brodmann's areas 10 and 13 seem to be different in ape brains. Allman et al. (2001) have proposed that those areas of the anterior cingulate cortex that contain von Economo neurons (VENs) are a phylogenetically new specialisation of the neocortex rather than a more primitive state of cortical evolution, as most other areas of the cingulate cortex are (Nimchinsky et al. 1995). Area 13 in the orbitofrontal cortex is similar qualitatively in apes and humans. As part of the limbic system (Heimer et al. 2008), it is heavily involved in social, motivational and emotional behaviour.

The human limbic system is larger in absolute size than that of the great apes, but not as a percentage of overall brain weight. The cingulate gyrus, the phylogenetically oldest part of the cortex, is a significant part of the limbic lobe and is of substantial size in humans (Mega and Cummings 1997). Other structures comprising an extended and in some ways unique human limbic system (Heimer et al. 2008) include the nucleus basalis of Meynert, the bed nucleus of the stria terminalis, the hippocampus and the amygdala, the septum, olfactory nucleus and the entorhinal cortex. The hippocampus receives its input from the entorhinal cortex, which does so from the neocortex, and the entorhinal cortex is involved in spatial orientation. The entorhinal and parahippocampal neocortices are located in the inferior and medial temporal lobe, adjacent to and richly interconnected with the hippocampus and the amygdala. Like these, the entorhinal cortices are involved in memory functioning (Gloor 1990; 1992). The cerebellum is smaller in humans than would be expected in an ape of human body size, and is larger than expected in the gorilla (Semendeferi 2001a). Presumably, the cerebrum became enlarged in humans at the expense of the cerebellum which subsumes such functions as fine motor tuning, balance and some aspects of cognition. In humans, the dentate nucleus of the cerebellum is more substantial than would be expected for an ape of human body size, which is striking given the smaller cerebellum in humans. The latter is implicated in cognitive tasks (Leiner et al. 1995) and contributes to the routinisation of complex cognitive procedures, leading thereby to mental agility. It is involved in error detection and language, performing functions such as prediction of others' actions and preparation for behavioural responses (Mueller and Courchesne 1998).

The parietal association area is larger in humans than apes at the expense of the occipital cortices (Gannon et al. 1998; Preuss and Kaas 1999; Holloway 2001), although the latter, like the frontal lobes, are not particularly larger in humans than in apes, based upon body weight (Semendeferi 2001b). In comparing the planum temporale in macaques, apes and humans, Gannon et al. (2001) propose that the distinguishable suite of neurological features related to receptive language began a gradual evolution 20 million years ago. It resulted in the appearance of asymmetrical anatomy for the planum temporale 10 million years later, in the common ancestor of gorillas and humans. If this is a structure shared by the great apes and humans, it challenges the notion of a simple relationship between the asymmetry in this part of the brain and language ability. It renders a homology in the hominoid and hominin branches far more likely, and the development of reorganisation in *Homo* as expressed in speech. Significantly, when the homologue of Broca's area in the ventral premotor region of the owl monkey is stimulated, it produces oral and laryngeal movements (Stepniewska et al. 1993).

Also relatively larger in humans than in apes appear to be the temporal lobes (Semendeferi and Damasio 2000). The insula, in the lateral fissure of the temporal lobe, is large in humans and seems to be involved in the processing of autonomic functions and internal and olfactory stimuli, constituting part of the extended limbic system (Heimer et al. 2008). The nucleus subputaminalis in the basal forebrain is unique in humans and provides cholinergic innervation to the inferior frontal gyrus where Broca's area is located. Broca's area, of course, is involved in speech production. In the human posterior temporal lobe, we find Wernicke's area, responsible for the reception of sounds, particularly for language comprehension. Buxhoeveden et al. (1996) report that human evolution has selected in favour of integrative functions of specific layers of this area of cortex, facilitating integrative language reception and comprehension. However, in some 11 per cent of linguistically competent humans, these areas are not present in the left hemisphere. In hemispherectomy patients, no matter which hemisphere is removed in childhood, speech and language comprehension are not affected (M. Henneberg, pers. comm. 2019).

The anterior principal nucleus and the mediodorsal nucleus appear to be unique in humans (Armstrong 1980; 1991). The first is well connected with the cingulate gyrus, hippocampus, subiculum, mammillary bodies; and the prefrontal, parietal and inferior parietal cortices. The anterior principal nucleus appears to be active in the encoding of information and sustaining attention to sensory stimuli. The mediodorsal nucleus is heavily connected with the dorsolateral prefrontal cortex, the last element to myelinate in the course of human development. It is believed to be involved with short-term working memory, helping to keep the subject focused on the task at hand. Other features unique in the human brain are structures in the extended limbic system involving neural circuits that regulate emotions and complex social behaviour (Heimer et al. 2008), and layers of primary visual cortex V1 or Brodmann's area 17.

An important factor is the role of von Economo neurons (VENs; also known as spindle cells; Nimchimsky et al. 1999; Watson et al. 2006), apparently participating in very rapid signal transmission. In humans, they occur in three particular regions: (1) in the anterior cingulate cortex (Allman et al. 2002; Hayashi 2006); (2) the frontoinsular cortex (Sridharan et al. 2008); and (3) the dorsolateral prefrontal cortex, Brodmann's area 9 (Fajardo et al. 2008). VENs can be found in various large mammalian species possessing large brains and extensive social networks, including besides primates also whales (sperm and beluga whale), dolphins (bottlenose and Risso's dolphin) and elephants (both African and Asian). However, they occur in far higher numbers in extant humans than in other species. VENs have massive cell bodies evolved for rapid transfer of information to various areas of the brain and spinal cord. Those of the frontoinsular cortex, especially on the right side, seem to be critical in switching between distinct brain networks across various tasks and stimulus modalities (Sridharan et al. 2008). These findings could have important implications for a unified view of network mechanisms that underlie both exogenous and endogenous cognitive control, certainly a high-level cognitive capacity. VENs seem to play a role in intuition, which, like perceptual recognition, involves immediate, effortless awareness rather than the activity of deliberative processes (Allman et al. 2005; 2010). They are also active in social contexts and Theory of Mind activities.

VENs are barely present in humans at birth but then increase rapidly in number until eight months after birth. This may account for postnatal pathogenesis through disruptions in development. The brain's right hemisphere comprises much higher numbers of VENs than the left (Allman et al. 2010). VENs are implicated in several brain illnesses of humans, such as frontal-temporal dementia (ibid.), Alzheimer's disease (Nimchinsky et al. 1999), depression (Drevets et al. 1997), schizophrenia and possibly bipolar illness. However, as they are not limited to human brains, they can only be a part of the explanation, and their role in human characteristics is one of quantity, not of distinction.

Similarly, the much-discussed issue of brain size (Jerison 1973; Aiello and Dean 1990; Changeux and Chavaillon 1996) is not a simple factor of discriminating between human and non-human animals; it is a rather complicated, multifaceted subject. Subspecies of *Homo sapiens*, such as *H. sapiens neanderthalensis*, have or had the largest absolute and relative brain size of any animal that existed on our planet. The human brain is three to four times larger than that of any of the great apes, and the human cortex is ten times larger than the macaque cortex. Large brains have cell bodies that are more scattered, leaving room for many more interconnections between areas (Semendeferi 2001b: 114). Not surprisingly, the incredible process of hominin encephalisation (the increase in brain volume) has dominated the discussion for over a century, but there are different facets of this topic. For instance, the dominant notion of encephalisation driving higher intelligence needs to be tempered by the observation that when the ratio of average brain diameter (calculated as a cube root of average cranial volume) to average body height is compared for various hominin species, there is no significant difference between *Australopithecus afarensis* and modern humans (Henneberg and Saniotis 2009: Figure 2). Of even more consequence to the entire notion of human evolution is the observation that about 40–50 ka ago, the endocranial volume of humans stopped expanding and instead began to contract (Bednarik 2014e). This brain atrophy has been known for several decades (Beals et al. 1984) but has been almost totally ignored by archaeology and palaeoanthropology. Henneberg (1988; 1990) has reported a Mesolithic mean brain volume of 1,567 ml ($n = 35$); for Neolithic/Eneolithic males, it is 1,496 ml ($n = 1,017$); and there are further reductions in Bronze and Iron Age specimens to 1,468 ml, in Roman times to 1,452 ml, early Middle Ages to 1,449 ml, late Middle Ages to 1,418 ml; while the males of post-medieval and recent times yielded a mean of 1,391 ml (1,241 ml for females). This reduction of 261 ml represents 37 times the rate of the long-term increase in brain size during the second half of the Pleistocene, and it follows previous brain atrophy since the Neanderthals, whose brains ranged in size up to 1,900 ml (Bednarik 2014e) (Figure 3.1).

Human brain atrophy presents a massive problem for the simplistic interpretation of the human past, which hinges on the correlation between encephalisation and increasing cognitive, intellectual and memory resources. It is universally assumed that the demand for neural resources has increased significantly since the Neanderthals, and yet this period has seen a significant reduction in brain size. Does this mean that brain size is not relevant to mental resources, or does it mean that we have less of them than our predecessors? To appreciate the significance of this question, it must be remembered that the relentless increase of

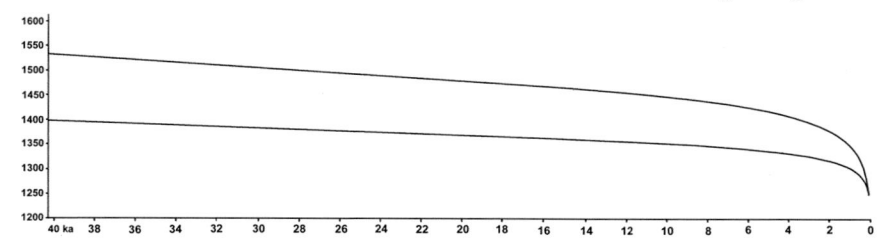

Figure 3.1 The atrophy of the human brain during the last 40,000 years: volume in ml versus age in 1,000 years; the upper curve represents males, the lower is of females

brain volume for some millions of years has involved enormous evolutionary and other costs (e.g. prolonged infant dependency, reduced fertility, obstetric costs; Joffe 1997; O'Connell et al. 1999; Bednarik 2011c) for the human line. These placed extraordinary constraints on human societies.

The encephalisation quotient (EQ) introduced by Jerison (1973) helps place the relative brain size into context. The EQ is the ratio of a species' brain size to 'expected brain size' based on average body size. The average mammal is the cat, with an EQ of 1.0. For the rhesus monkey with a brain volume of 196 ml, it is 2.09, and the EQ of the chimpanzee (440 ml) is 2.48. The present human brain's average of 1,316 ml scores a massive EQ of 7.30. However, it would still have been dwarfed by that of *Homo sapiens neanderthalensis* of 50 ka ago.

The processes of enhanced encephalisation also deserve some consideration. The various neural areas differ in the length of the embryonic period of neuronal cytogenesis (cell birth) of precursor cells. The most extended periods of cytogenesis apply to those portions of the brain that have experienced the highest degree of encephalisation (e.g. the cortex), while the least enlarged (e.g. the medulla) have the shortest cytogenesis (Kaskan and Finley 2001). Therefore, heterochrony (changes in developmental timing) could account for species variations in both absolute brain size and in differential size of various structures.

This summary of the main differences between the brain of the sole surviving member of *Homo* and that of the extant pongids or other primates collectively shows that, although there are dissimilarities in their brain structures, these are perhaps not as great as one would suspect based on behavioural differences. The issue of brain atrophy from robust *Homo sapiens* subspecies to purported anatomically modern humans is very troubling because it seems to contradict the central premise defining human evolution squarely. This also applies in several other respects, for instance, it is evident that modern humans suffer from thousands of unfavourable genetic conditions absent in other extant primates, and yet natural selection seems to have failed to eradicate these from the human genome. In considering such fundamental contradictions to the widely preferred narratives, we are at once reminded of the narrative of how modern humans arose by speciation in Africa and replaced all humans in the world. These simplistic origins stories explain literally nothing about how humans became what they are today: they cannot even account for our atrophied brains or our troubled genome; in fact, they discourage investigation of such inconvenient truths. Let us look at such issues more closely.

The rise of pathologies

There is one other significant disparity between apes and us: humans are subject to a vast number of pathologies that other animals, including other primates, seem mostly to be free of. Such disadvantages include numerous genetically derived susceptibilities or defects of many kinds which humans are saddled with, as well as neurodegenerative diseases, but which are either absent in the extant ape species or have not, at this point, been detected in them (Rubinsztein et al. 1994; Walker and Cork 1999; Olson and Varki 2003; Bailey 2006). This is not entirely unexpected: relevant brain illnesses seem to involve precisely those areas of the human brain that are the phylogenetically most recent and are therefore those where we differ neurologically most from the other primates (see the previous section). Those newest structures in the human brain are the ones that have not been 'tested' as much by selection as the more ancient ones have (Horrobin 1998). Selection for the traits that have 'made us human', especially the neural systems underlying language and social cognition, has led to psychosis as a secondary result. This hypothesis can be tested directly, using the human haplotype map to assay for selective sweeps in regions associated with psychosis in genome scans, such as 1q21 (Voight et al. 2006). Many of the selective sweeps inferred from such data are remarkably recent (less than 20,000 years old).

Moreover, the most recent history of humans involved new selection agents that never affected any animal's evolution before. Being of the 'cultural' kind or related to niche construction, they quite probably also interfered with the processes of natural selection. Therefore, they could easily have helped preserve new features of the brain that purely 'natural' effects would have selected against. Nevertheless, it is these latest modifications to the human brain that support our advanced cognition and intelligence. To attain them we seem to have been saddled with a brain that developed too fast for its own good (Damasio et al. 1990; Crow 1997a; Randall 1998; Keller and Miller 2006; Ghika 2008; Bednarik and Helvenston 2012). A good example is provided by schizophrenia which has been proposed to be best understood as a deleterious side effect of the newly expanded social brain areas during hominisation (Farley 1976; Crow 1995a; 2000; 2002; Cosmides and Tooby 1999; Burns 200; 2006; Brüne and Brüne-Cohrs 2007).

The neural developments in our brain underwriting our mental and cognitive capacities have rendered the brain vulnerable to numerous neurodegenerative and mental diseases as well as frontal lobe connectivity issues, demyelination or dysmyelination. They may account not only for schizophrenia but also bipolar or manic-depressive illness, autism spectrum disorder, obsessive-compulsive disorder, sociopathic or antisocial personality disorders, chronic fatigue syndrome, Rett or Down syndrome and thousands of other impairments endemic to humans. On the neurodegenerative side the better known are Alzheimer's, Huntington's and Parkinson's diseases, frontotemporal dementia and behavioural variant FTD; but amyotrophic lateral and diffuse myelinoclastic sclerosis, AIDS dementia, Batten disease and neuronal ceroid lipofuscinosis, Creutzfeldt-Jakob disease, multiple sclerosis, temporal lobe epilepsy, middle cerebral artery stroke and many others may also pertain. Then there are the many thousands of Mendelian disorders. What all these human conditions have in common is that for most of human history, their sufferers are not very likely to have contributed significantly to the gene pool – or even survived long enough to have had the opportunity to do so. Which raises a fundamental question concerning the rise of these conditions: how do we account for it in the face of evolution's uncompromising natural selection? Surely the establishment of these thousands of deleterious conditions does not prove that selection against them is a mirage, just as we have seen the postulated role of encephalisation dissolving? Let us briefly review the perhaps most important of these conditions that extant hominins suffer from before developing this reasoning further.

Schizophrenia

Schizophrenia is a polygenetic neuropsychiatric disease (Cardno and Gottesman 2000; Kennedy et al. 2003; Riley and Kendler 2006; Van Os and Kapur 2009) that afflicts the frontal lobes with connectivity problems (Mathalon and Ford 2008), contributing to the appearance of atrophy. It also affects the cingulate cortex, temporal lobes and hippocampus adversely and it involves volumetric changes of grey matter in the right and left middle and inferior temporal gyrus (Hershfield et al. 2006; Kuroki et al. 2006). The hippocampal volume is reduced, and lateral ventricular enlargement occurs (Harrison 1999). These morphometric changes are suggestive of alterations in the synaptic, dendritic and axonal organisation, which is supported by immunocytochemical and ultrastructural findings. However, a properly integrated aetiology and pathophysiological model of schizophrenia remain elusive, because the underlying abnormalities and contributing environmental factors remain inadequately understood (Ayalew et al. 2012). Nevertheless, numerous schizophrenia susceptibility genes have been identified or proposed, including, but certainly not limited to NRG1, NRG3, DTNBP1, COMT, CHRNA-7, SLC6A4, IMPA2, HOPA12bp, DISC1, TCF4, MBP, MOBP, NCAM1, NRCAM, NDUFV2, RAB18, ADCYAP1, BDNF, CNR1, DRD2, DRD4, GAD1, GRIA1, GRIA4, GRIN2B, HTR2A, RELN, SNAP-25, TNIK, HSPA1B, ALDH1A1, ANK3, CD9, CPLX2, FABP7, GABRB3, GNB1L, GRMS, GSN, HINT1, KALRN, KIF2A, NR4A2, PDE4B, PRKCA, RGS4, SLC1A2, SLC6A4, SYN2, APOL1, APOE, FOXP2, GRM3, MAOA, MAOB and SYNJ1 (Yoshikawa et al. 2001; Spinks et al. 2004; Cho et al. 2005; Li et al. 2006; Xu et al. 2006; Ayalew et al. 2012).

These susceptibility alleles are all of small or non-detrimental individual effects but constitute an increased risk for schizophrenia through aggregation (Cannon 2005). Carriers of small numbers of susceptibility genes account for about 15 per cent of any population, yet the disorder is fully expressed in only about 1 per cent of people worldwide (Jablensky et al. 1992). That percentage is relatively stable across all societies, and all cultures perceive the illness as a severe maladaptive dysfunction (Pearlson and Folley 2008). However, there can be significant variations in frequency; for example, the population of Kuusamo in north-eastern Finland has triple the risk of schizophrenia relative to Finland's national population, yet with the comparable clinical phenotype (Hovatta et al. 1999; Arajärvi et al. 2004). If schizophrenia in Finland has been as disadvantageous over the last twenty generations as it appears today (selection coefficient $s \sim .50$), and is caused by a single recessive allele with $p = .10$ (explaining the current disease prevalence of 1 per cent), it would follow from standard evolutionary genetics that 42 per cent of Finns were schizophrenic in 1600 – which we presume is a nonsensical result (Keller and Miller 2006).

Among the variations in the incidence of schizophrenia are those affecting groups that are at increased risk, such as men, first-generation and second-generation migrants, those born and/or raised in cities, the offspring of older fathers, and those born in winter and spring (McGrath et al. 2004); as well as those of low birth weight, maternal infection during pregnancy and exposure to house cats. Patients later diagnosed with schizophrenia had a persistent reading impairment and low IQ scores in their childhood (Karlsson 1984; Crow et al. 1995). Schizophrenia and language seem to be linked to cerebral asymmetry (Crow 1997b) and the hemispherical dominance for language is thought to have led to the collateral hemispheric lateralisation (Crow 1995a; 1995b). However, this notion is countered by several indices, not only the apparent error of linking language origins with the falsity of speciation of Graciles (the replacement hypothesis, see above, and Falk 2009; Bickerton 2010). For instance, the planum temporale, presenting a left-right asymmetry favouring the left (Geschwind and Levitsky 1968), presumably related to language reception, is also present in great apes (Gannon et al. 1998; 2001). Moreover, the detection of the FOXP2 gene on chromosome 7 of Robusts from El Sidrón in Spain (Krause et al. 2007; cf. Enard et al. 2002; Zhang et al. 2002; Sanjuan et al. 2006), but absence of such distinct schizophrenia susceptibility alleles as NRG3 and NRG1 in Robusts refutes the idea (in fact, schizophrenia may have appeared much later than Graciles; Bednarik and Helvenston 2012). Genomic regions that have undergone positive selection in 'modern humans' are enriched in gene loci associated with schizophrenia (Srinivasan et al. 2016; 2017). This supports the possibility that most of the many schizophrenia risk alleles appeared recently in human evolution, and they are absent in apes (Brüne 2004; Pearlson and Folley 2008).

As in autism, there is a schizotypy spectrum within which schizophrenia is merely the extreme form. First-degree relatives of psychotic patients are notably creative (Heston 1966; Karlsson 1970; Horrobin 2002). Elevated levels of some of the schizotypal or schizoaffective personality traits are commonly observed in individuals who are active in the creative arts (Schuldberg 1988; 2000; Brod 1997; Nettle 2001). Schizotypal diathesis, which may lead to actual illness under specific environmental factors (Tsuang et al. 2001) but in most cases does not, is, therefore, more convincingly implicated in creativity, much in the same way as mild forms of autism can yield high-performing individuals (see below). Moreover, introvertive anhedonia, a typical symptom of schizophrenia (Schuldberg 2000), decreases creative activity significantly, thus providing a clear separation between creative and clinical cohorts.

The involvement of schizotypy in shamanism deserves further examination (Bednarik 2013g). The discovery of the rubber hand illusion (RHI) in schizophrenic patients (Peled et al. 2003) has implications for the notion of out-of-body experiences (Thakkar et al. 2011). It has been suggested that a weakened sense of the self may contribute to psychotic experiences. The RHI illustrates proprioceptive drift, which is observed to be significantly higher in schizophrenia patients than in a control sample and can even lead to an out-of-body experience, linking 'body disownership' and psychotic experiences. However, there is no credible empirical evidence linking schizophrenia with palaeoart production, just as there is none linking shamanism with it or with schizophrenia (Bednarik 2013g), contrary to various claims. Nevertheless, susceptibility to proprioceptive drift can be shown to be linked to schizotypy, and may thus well account for certain experiences of shamans.

Bipolar or manic–depressive disorder

Bipolar or manic–depressive disorder (BD) is another grave mental illness that can be found in a range of severity. Its milder version is called cyclothymic disorder (Goodwin and Jamison 1990), which affects close to 4 per cent of the population. Between 1 per cent and 1.6 per cent of a population suffers fully developed bipolar disorder, characterised by extreme mood swings between alternatively depressed and euphoric states. Several genetic regions have been implicated in these conditions, including six specific chromosomes: 18, 4p16, 12q23-q24, 16p13, 21q22 and Xq24-q26 (Craddock and Jones 1999; Craddock et al. 2005). Since the involvement of the neurotransmitters norepinephrine and serotonin has been suggested (Schildkraut 1965), and with the advent of neuroimaging, several brain areas have been implicated in BD. This includes the observed reduction of grey matter in the left subgenual prefrontal cortex and amygdala enlargement (Vawter et al. 2000); decreased neuronal and glial density in association with glial hypertrophy (Rajkowska 2009); and significant shape differences in caudate and putamen, thus implicating the basal ganglia (Hwang et al. 2006).

In contrast to schizophrenia, BD is not accompanied by progressive dementia and the periods of 'normal' personality between alternatively euphoric (manic) and depressed mood episodes may be relatively creative and productive (Goodwin and Jamison 1990). Twin studies, family studies, adoption studies and the higher concordance rate for monozygotic twins than for fraternal twins all point to a high degree of heritability. Although there are several similarities with schizophrenia, that illness involves increased neuronal density in the prefrontal cortex, whereas, in BD, there is decreased neuronal and glial density in association with glial hypertrophy (Rajkowska 2009). Ongur et al. (1998) observed a significant reduction in glial density in the subgenual prefrontal cortex but did not identify which type of glial cell, astrocytes, oligodendroglia or microglia were involved. Decreased cortical thickness in both bipolar and depressive disorders has been found in the rostral orbitofrontal and middle orbitofrontal cortex, whereas caudal orbitofrontal and dorsolateral prefrontal cortex appeared normal. Areas of neuronal densities were significantly decreased in the CA2 sector of the hippocampus of patients with both BD and schizophrenia, compared to control subjects.

The symptoms of bipolar illness are very dramatic and noticeable, and the sufferers are often highly intelligent, creative and verbally communicative people. Therefore, these mood disorders have been recognised, described and linked together since at least the Hippocratic Corpus in Western civilisation (Helvenston 1999), whereas schizophrenia has been described only in recent centuries.

Autism spectrum disorders

Autism spectrum disorders are human brain conditions manifesting themselves in early childhood and their aetiology remains virtually unknown (Hermelin and O'Connor 1970; Frith 1989; Hughes et al. 1997; Baron-Cohen 2002; 2006; Allman et al. 2005; Grinker 2007; Balter 2007; Burack et al. 2009; Helvenston and Bednarik 2011; Bednarik and Helvenston 2012). Autism has recently become a widespread illness, affecting one in 110 children according to one study (Weintraub 2011), and one in 38 as demonstrated in a more reliable analysis (Kim et al. 2011). The epidemic increase in this diagnosis, from one in 5000 children in 1975, cannot be entirely attributed to changing diagnostic criteria (cf. Buchen 2011). The explanation offered by Bednarik (2011a) is perhaps more relevant: an exponential increase in human neuropathology due to suspension of natural selection.

Abnormalities have been detected in autism patients' frontal and temporal lobes, the cerebellum, the amygdala and the hippocampus. VENs have also been implicated in autism (Allman et al. 2005: 367; challenged by Kennedy et al. 2007). Under-connectivity in the brains of children with autism (Hughes et al. 1997) offers a basis for further investigation of this and other pervasive developmental disorders. In some subgroups, cerebellar dysfunction may occur; in others, there is dysfunction of the prefrontal cortex and connections to the parietal lobe. Reduced activation in the fusiform gyrus, the portion of the brain associated with facial recognition, and increased activation of adjacent portions of the brain associated with the recognition of objects have been observed. The amygdalas of patients with autism have fewer nerve cells, especially in a subdivision called the lateral nucleus of the amygdala (Balter 2007).

Several authors have proposed autism to have been instrumental in introducing Pleistocene palaeoart (Humphrey 1998; Kellman 1998; 1999; Haworth 2006; Spikins 2009; Bogdashina 2010: 159–160), typically without realising that their proposals coincided with those of others. The most comprehensive attempt of this nature has been that of Humphrey, but even he fails to present evidence in support of his case, for which he offers a single example (Selfe 1977). Although there are several other cases known of exceptional artistic skills in autistic savants that he is not aware of (Waterhouse 1988; Mottron and Belleville 1993, 1995; Pring and Hermelin 1993; Kellman 1998, 1999; Mottron et al. 1999; Happé and Vital 2009; Happé and Frith 2010), they prove no nexus with the abilities of Upper Palaeolithic graffitists. Humphrey and others pursuing this line of reasoning also seem unaware of the phenomenon of 'precocious realism' in the art of non-autistic children (Selfe 1983; Drake and Winner 2009; O'Connor and Hermelin 1987, 1990). The absurdity of his reasoning that the cave art's naturalism implies an absence of language becomes apparent when it is extended to the similarly realistic rock art of recent groups such as the San Bushmen of South Africa or the Aborigines who created the Gwion rock art in Australia. Moreover, the extremely rare occurrence of autistics of exceptional depictive abilities does not explain why 99.99 per cent of autistic spectrum disorder patients lack them. Similar proposals that the first art-like productions by humans derive from schizophrenia or bipolar disorder (Whitley 2009) have been soundly refuted, as have been many claims that shamanism is attributable to one of several mental illnesses (Bednarik 2013g).

Another mental illness credited with the introduction of palaeoart is *Asperger's syndrome*, a variant of autistic spectrum disorder that lacks the aberrations or delays in language development or cognitive development that are typical for autism. Patients may also have average or even superior intelligence, in contrast to the low IQ typically associated with autism (Rodman 2003; Bednarik and Helvenston 2012). They share social insensitivity and other characteristics with autism patients, and Asperger's also develops in early childhood, generally after the age of 3. As in autism, patients have cognitively based deficiencies in the Theory of Mind, the ability to attribute mental states to oneself and others; and to understand that others have beliefs, desires and intentions that are different from one's own.

Spikins' (2009) 'different minds theory' endeavours to explain 'modern behaviour' as a rise in cognitive variation when autistic modes of thinking were integrated into the practices of human societies in the Early Upper Palaeolithic. Spikins cites projectile weapons, bladelets, bone artefacts, hafting, "elaborate fire use", exploitation of marine resources and large game as being attributable to the "attention to detail, exceptional memory, a thirst for knowledge and narrow obsessive focus" of autistics, particularly those with Asperger's syndrome. This suggests that she is unaware, as an archaeologist, that all these practices have long been demonstrated from the Lower and Middle Palaeolithic periods, together with palaeoart and personal ornamentation. Asperger's and autism include diagnostic characteristics such as inflexibility in thinking, difficulty with planning and organisation, and a rigorous adherence to routine (Pickard et al. 2011), which impede originality and innovative thought. The creativity Spikins invokes is impoverished in such patients (Frith 1972; Craig and Baron-Cohen 1999; Turner 1999), unless fostered. The savant skills ascribed to a very few of them need to be nurtured and are specific to the ordered cultural context of modern life (Baron-Cohen 2000; Folstein and Rosen-Sheidley 2001; Thioux et al. 2006).

As in all cases of hypotheses linking mental illnesses with the perceived modernity of hominins (Bednarik 2012c), the neuropsychiatric disorders of humans, absent in other extant primates (Rubinsztein et al. 1994; Walker and Cork 1999; Enard et al. 2002; Marvanová et al. 2003; Olson and Varki 2003; Sherwood et al. 2011; Bednarik and Helvenston 2012), needed to exist before they could have had any of the proposed effects. However, to exist in a population, the selective processes acting on that population first had to tolerate these disorders. The lack of social skills typical of autism or Asperger's in societies heavily reliant upon social dynamics, such as those of Pleistocene hominins, would tend to select against them, socially, reproductively and genetically. Thus all of these hypotheses run up against the classical Keller and Miller paradox, which we will soon consider in detail.

Obsessive-compulsive disorder

Obsessive-compulsive disorder (OCD) refers to persistent thoughts, feelings and impulses of an overactive inferior prefrontal cortex. This anxiety disorder first appears typically in childhood to early adulthood, affecting areas of the brain called the 'worry circuit' and it is connected with an imbalance of the neurotransmitter serotonin (Schwartz and Begley 2002). Excessive activity in the inferior prefrontal cortex leads to the development of obsessive stereotypical behaviours, and the striatum (caudate nucleus and putamen) is also excessively active in OCD sufferers. Projecting to the striatum, the inferior prefrontal cortex, orbitofrontal cortex and the cingulate cortex cause the caudate to be overactive in the striosome area, thus bringing emotional tones and valences into the experience via the amygdala because it also projects into this same striosome area. Between the matrisome and the striosome areas are the tonically active neurons (TANs), which integrate the input from the inferior orbital frontal cortex via the striosomes with the input from the amygdala and orbital-frontal region, also via the striosomes. The TANs thus function as a gating mechanism between the matrisome and the striosome regions. The direct route by which the striatum projects to the cortex is via the globus pallidus, thalamus and motor and premotor cortex. The indirect route neurons project from the striatum to the globus pallidus, the subthalamic nucleus to the thalamus and then to the motor and premotor cortex. The direct pathway from the striatum calms the cortex, whereas the indirect stimulates it, and the gating TANs determine which pathway is taken.

When working effectively, the TANs modulate the orbital frontal cortex and anterior cingulate by adjusting the degree to which the thalamus drives both areas. However, in OCD, the error detector centred in the orbital frontal cortex and anterior cingulate can be locked into a pattern of repetitive firing. This triggers in the patient an overpowering feeling that something is 'wrong', eliciting compulsive thoughts or behavioural patterns intended to make it somehow 'right'. This leads to the anxiety disorder in which people have unwanted and repeated thoughts, feelings, images and sensations (obsessions) and engage in ritualised behaviours in response to these thoughts or obsessions. OCD affects about 2 per cent of the population.

Multiple sclerosis

Multiple sclerosis (MS) is apparently (but not conclusively) an autoimmune inflammatory disease of the central nervous system, causing demyelination of axons (Sailor et al. 2003). This inflammation of the fatty sheaths of nerve fibres leaves scar tissue, and the affected nerve cells cannot transmit or receive impulses. The outcome of MS is, therefore, a great variety of effects on muscle function and sensation. Activated mononuclear cells (white blood cells called T-cells) are thought to destroy myelin and to some degree oligodendrocytes, the glial cells that produce the myelin in which axons are wrapped. Remaining oligodendrocytes then attempt to produce new myelin, but this pattern of inflammatory reaction tends to subside locally only to appear at another location or at another time. The symptoms may, therefore, come and go, with relapses and remissions lasting from days to months. The pattern of progression suggests a relative hierarchy of changes over time, involving first frontal and temporal regions (Lumsden 1970) and later the pre-central gyrus (Wegner and Mathews 2003). MS can impact on any area in the central nervous system, although visual areas are commonly affected. The onset of MS occurs most often in young adults, 20–40 years of age, but can even occur before age 15 or after age 50 (Compston and Coles 2008).

Alzheimer's disease

Alzheimer's disease represents 60–70 per cent of all dementia cases. Although primarily an illness of ageing, it can be diagnosed in people from their thirties onwards. It derives from the deposition of extracellular plaque of beta-amyloid and intracellular accumulation of tau, a protein and a component of intracellular neurofibrillary tangles. These plaques and tangles are visible with MRI, and are initially found primarily in the hippocampus and entorhinal cortex, later in some areas of the frontal cortex and temporal (medial temporal lobe) and parietal association cortex. Alzheimer's targets the limbic structures (Hyman et al. 1984), including the amygdala, the locus coeruleus and the cholinergic neurons of the nucleus basalis of Meynert. As a result of the plaque deposition, neurons and synapses die, axons degenerate and connections are lost. General atrophy of the cortex and brain shrinkage occur (Smith 2002). VENs are particularly vulnerable to Alzheimer's; about 60 per cent of them may be lost in the anterior cingulate cortex (Seeley et al. 2006).

Frontotemporal dementia

Similarly, VENs are also implicated in frontotemporal dementia (Pick's disease). This group of relatively rare disorders affects the frontal and temporal lobes primarily. In one variant of frontotemporal dementia known as behavioural variant bvFTD, the anterior cingulate cortex and the orbital frontoinsula both show marked signs of focal degeneration, which is prominent in the right hemisphere (Seeley et al. 2007).

Huntington's disease

Huntington's disease derives from cell loss in the basal nuclei and cortex. This movement, cognitive and behavioural disorder is also not limited to older cohorts but can affect most age groups. It occurs initially in the neostriatum, where marked atrophy of the caudate nucleus and putamen is accompanied by selective neuronal loss and astrogliosis. This is followed by atrophy in the globus pallidus, thalamus, subthalamic nucleus, substantia nigra and cerebellum as the disease progresses. Its genetic basis involves the expansion of a cysteine-adenosine-guanine (CAG) repeat encoding, a polyglutamine tract in the N-terminus of the protein product called Huntington.

Parkinson's disease

Parkinson's disease is characterised by a loss of dopaminergic nigrostriatal neurons in the substantia nigra of the midbrain. By the time a patient is diagnosed with Parkinson's, about 60–70 per cent of the substantia nigra dopamine cells are already lost. By that time the substantia nigra, basal nuclei, amygdala, part of the limbic system, nucleus basalis of Meynert and part of the extended amygdala have all been affected. Recent research suggests that *Helicobacter pylori*, a bacterium implicated in ulcers and stomach cancer, may help trigger Parkinson's. There is also a juvenile-onset Parkinson's disease that is a Mendelian disorder, being a recessive mutation at 1p and 6q26 and, like all Mendelian disorders, very rare.

Prion diseases

Prion diseases (transmissible spongiform encephalopathies) are progressive neurodegenerative conditions that can also be found in animals other than humans. Scrapie is such a disease when found in domesticated ovicaprines and was the first prion disease to be identified. Other neurodegenerative diseases specifically in humans are *motor neurone diseases, spinocerebellar ataxia* and *spinal muscular atrophy*.

Other conditions

Already at this point, it becomes clear that our extant species suffers from a significant number of neuropathological conditions that are necessarily of genetic aetiology and that should not be tolerated by unfettered natural selection. However, we have almost countless other conditions, endemic to 'anatomically modern humans', that challenge the 'survival of the fittest' notion of evolution. Among them are, for instance, *cleidocranial dysplasia* or delayed closure of cranial sutures, certainly endemic to our subspecies. There are *malformed clavicles* and *dental abnormalities* (including anodontia), for which genes RUNX2 and CBRA1 are responsible. The mutation THADA relates to *type 2 diabetes*. Then there is the microcephalin D allele, introduced approximately 37,000 years ago through a single progenitor copy (but which could be as recent as 14,000 years ago, at 95 per cent confidence interval; Evans et al. 2005). Another contributor to *microcephaly*, the ASPM allele, reportedly appeared only around 5,800 years ago (ibid.; Mekel-Bobrov et al. 2005). Nor can natural selection be responsible for our susceptibility to the leading neurological illness, *middle cerebral artery stroke*. The middle cerebral artery is the largest cerebral artery and also the most commonly affected by cerebrovascular incidents. It supplies most of the outer convex brain surface, nearly all the basal ganglia, and the posterior and anterior internal capsules. Infarcts can lead to diverse conditions, including apraxia (inability to perform previously learned physical tasks) and dyspraxia (inability to perform a physical task), and Broca's and Wernicke's aphasia (expressive and receptive language deficits).

Mendelian disorders

Still, many thousands more genetically based neuropathologies need to be listed, including the *Mendelian disorders*. In 2011, we reported that for 1,700 of them, the precise gene responsible had been identified (Bednarik 2011a; based on 2006 data). At that time, their total number was estimated at 6,000. Since then, the numbers have steadily grown as research continued. As of January 2019, the *Online Mendelian Inheritance in Man* listed 6,328 Mendelian disorder phenotypes for which the molecular basis is known and 4,017 genes with phenotype-causing mutation. Despite their large number, Mendelian disorders account for only around 2 per cent of all births (Sankaranarayanan 2001), which illustrates their in most cases extremely low individual frequencies. They are rare precisely because selection keeps harmful mutations very rare. Mendelian disorders have frequencies consistent with the mutation-selection balance – a balance between genetic copying errors that turn normal alleles into harmful mutations, and selection that eliminates these mutations (Falconer and Mackay 1996). Mutations that affect the phenotype are nearly always harmful: entropy erodes functional complexity (Ridley 2000). Selection removes these mutations at a rate proportional to the fitness cost of the mutation. If the selection coefficient against the mutation is reproductively lethal, the newly arisen mutation exists in only one body before being eliminated from the population, but if it is relatively small, the mutation may pass through and affect many individuals through many generations before being removed by selection.

Summary

These are the major neuropathologies of present-day humans, and some of them have increased to epidemic proportions in recent times. Of schizophrenia, there does not even appear to be any recorded historical evidence until just a few centuries ago, and the proliferation of autism spectrum disorders, obsessive-compulsive disorder and other compulsive syndromes is of considerable concern. In the case of the neurodegenerative diseases, the increased incidence can perhaps be ascribed to a tendency of greater longevity in recent times. However, even here, it needs to be asked why natural selection tolerated the relevant predispositions in the first place. Neurodegenerative diseases affect the neurons primarily in the human brain, and they are incurable and debilitating. Where they affect movement, they cause ataxias, whereas those affecting mental functioning are called dementias. While they tend to affect older people more than younger ones, this is undoubtedly not true universally. Most importantly, other animals, including other primates, seem to be free of neurodegenerative diseases. Some of these disorders, such as Parkinson's, Alzheimer's and (infectious) prion diseases or amyotrophic lateral sclerosis can sometimes be caused genetically, whereas others, especially Huntington's disease, are passed on entirely by inheritance of gene mutations.

The Keller and Miller paradox

This brief review of some of the pathologies *Homo sapiens sapiens* suffers from shows that many of them affect the regions generally defined as supporting the 'higher cognitive functions' of the human brain (Damasio et al. 1990). In all probability, this neurological susceptibility is directly linked to the complexity of our once burgeoning brain. We have already visited one relevant paradox: as the assumed demand on the higher cognitive functions became more pressing with the ever-increasing complexity of culture, during the last 40 or 50 millennia the human brain not only stopped expanding; it began contracting at an alarming rate. The paradox is that the previous expansion, called encephalisation, is supposed to have underwritten those higher cognitive functions. If that had been the case, how do the palaeoanthropologists explain that our ancestors of the final Pleistocene and the subsequent Holocene managed apparently quite well with a smaller brain? The fact that natural selection tolerated the extremely high costs of brain expansion seems to indicate that the benefits of a larger brain outweighed the significant disadvantages it involved. Those disadvantages include obstetric demands and complications (including increased rates of death in childbirth), the foetal stage of the new-born, greatly prolonged childcare period and consequently reduced fertility of populations. They can reasonably be assumed to have been of enormous effects for the societies concerned, determining their social structures and significantly increasing their investment in the next generation. The only evolutionary benefit of this investment was a larger brain, but if during a phase of increasing demands made on that brain our species managed with less brain tissue (Bednarik 2014e), it seems that the traditional explanation for encephalisation needs to be abandoned. The rationalisation that perhaps a more efficient way to pack the tissue was found by evolutionary processes does not stand up to scrutiny. If such better neuronal efficiency were possible, the relevant mutations would have been strongly favoured by natural selection all along, which instead preferred encephalisation for millions of years.

So here we have one paradox that has never been resolved – or even investigated in any consequential manner. It defines a monumental problem in hominin development, but here is another, even greater conundrum we must address. We have in the previous sections considered the differences between humans and other primates; we have discovered that humans suffer from many thousands of detrimental or at least disadvantageous conditions of which other primates seem to be largely if not entirely free. We have also noted that as far as the human brain is concerned, those regions that suffer the most from genetically imposed impairments seem to be among the ones most recently developed or enlarged – which corroborates the absence of intrinsically maladaptive conditions in other primates. This brings us to a core issue considered in this volume: the disadvantageous conditions encoded in the human genome pose a fundamental evolutionary paradox because natural selection is expected to make harmful, heritable traits very uncommon over the course of generations. Low-fitness alleles, even those with minor adverse effects, tend to go extinct reasonably quickly. Alleles that reach fixation or extinction cause no genetic variation, and therefore cannot contribute to heritable variation in traits. So, for instance, if mental disorders existed in ancestral environments in much the same form as they do now, it is reasonable to assume that they would have resulted in lower ancestral fitness.

A critical discussion paper (Keller and Miller 2006) explores why, given that natural selection is so effective in optimising complex adaptations, it seems unable to eliminate genes that predispose humans to common, harmful and heritable mental disorders. These authors considered such disorders as the very embodiment of maladaptive traits. They reviewed three potential explanations for the paradox: (1) ancestral neutrality (susceptibility alleles were not harmful among ancestors); (2) balancing selection (susceptibility alleles sometimes increased fitness); and (3) polygenic mutation-selection balance (mental disorders reflect the inevitable mutational load on the thousands of genes underlying human behaviour). There are no good explanations, Keller and Miller noted, of why human brains seem to malfunction so often today, and why these malfunctions are both heritable and disastrous to survival and reproduction.

The relatively high frequency of mental disorders, hundreds or even thousands of times more common than nearly all of the multitude of single-gene mutations, renders the mutation-selection balance explanation unlikely (Wilson 1998: 390; see also Wilson 2006; Keller and Miller 2006: 440). Keller and Miller favoured that explanation over the two alternatives, although they admitted that the roles of other evolutionary genetic mechanisms could not be precluded. In the end, they explicitly accepted that they had not resolved the paradox they were trying to address. Their preference reflects their firm belief that mental disorder susceptibility alleles were ancestrally maladaptive. If mental disorders were debilitating in ancestral conditions, Keller and Miller argue, their developmental timing would have harmed fitness, given any reasonable model of ancestral life-history profiles.

Keller and Miller (2006) perceive natural selection as a purely mechanistic, iterative process whereby alleles from one generation have a non-random probability of being represented in subsequent generations. Thus natural selection, being an entirely dysteleological process, cannot "hedge bets by stockpiling genetic variation in the hope that currently maladaptive alleles might become adaptive in the future". Neutral mutations are not thought to be the main source of phenotypically expressed variation (Ridley 1996), as neutral alleles have no fitness effects. Keller and Miller all but rule out ancestral neutrality as an explanation for their paradox. They find it hard to believe that phenotypically expressed alleles associated with mental disorders that have such harmful effects in modern environments would have been precisely neutral (selection coefficient $s < 1/40{,}000$) across all ancestral environments. The most harmful are relatively rare, in an absolute sense; if neutral evolution were the right reason, one would have to explain why the most harmful mental disorders are so consistently rare. At $s = 0.05$, an allele frequency of 0.5 will approach 0 after 40 generations, ignoring effects of genetic drift; at $s = 0.1$ it will be 200 generations, and at $s = 0.01$ after c. 1,400 generations.

Of course, alleles affecting the heritable disorders might have been closer to being neutral in ancestral populations of the Pleistocene, while extant prevalence and heritability rates may be higher than predicted. This might be plausible for disorders implying G–E (gene-by-environment) interactions, showing wide cross-cultural variation in prevalence rates, increased or decreased rates in recent historical time as environments change, and a credible mismatch between ancestral and modern conditions that affect the mental disorder in question.

We share with Keller and Miller their rejection of the possibility that perhaps alleles which "increase the risk of mental disorders today had no such effect in ancestral environments". The 35 outstanding psychiatric, psychological and neuroscientific authorities who discussed their paper lacked the relevant information to adequately consider what we might or might not know about the psychopathologies of ancestral populations, or even about their cognition (Bednarik 2011c; 2012a; 2012c; 2013d; 2014e; 2015a; 2015c; 2016a; Helvenston and Bednarik 2011; Bednarik and Helvenston 2012). Many authors have favoured balancing selection as explaining the persistence of harmful and heritable mental disorders (Karlsson 1974; Allen and Sarich 1988; Mealey 1995; Wilson 1998; Longley 2001; Barrantes-Vidal 2004). Essentially, in balancing selection, two or more alternative alleles are preserved because their net fitness effects balance each other out. These effects are positive in some genetic or environmental contexts, and detrimental in others, but their potential availability determines their efficacy. While not dismissed by Keller and Miller as a possible explanation, they expressed a preference for their third alternative, polygenic mutation-selection balance. This model suggests that susceptibility alleles with the largest effect sizes may also be the rarest, the most recent, and the most population-specific. This has not only implications for the methods most likely to locate mental disorder susceptibility alleles (Wright et al. 2003); it provides valuable direction for the search for a resolution to the paradox, and psychiatric genetics and evolutionary genetics generally. Common mental disorders, as Keller and Miller note, are probably fundamentally different from Mendelian disorders in that the former are influenced by a much larger number of environmental and genetic factors, most of which have only minor influences on overall population risk. Indeed, the contention "that mental disorders usually or always represented the proper functioning of universal adaptations, and only seemed to be maladaptive because of the modern environment, or medical stigmatisation" (Nettle 2006; see also Nettle 2004) needs to be abandoned. This has been negated for some time by the discovery that ionising radiation can cause both parental germ-line cell and somatic mutations, increasing the risk of mental retardation and schizophrenia (Otake 1996; Loganovsky and Loganovskaja 2000). This evidence seems to reinforce the notion that harmful mutations increase mental disorder risk.

A significant constraint on the Keller and Miller paper is that it deals mostly just with 'mental disorders', which constitute not only a poorly definable taxonomic entity (Polimeni 2006; Preti and Miotto 2006; Wilson 2006), but also do not exist in an epistemological vacuum. A vast number of debilitating heritable conditions are detrimental or disadvantageous to the 'modern' human host, impact negatively on his fitness and are selected against in unencumbered evolutionary processes. The conditions generally subsumed under the label of 'mental disorders' are only a small part of the puzzle of why natural selection seems to tolerate so many genetic impairments in 'anatomically modern humans' (for critiques of this term, see Latour 1993; Tobias 1995; Bednarik 2008a). Keller and Miller have, therefore, posed the paradox quite correctly but applied it in a limiting way. They end their 2006 paper with this optimistic statement: "It is exciting, and humbling, to realize that, within the next 50 years, we will probably know why alleles that predispose people to common, harmful, heritable mental disorders have persisted over evolutionary time."

It took only a year or two to clarify the most fundamental issue, of which the discussed psychopathologies represent only a small part. The domestication hypothesis was first mentioned in the year following the landmark Keller and Miller paper (Bednarik 2007), in the context of the origins of human modernity. This revolutionary hypothesis was fully defined the following year, in most of its key elements (Bednarik 2008b). The right questions had been asked by psychiatric genetics and evolutionary psychology, but ultimately there had to be broadly based input, unencumbered by dogmatic views, from the empirical fields of palaeoanthropology and Pleistocene archaeology to solve the paradox in an all-encompassing explanation of the 'human condition' (Bednarik 2011c).

The observation that many of the pathologies discussed affect the regions generally defined as supporting the 'higher cognitive functions' of the human brain already provided an important clue. These functions are generally assumed to be relatively recent, although it needs to be cautioned that just how recent they are remains unknown. Nevertheless, the evidence as it stands suggests that the many disorders we are concerned with also must be recent developments in the development of humans.

There are anecdotal reports of animals mourning the loss of their mothers and becoming so dysphoric they died from grief. Goodall documents dysphoric behaviour and behavioural disorders and death in young chimps that have lost their mother (1986: 101–104). No doubt some animals in restricted environments, isolated from conspecifics appear dysphoric, although their overt behaviour more closely resembles obsessive-compulsive disorder. It is also possible to create animal models of bipolar disorder using surgical lesions or assorted neuroleptic drugs, but no spontaneous illness such as bipolar disorder is known in captive chimpanzees or other apes.

Therefore, one of the most burning issues to emerge from this chapter is the need to explain why natural selection apparently failed to do its job and select against the thousands of deleterious or disadvantageous genetic mutations that define present-day humans. Our solution to this rather major paradox is incredibly simple: it is just one of the very many effects that define the differences between our robust ancestors of 40,000 years ago from us, all of which are a direct or indirect result of the auto-domestication of humans that is the very subject of this book. It will be examined in Chapter 4.

4 Human self-domestication

Domestication

The underlying idea of the auto-domestication hypothesis has been said to have been expressed, however fleetingly, in Darwin's seminal work on domestication (Darwin 1868). However, Darwin regarded domestication as a deliberate intervention as well as being only by human domesticators, stating that "the will of man thus comes into play" (ibid., Vol. 1: 4). However, Darwin's ambivalence on this crucial issue is illustrated by his previous remark that the human domesticator "unintentionally exposes his animals and plants to various conditions of life, and variability supervenes, which he cannot even prevent or check" (ibid., Vol. 1: 2). Moreover, Darwin lacked today's understanding of domestication as a biological concept of wide variation and significance. Nowadays, we know of countless examples of non-human domesticators, and we have some level of comprehension of the symbiotic (or mutualistic) nature of domestication.

A few German researchers considered the idea that humans might, in some way, have become domesticated during the late nineteenth and the early twentieth centuries. Among them were the ethnologist Eduard Hahn (1856–1928), who specialised in a geographical study of the origins and distribution of domestic animals. As a sceptical cultural critic and outspoken opponent of colonialism, he mentioned the notion that some aspects of humans are reminiscent of domesticates (Hahn 1896) without pursuing it further. In this, he had probably been influenced by Johannes Ranke (1836–1916), a physiologist and anthropologist who eventually focused mainly on human variation and pre-History. Of particular relevance is his work on individual and group variability of the human skull. He rejected race-ideological theories most energetically and paid particular attention to the effects of social conditions and environmental factors on the historical development of populations. The issue of human domestication, however, was more directly addressed by Eugen Fischer (1874–1967), professor of medicine, anthropology and eugenics, and a member of the German Nazi Party. It was included in his objectionable proposal that 'races' may have been subjected to 'breeding' (Fischer 1914; cf. Schwalbe and Fischer 1923; not to be confused with evolutionary statistician R. A. Fisher). Both his racialism and the offensive connection he made between radical eugenics and genetics, not to mention his association with a fascist regime, contrasted sharply with the exemplary position taken by his two German predecessors. The entire notion of racialism is a misguided and misleading program (Brace 1996; 1999) and this predilection contributed undoubtedly to the eventual ruin of the once highly regarded German school of anthropology. These circumstances and the distortion of the issue for sinister political ends may also have prevented scientific advancement on the question of self-domestication, which for the following century experienced practically no progress. Bagehot (1905: 51) proposed more than a century ago that it was human self-domestication that led to civilisation, and although this is not the case, that century failed to see the claim tested.

The work of two more researchers who visited the subject, Boas and Eibl-Eibesfeldt, also suffered from Fischer's influence on them, and from the particular spin that had been put on the topic by that time. This slant focused on the assumed changes to humans due to the introduction of cooking and apparel, thought to have occurred in the last 50,000 years (Boas 1938). Today we know that the controlled use of fire was mastered at least 1.7 Ma ago (Beaumont 2011; Bednarik 2011c: Figures 6.1 and 6.2) and clothing must have been available in the Middle Pleistocene. While both factors likely had some minor effects on genetics, they are certainly not causes of human domestication. This preoccupation with a tangential issue seems to affect all subsequent considerations of that subject for the twentieth century, rather than helping to clarify it. Boas did consider some effects of domestication possibly evident in humans, but he attributed them to changes in lifestyle, nutrition and the use of implements. He also noted changes in sexual behaviour, lactation, body pigmentation, hair types and facial structures. Eibl-Eibesfeldt (1970: 197) recognised a significant aspect in commenting on the "deficiencies" induced by domestication. Wilson (1988) addressed the domestication of humans in a major work, but he too opted for equating that process with the Neolithic revolution and with "economic domestication". He attributed the change to assumed associated developments, such as the built environment, establishment of boundaries and spatial divisions, which is reminiscent of the suggestion that Mesolithic people counting shells developed numbering. After all, dwellings have been built since the Acheulian technocomplex of the Lower Palaeolithic (Bednarik 2014d: 87, Figure 15) and are not an innovation of the Holocene.

In Hodder's (1990) subsequent work, this emphasis on domestication is again entirely linked to the Neolithic. He completes the correlation of domestication with *domos* and the domicile and thus deflects the scientific approach to the general issue. Perceiving the human ape as the sole domesticator is itself extremely limiting and misleading. Humans have domesticated a good number of plant and animal species, some deliberately, other unintentionally (e.g. the house mouse and the sparrow); but they have also been domesticated themselves (e.g.

by the dog; Groves 1999). More importantly, hundreds of other animal species ranging from mammals to insects have domesticated other animal as well as plant and fungi species. Ultimately there is no hard and fast separation of domestication and symbiosis: one grades into the other. Therefore, domestication is a phenomenon not adequately understood by most humanists and has been misconstrued by them as something to do with domesticity. After the damage done to this topic by the earlier endeavours to connect it with eugenics, this misunderstanding of humanists unschooled in biological sciences and genetics prevented progress in exploring these issues. An approach informed by biology and genetics, i.e. by the sciences, was needed. The framework casting the question of domestication within an anthropocentric agenda, in which the human is the originator and manager of the process needs to be replaced with one perceiving it as genetic, in which Mendel trumps Darwin.

The first comprehensive review of human domestication in the new millennium (Leach 2003) offered no improvement. It also argued that the selective pressures effecting animal domestication applied to humans adopting sedentary ecologies, which involved changes in diet and mobility, among others. Perhaps most importantly, this reasoning implies that human domestication is a development of the Holocene, intrinsically tied to the Neolithic revolution. This is a fundamental falsity running like a red thread through the entire twentieth century, and no attempt was made to ascertain when signs of domestication begin to appear in hominin history. To give just one example: the dental and craniofacial reductions in humans are clearly not related to Holocene changes in lifestyle; they commenced tens of thousands of years earlier.

A significant improvement in understanding the nature of domestication as a universal phenomenon is attributable to the late David Rindos (1984), whose premature death in 1996 is attributable to vicious archaeologists at the University of Western Australia (Bednarik 2013a: 60). His reinterpretation of domestication as a co-evolutionary process represents a clear departure from the anthropocentrisms of the humanities, especially archaeology and anthropology. In introducing the concept of mutualism, which is a symbiosis beneficial to both the domesticator and the domesticate, Rindos's insight anticipated the gene-culture co-evolutionary model developed over the following years (Cavalli-Sforza and Feldman 1973; Feldman and Cavalli-Sforza 1989; Aoki and Feldman 1991; Durham 1991; Boyd and Richerson 2005; Richerson and Boyd 2005). The idea that evolution is not a purely biological process was not new (Dobzhansky 1962: 18; 1972) but these recent developments helped to extract domestication from the simplistic construals created around it. Another relevant brief comment states: "One might seriously entertain the hypothesis that an important first step in the evolution of modern human societies was a kind of self-domestication (selection on systems controlling emotional reactivity) in which human temperament was selected" (Hare and Tomasello 2005: 443).

These new currents provided a revolutionary new perspective on the operation of domestication, presenting it as a complex, multifaceted and generic phenomenon that has presumably always existed in nature, affecting numerous species well before humans even came into existence. Generically the domestication of animals, plants and fungi consists of the collective genetic alteration of their physiology, behaviour, appearance or life cycle through selective breeding. Typically, one organism assumes a significant degree of influence over the reproduction and care of another organism in order to secure a more predictable supply of a resource or characteristic of benefit. However, this tends to increase the fitness of both the domesticator and the domesticate (Zeder 2012a; 2014). For instance, most species bred by humans, be they plant or animal, have increased their geographical range and numbers greatly as a result, far beyond any natural potential. Many animal species, vertebrate and invertebrate, from mammals to social insects, have domesticated others, for example, to serve as a staple food source, to modify foods indigestible by the domesticators, or for their labour. Insect domesticators are well known for their animal and fungi domesticates. Examples are provided by the mutualistic agriculture involving leafcutter ants or certain African termites and specific fungi, rendering both the insects and the fungi dependent on each other for their symbiotic survival (Hölldobler and Wilson 1990; Munkacsi et al. 2004; Mueller et al. 2005).

In considering the extent of domestication, most discussions have focused on mammals and birds domesticated by humans, but it needs to be remembered that among the species selected by humans are also insects (e.g. the silkworm, the honey bee), ornamental plants of a wide range, several invertebrates used in biological pest control, and, of course, a wide range of plants as well as numerous fungi. These and many other factors need to be considered in any comprehensive review of the subject. The humanistic discussion of domestication from the nineteenth century onwards and right through to the early twenty-first century has thus completely ignored the biological nature of the phenomenon and replaced it with an anthropocentric construct, a simplistic notion that equated human domestication with an archaeologically perceived notion of domesticity. It has therefore been entirely unable to address the question of human self-domestication in any consequent form or fashion.

The idea of a 'domestication syndrome' (Hammer 1984; Brown et al. 2008), a suite of phenotypic traits arising from domestication has been known for about a century, but its systematic exploration is a quite recent development (Wilkins et al. 2014). When applied to vertebrate animals, the domestication syndrome includes changes in craniofacial morphology (shortened muzzle) and reductions in tooth size; depigmentation; alterations in ear and tail form; shortening of the spine; more frequent and non-seasonal oestrus cycles, ranging to their complete elimination; alterations in adrenocorticotropic hormone levels; changed concentrations of several neurotransmitters; prolongations in juvenile behaviour and several other neotenous effects; increased docility and tameness; and reductions in both total brain size and of particular brain regions (Price 2002).

In examining the last-named factor of the domestication syndrome, the profound changes in brain form, size and function due to sustained selection for lowered reactivity, we note that the reduction is highest in those wild progenitor species that possess the greatest brain volume as well as maximal degree of folding (Kruska 1988a; 1988b; Plogmann and Kruska 1990; Trut 1999; Zeder 2012a). Brain volume reduction has affected nearly all domesticated animal species and can range from 6 per cent to as high as 30–40 per cent in some species, with pigs, sheep and dogs occupying the higher range. "However, the methods used in virtually all studies on brain size reduction in domesticated animals confound the effect of brain reduction with the often substantial body size augmentation that has occurred during domestication" (Henriksen et al. 2016). Not surprisingly, it tends to be the limbic system, which regulates endocrine functions and thus behaviour, that is the most affected part of the brain. The size reduction is not limited to mammalian species; it can also be observed in rainbow trout (Marchetti and Nevitt 2003) and birds.

The changes caused by domestication are essentially genetic and seem to be irreversible. This belief is based on the observation that feral animals, i.e. those that were domesticated but have reverted to a wild state, retain their smaller brains and many of their other characteristics of domestication (Birks and Kitchener 1999; Zeder 2012a). It has been suggested that the time required to experience such reversal may not have been adequate to detect it, but there are at least two arguments against this view. First, it can take only a very short time to initiate domestication effects (Trut et al. 2009), and some species, such as the rabbit, have been domesticated for only relatively short times. Therefore, the argument goes, the reversal should be similarly swift. Second, the dingo must have been domesticated when it was brought to Australia on rafts in mid-Holocene times, yet despite having since then been partially feral for millennia, it retains the brain size of domestic dogs (Schultz 1969; Boitani and Ciucci 1995). One reason for the irreversibility of domestication effects may be the mechanism of pleiotropy, an essential principle in domestication that describes when one gene affects two or even more, apparently unrelated traits. This occurs when selection for a particular trait or set of traits, such as docility, consequentially prompts changes in other phenotypic traits, for instance, in facial architecture (Trut et al. 2009). It has been suggested that genetic changes in upstream regulators of the network affect downstream systems and that deficits in neural crest cells (NCCs) are related to adrenal gland function (Wilkins et al. 2014). The neural crest is a multipotent population of migratory cells unique to the vertebrate embryo, arising from the embryonic ectoderm cell layer (Xiao and Saint-Jeannet 2004) and developing a wide variety of cell types that define many features of the vertebrate clade (Meulemans and Bronner-Fraser 2004). NCCs first appear during early embryogenesis at the dorsal edge (hence the term 'crest') of the neural tube and then migrate ventrally in both the cranium and the trunk. They give rise to the cellular precursors of many cell and tissue types and indirectly promote the development of others. Among this vertebrate-specific class of stem cells are peripheral and enteric neurons and glia, craniofacial cartilage and bone, smooth muscle and pigment cells. Defining the mechanisms of the gene-regulatory network controlling neural crest development is key to insights into vertebrate evolution.

The gene-regulatory network can be divided into four systems establishing the migratory and multipotent characteristics of NCCs. *Inductive signals* separate the ectoderm from the neural plate by various pathways, but these remain incompletely understood; *neural plate border specifiers* are molecules that influence the inductive signals; *neural crest specifiers* are a suite of genes activated in emergent NCCs; *neural crest effector genes* confer properties such as migration and multipotency. The cascade of events constituting the migration of NCCs begins with the closure of the dorsal neural tube that is created by a fusion of the neural fold (Sanes 2012: 70–72). Delamination follows, in which the NCCs separate from surrounding tissue (Theveneau 2012). The NCCs then develop into various tissues that can be divided into four functional domains: cranial neural crest, trunk neural crest, vagal and sacral neural crest, and cardiac neural crest. The evolutionary relevance of NCCs derives from their roles in determining specific vertebrate features, including sensory ganglia, cranial skeleton and predatory characteristics (Gans and Northcutt 1983).

Human auto-domestication

The role of the NCCs in the domestication syndrome will be revisited further below. At this point, it is relevant to observe that with the beginning of the twenty-first century, a degree of consensus had emerged concerning the species-specific brain growth trajectory in primates, especially hominins. This agreement emphasises the roles of the parietal and frontal lobes, the cerebellum, the olfactory bulb and possibly the temporal lobe (Neubauer et al. 2010; Gunz et al. 2010; 2012; Bastir et al. 2011; Scott et al. 2014). However, what drove these developments remained as unclear as it had been since Darwin's time. The gracilisation or globularisation of the human cranium, too, was well known but unexplained, as were indeed all of the many differences between 'anatomically modern humans' and their archaic, robust predecessors. It was precisely these differences and their rapid introduction that persuaded the discipline in the 1980s that there could only be one explanation for the relatively sudden – on an evolutionary scale – appearance of gracile humans. The change occurred in the course of just a few tens of millennia, ending millions of years of slow and gradual evolution. Because of the punctuated nature of archaeological and palaeoanthropological insights, which can never be more than isolated tiny peepholes into our past, it is challenging to perceive continuities and processes; empirical observations tend to capture just collections of instants in history. Therefore, the record is somewhat like a series of still photographs of various events, not like a continuous video of any event or development. This factor has contributed to archaeology's judgement that only a catastrophist process such as sudden speciation could explain what is, in reality, a very scattered record.

Nevertheless, there were other factors involved in the preference for catastrophism over gradualism, though they are subtler. Archaeology, like all anthropocentric humanities, subscribes to an ideology ultimately derived from religion. The finest expression of this subliminal predisposition is the discipline's perception of evolution as a teleological process, leading almost inevitably to the crown of evolution, defined as the anatomically modern human. This is the precise opposite of the biological concept of the process, which sees evolution as dysteleological and thus stochastic: it yields no superior species, and least of all one that resembles a deity. This is one of many examples showing that archaeology is not a science – cannot be a science. The concept of 'cultural evolution', so central to the project of archaeology, is simply a scientific oxymoron: there can be no such thing, as we have already noted in Chapter 1. It is not just a terminological transgression; it illustrates what happens when the humanities corrupt scientific concepts.

Moreover, science can never be a part of a self-serving agenda of one self-important subspecies that considers it was created in the image of a god, and it cannot share the narcissistic conceit of some sections of humanity. A value judgement would be that our rather 'imperfect' subspecies has been damaging to this planet, and proper science might find that most of the characteristics distinguishing us from our robust ancestors offer no teleological advantages. The mantra of archaeology, in contrast, *is that the existence of phenomena may be explained with reference to the purpose they serve*. In this ideology, human ascent serves the purpose of yielding superior people and cultures. In short, the discipline is part of a self-delusional project about the human condition (Bednarik 2011c).

With the benefits of lacking humanistic conditioning, we arrived at the topic of human evolution from a very different perspective. During the late 1980s, as the replacement hypothesis rapidly gained wide acceptance, we became one of a dwindling number of its opponents. African Eve was marketed very effectively, by the "mastery and shameless willingness to use various online social networking platforms, publishers and all manners of other deafening self-promotion", not to mention manipulation of the peer review system, and archaeology and palaeoanthropology began to "resemble poorly managed but well-advertised corporations" (Thompson 2014). The popular science writers, science's "lumpenproletariat" (Rose 2016), played a vital role in this campaign to neutralise opposition to African Eve. As the "high priests" (Thompson 2014) of African Eve's crusade gained control in the debate in the mainstream journals, dissenting views were spurned and effectively censured. We responded with counterarguments addressing specific central tenets of African Eve. First, we refuted the notion that the advent of palaeoart coincided with the appearance of Graciles towards the end of the Pleistocene, showing that it extends hundreds of thousands of years into the deep past of hominins (e.g. Bednarik 1992a). Next, we stunned some of African Eve's advocates by showing that there was no evidence that the people of the Early Aurignacian tool industries were anatomically modern (Bednarik 1995a); apparently, they had not noticed this. Indeed, as we have seen, the hominins of the last half of the Late Pleistocene show a distinctly clinal development of skeletal gracility. At no point did a significantly different kind of people suddenly appear in Eurasia, as the African Eve scenario would demand.

On the contrary, late Neanderthals tend to be much more gracile than their predecessors, and from about 35–25 ka ago, there is a succession of increasingly gracile humans (see Chapter 2), and a co-existence of robust and gracile forms can be found in various parts of the world. By 2007, we had challenged the most fundamental tenets of the replacement hypothesis, proposing that not only the Aurignacian, but all of the Early Upper Palaeolithic traditions of Europe were the work not of 'modern' humans, but Robusts or their progenies (Bednarik 2007). At the same time, we attributed the neotenous changes during that period to sexual selection based on culturally determined somatic features, promising to present the details of our 'domestication hypothesis' shortly. Those details were submitted the following year, describing the main features of this hypothesis that was offered as an alternative to the replacement hypothesis (Bednarik 2008b). The present book is an expanded version of that theory.

The domestication hypothesis was thus presented in 2008, attempting to explain a whole raft of hitherto unexplained or inadequately explained aspects about the origins of present-day humans. One of the most consequential questions to ask concerning those origins is: what could have caused the change from the dysteleological progress of evolution to the apparent teleology of cultural development? What could have suspended the inherent laws of biological evolution? However, this question had never been asked by the mainstream. Nor had the question why evolutionary natural selection failed to select against thousands of deleterious genetic predispositions and defects ever been asked by those concerned with the human past. Neuroscientists and cognitive scientists have prominently debated this issue (culminating in Keller and Miller 2006), yet they have been unable to arrive at any credible answers because the disciplines taking care of the human past are lagging decades behind those dealing with the workings of our brain or genes. Nor has there been a sustained attempt to deal with such issues as to why the aetiologies of brain illnesses suggest that they involve primarily the same areas of the brain that are the phylogenetically most recent; or why it should be that other extant primates are largely if not wholly free of such pathologies. The apparently quite recently developed toleration of maladaptive traits, which range from physical features universally related to neotenisation to mental disorder susceptibility alleles, and to almost countless other detrimental susceptibilities (e.g. thousands of Mendelian disorders), remained supremely unexplored. The preservation of the mutations deriving from multiple mutant alleles at different genetic loci involved in the major deleterious aetiologies had remained entirely unexplained until 2008. Instead, we had a dominant and aggressively promoted hypothesis proclaiming the replacement of all hominins by a new African species.

Not only does the alternative domestication hypothesis offer a plausible explanation for the toleration of human neuropathologies by natural selection. It also provides a credible elucidation for the extraordinary and relatively sudden changes that led to 'anatomically modern' people. These include the reduction in both brain size and somatic robusticity, as well as the loss of oestrus, and many other features so crucial to appreciating what it is that made us what we are today. It explains why the aetiologies of neuropathologies suggest that the phylogenetically most recent areas of the brain are affected, which are the very same areas that underwrite our advanced cognitive abilities. The domestication hypothesis also explains the absence of neurodegenerative diseases in other primates, and why human males today strongly prefer females presenting neotenous appearance and other features. It even offers an explanation for how modern humans coped with the atrophy of their brains, a topic not considered by the mainstream. Without some appreciation of these issues, human modernity has no causal context or explanation. Science, we would argue, expects some level of causal reasoning from us, and the failed replacement hypothesis never offered that.

The main points of our original paper (Bednarik 2008b), amplified and explained in more detail in Bednarik (2011c), were the following. The 'explosion' of the Upper Palaeolithic in Europe is a myth: there is no evidence that the rate of technological development between 45 ka and 28 ka, when Neanderthals lived, was higher than that of the second half of the Upper Palaeolithic. The assumption that the people of the Aurignacian are 'indistinguishable' from us in terms of cognition, anatomy, behaviour or cultural potential is a complete fallacy. No humans free of any robust features existed in the Early Upper Palaeolithic of Eurasia, which is a period of intermediate forms ranging from very gracile Robusts to still robust Graciles. Therefore, beginning about 45 ka ago, robust *Homo sapiens* types experienced rapid gracilisation. However, that process occurred much faster in females than in males. By 30 ka ago, the females were as gracile as the males only became 20,000 years later, with the end of the Pleistocene (Figure 2.13). These significant skeletal changes are not limited to Europe; they occur in all four continents then occupied. Food-processing techniques or lifestyles remained much the same well into the Holocene, so they cannot account for these significant skeletal revisions. The archaeological record does not indicate that the diet of humans changed so significantly between 45 ka and 30 ka ago that this could have become expressed in domestication traits. The reversal from encephalisation to brain atrophy also lacks an explanation. Dimorphism in mammals generally reflects one of two selection pressures: competition between males for access to females, or male-female differences in food-procuring strategies, with males provisioning females with fat and protein (Biesele 1993; Aiello and Wheeler 1995; Deacon 1997). Around 40 ka ago, cultural practices had become so dominating that mate choice was influenced by factors attributable to learned behaviour. In all animals, the phylogenetic direction is determined by reproductive success. Since culturally governed mating imperatives undeniably dominate sexual choice in present humans (Buss and Barnes 1986), it is inescapable that they must have been introduced at some point in the past, so the question is: when? We could look for signs that the attributes of natural fitness were replaced by characteristics that confer no Darwinian survival benefits. Alternatively, we could look for indications of a culturally mediated preoccupation with female sexuality. We find both factors between 40 ka and 30 ka ago, the latter in the form of frequent depiction of female attributes in the Aurignacian, ranging from vulvae (Delluc and Delluc 1978) to figurines.

Similarly, *Homo sapiens sapiens* is the only animal that shows distinctive *cultural* mating preferences concerning personality or anatomical traits, such as 'attractiveness' (Laland 1994). *It is self-evident that in a population whose males consistently favour neotenous females because of culturally transmitted preferences, neotenous genes will progressively become dominant.* It is equally self-evident that such a process resembles domestication: it partially replaces natural selection. Individuals considered attractive had more offspring, and it was they who replaced the robust genes, not a purported intrusive population from Africa:

> We have three basic hypotheses to account for the universal change from Robusts to Graciles: replacement by an invading population in four continents (for which we lack any evidence, be it skeletal, cultural or genetic); gene flow and introgression without any mass movement of population (which is somewhat more plausible, but fails to explain the apparent suspension of evolutionary canons); or cultural moderation of breeding patterns (i.e. domestication). Only the last-named option can account for all the hard evidence as it currently stands.
>
> (Bednarik 2008b: 12)

Individual physical attractiveness is a culturally determined construct; awareness of it does not seem to exist in any other species, in the sentient sense present-day humans perceive it. To the best of our knowledge, our closest living relatives, the bonobos, show in their mating behaviour no preference for youth, skin or hair colour, body ratio (e.g. hips to waist; Singh 1993), cephalofacial appearance (cardioidal strain), facial symmetry, or most especially neoteny in females. The same applies to any other primate – or any other species, for that matter. In extant humans, these preferences are strong and universal, and they are independent of cultural, social or religious customs and found in all societies investigated (Buss et al. 1990; Grammer and Thornhill 1994; Jones 1995; 1996; Shackelford and Larsen 1997; Barkow 2001). They are deeply ingrained in the neural systems of all present-day humans: Buss et al. (1990) reported this after investigating 37 population samples in 33 countries, and after reviewing almost 200 societies. Ford and Beach (1951: 94) concluded that female attractiveness is universally more important than male; while Gregersen (1983) arrives at the same conclusion based on nearly 300 mostly non-Western groups. This is another fundamental difference between present-day humans and other animals: the human emphasis on female attractiveness in mate selection is fundamentally different from all other animals, where almost without exception male attractiveness is more critical (leading to Darwin's exasperation with the male peacock). The shift from the natural and universal emphasis on male attractiveness in mate selection in which the female selects, to a human emphasis on female attributes in which the male selects, is another crucial difference not explained hitherto. This deficit of viable explanations of crucial differences between humans and other animals, or humans and other primates, runs like a red thread through the discipline. It seems almost deliberate but is more likely just the outcome of flawed models and distorted expectations deriving from systematic biases.

Most of the universal preferences observed in mate choice of modern humans seem to serve no processes of natural selection. The obvious exception is youth because it offers greater procreative potential through longer fecundity. Since human female fecundity ends with the menopause, this may explain the importance attached to feminine youthful or neonate appearance (Jones 1995). Humans are almost the only mammalian species that experiences the menopause, in the sense that the latter generally reproduce until they die; apparently, the only other menopausal species are short-finned pilot whales and killer whales. Therefore, one might be tempted to predict a connection between the early menopause of human females and the male preference for youth. Another proposed explanation for the universal male favouring of young females is that many men seek to dominate their partners, which is perceived to be easier with young females (Gowaty 1992). This idea may be negated by Berry and McArthur (1986) and does not seem to reflect the real world, in which far more complex processes negotiate relationships. Another variable, besides youth, of possibly Darwinian value could be facial symmetry, having been suggested to indicate elevated immunocompetence. However, most of the factors of male preference provide no survival or selective benefit whatsoever, especially the main elements of 'attractiveness' (an utterly arbitrary circumstance; Barkow 2001) and neotenous features. The neoteny that separates robust from gracile *Homo sapiens* representatives is decidedly detrimental to the species in its effects. For instance, massive reduction in skeletal robusticity (especially in the cranium) and physical strength both reduce rather than increase survival chances. Significant reduction in brain size (see Figure 3.1) at a rate 37 times greater than the previous long-term encephalisation severely challenges all previous palaeoanthropological reasoning concerning the role of encephalisation and plunges the traditional explanations of human evolution into chaos (Bednarik 2014e).

Gradual brain enlargement has long been the guiding light of the discipline, with the mantra of brain prevailing over brawn explaining why evolution tolerated the development of gracility. Humans more than compensated for their loss of robusticity by becoming smarter through their larger brains. In this, it was completely ignored that we have no evidence whatsoever that the gracile humans were more intelligent or had better communication or cognition than their more robust predecessors. If we review the limited evidence we currently have of the technology and cultural complexity of the 'Denisovans', for instance, we find that at least in some respects they had already attained Neolithic technologies. The green chlorite bracelet made by them perhaps 50 ka ago must have been created by the intricate method described by Semenov (1964: 74–83), which until now has been seen as a typical Neolithic invention (Figure 4.1 (i)). This is not the only indication suggesting that the 'smart

110 *Human self-domestication*

graciles' lacked the superiority attributed to them, relative to their robust predecessors. Many indicators suggest that the sophistication of palaeoart of robust or intermediate humans in Europe in the Early Upper Palaeolithic was higher than the more regimented and standardised production in the second half of the Upper Palaeolithic. Examples are the Aurignacian figurines of south-western Germany and Lower Austria, or the most elaborate cave art in the world, especially that of Chauvet Cave in France, all created by people more robust (Neanderthaloid) than present-day humans (Bednarik 2007; 2008c). When we add to this the significantly larger brain volume of 'Neanderthals' (between 13 per cent and 15 per cent greater than that of present-day humans), it becomes evident that the arguments citing our intellectual, cognitive, social or cultural superiority lack empirical evidence. If it were true that Siberian Denisovans had begun to reach Neolithic technology tens of thousands of years before Graciles 're-discovered' it in Neolithic times, another long-standing archaeological belief would hit the dust.

When we look more closely at neoteny (the attainment of sexual maturity before full somatic development) as explicitly expressed in human females, we find that the universal facial aspects are relatively large eyes, small nose, high forehead and reduced lower face (Jones 1995). Once again, they provide no prospects for natural selection, and like most other characteristics separating us from our robust ancestors, they are also typical features we share with mammalian domesticates. Neoteny is also called foetalisation, and it has long been

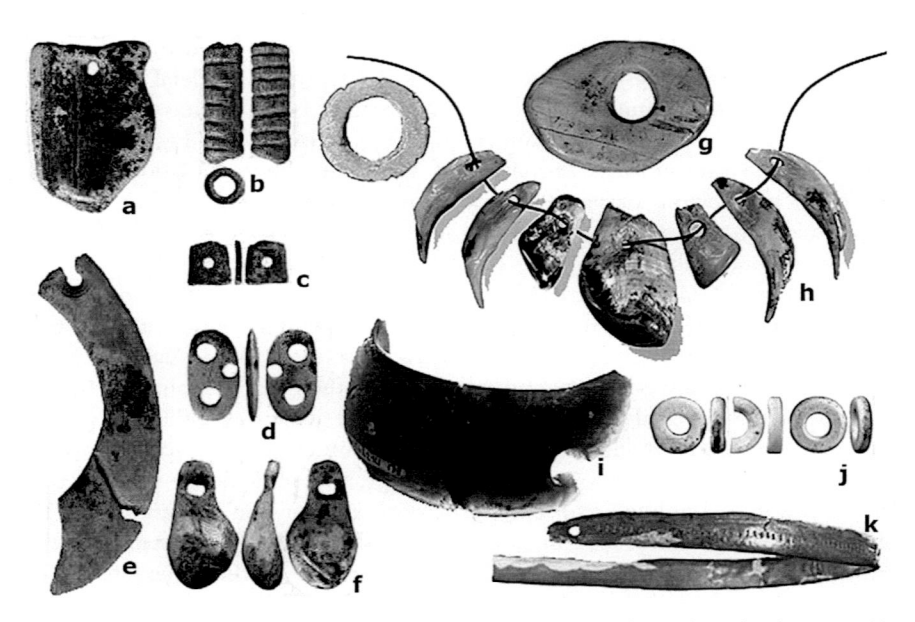

Figure 4.1 Presumed decorative objects, including a green chlorite bracelet fragment (i), made by Denisovans from Denisova Cave, Siberia

known that of all the animals in the world, modern humans are most like foetal chimpanzees (see Chapter 2). The neotenous traits characterising present-day humans include smaller body size, more delicate skin and skeleton, smaller mastoid features, thinness of cranium bones, flattened and broadened face, significantly reduced or absent tori, relatively large eyes, smallish nose, small teeth, less hair but retention of foetal hair, higher pitch of voice, more forward tilt of head but more backward tilt of pelvis, smoother ligament attachments and narrower joints, and limbs that are proportionally short relative to the torso, especially the arms; but they include also increased longevity, lower amount of energy expended at rest, faster heartbeat and prolonged development period (Ashley Montagu 1960). These developments, we have noted, need to be explained because many of them contradict the principles of evolution.

The lack of oestrus and the periodicity of libido in the female are other features of modern humans the origins of which anthropologists have not explained. Indeed, the causes of none of the traits characterising us as a subspecies (i.e. distinguishing us from other members of the *Homo sapiens* family) have been adequately elucidated, including our appearance, neoteny, intellect, language ability, cognition, exogram use, exclusive homosexuality, ability to form constructs of reality, high susceptibility to neuropathologies, and so forth. In the case of sexual receptivity of modern human females, they share the lack of oestrus with very few other animals. Two potential explanations offer themselves. One involves the need for frequently pregnant hominin females to have proper access to fat and protein for the growth of their unborn. If they could not obtain such foods themselves because of dependent children, they would have had to secure male support, i.e. provisioning by a male consort. Such bonding would have favoured the selection of female mutations allowing extended periods of sexual receptivity that could have led to the abandonment of oestrus (Lovejoy 1981; Leonard and Robertson 1992; 1994; 1997; Biesele 1993; Aiello and Wheeler 1995; Deacon 1997; Leonard 2002). Female chimps are known to only secure meat from a kill after first copulating with the successful hunter, and it has been suggested that human females who were longer or always receptive were similarly better supplied with the animal protein needed by their unborn.

The alternative second explanation for the loss of oestrus in humans is more straightforward and would favour a relatively late introduction. Most domesticated animal species lack the seasonal reproduction of their wild ancestors, and many can reproduce themselves at almost any time of the year. Given the extensive other evidence of rapid domestication-driven neotenisation in humans towards the end of the Pleistocene, the abolition of oestrus may well be attributable to the same cause. Selection in favour of infantile physiology has occurred at various times in hominin evolution, but at the start of the last quarter of the Late Pleistocene, effectively around 40 ka ago, this human foetalisation or paedomorphosis (Shea 1989) accelerated at such an unprecedented rate that palaeoanthropologists mistakenly perceived only one possible explanation: a gracile new species replaced all robust populations of the world.

Therefore, the former must have evolved in some specific region. The alternative explanation for this rapid change is presented here: not by evolution, but by domestication. All robust skeletal features were shed in a relatively short time, but the neotenisation process is continuing today. Muscle bulk waned significantly, as did brain volume. Around 35 ka ago, partially gracile specimens occur in Australia and Europe. A distinctive sexual dimorphism becomes evident in the skeletal remains, with gracilisation occurring much faster in the presumed females than the males (Bednarik 2008b). This includes the globularisation of crania, the rapid reduction of supraorbital tori, occipital projections and cranial bone thickness, the decline in prognathism and reduction of tooth sizes. Ten thousand years later, female skeletal architecture has become significantly more gracile, but contrary to the claims of the African Eve advocates, it is still much more robust than at the end of the Pleistocene – or, for that matter, today. This process is much slower in the males, who at 35 ka before the present day remained almost as robust as late, 'gracile Neanderthals'. By the beginning of the Holocene, the dimorphic gap has become markedly smaller, but gracilisation continues right up to the present time (Figure 2.13).

Bearing in mind that the skeletal and genetic changes – the only ones ever considered by the replacement advocates – towards the end of the Pleistocene are only a small part of the change from robust to gracile *Homo sapiens* types, the futility of the African Eve hypothesis becomes starkly apparent. Its most prominent feature by far is that it explains virtually nothing. Its keystone principle that Graciles are a separate species has been falsified by the evidence that they could breed with Robusts such as 'Neanderthals' and 'Denisovans' (and quite probably even with earlier hominins). Therefore, the hypothesis has no longer much credence in any case, and yet its principal advocates continue to support it (Stringer 2014). One of them, Chris Stringer, has proudly informed us that he rejected the publication of the alternative domestication hypothesis when asked to referee our paper submitted to an establishment journal. Not only do these detractors of open debate continue to censure any opposing propositions, they even seem to be proud of their censorship. It does not even seem to occur to them that a just-so hypothesis which, in fact, does not explain or elucidate anything, is not particularly useful. The replacement hypothesis does not explicate the reduction in cranial volume, which is indirectly contradicted by the proposition that the larger brains of the 'anatomically modern humans' helped them defeat the 'inferior' Robusts. The hypothesis does not explain the preservation of thousands of genetic disorder alleles. It offers no rationalisation for the appearance of neurogenerative diseases, of which other animals are free. It fails to account for the significant and accelerating reduction in human dentition size (Brace et al. 1987). Why brain illnesses involve mostly those areas of the brain that are the phylogenetically most recent remains unexplored by it. Nor does it attempt to explain the human menopause or, for that matter, the abolition of oestrus in humans. The replacement hypothesis is silent on the dramatic neotenisation of humans in the last few tens of millennia; it explains it away with some unlikely Biblical narrative about Exoduses and

Eves and Edens. The same applies to the loss of bone robustness and of muscle bulk, and thus physical strength. This concocted hypothesis does not explain exclusive homosexuality, Mendelian disorders, or pronounced male preference of neotenous females. Nor does it explain why the extant human is the only mammal whose males select females based on cultural variables. The domestication hypothesis explains all of these factors, and many more still.

For instance, the most disturbing aspect of the replacement hypothesis is its demand that all pre-Eve humans were too primitive to master symbolic behaviour, language or art-like production. In the quest to sustain this assertion, servile followers have systematically and categorically rejected any such empirical evidence predating African Eve's progeny. This campaign of discrediting any finds contradicting the replacement dogma, that earlier humans lacked any cognitive sophistication, has now been conducted for several decades. It is profoundly unscientific and intellectually corrupt. Most of its academic minions have not seen any of the relevant finds or find sites, relying on and misinterpreting the accounts of others (e.g. Langley et al. 2008). The wholesale dismissal of all palaeoart finds predating Mode 4 (Upper Palaeolithic) industries and 'anatomically modern humans' has shown the proselytising traits of this programme. This is partly because it is in this area that African Eve's advocates encountered the fiercest opposition, partly it is where their boat best revealed its leaks early on. After all, there is far more Pleistocene rock art in Australia than there is in Europe (Bednarik 2014a; 2017a), yet all of that in Australia is of Mode 3 industries, while in Europe most of it is of Mode 4 traditions. The obviousness of this one factor alone suffices to expose the ignorance of the replacement archaeologists because it negates their core belief that sophisticated cognition was introduced together with Mode 4 technologies. Not only is it evident that Upper Palaeolithic-type technologies were not introduced in Sub-Saharan Africa, African Eve's purported homeland, any earlier than elsewhere; they developed very slowly *in situ*, over a transitional period of tens of thousands of years, across Eurasia (Camps and Chauhan 2009). The massive corpus of Australian palaeoart of the Pleistocene and the early Holocene was produced by people of the 'core and scraper' technology, a typical Mode 3 industry. How could the replacement camp have been so oblivious to this factor? Today we know that the Robusts who lived in Denisova Cave in southern Siberia between 52 ka and 195 ka ago (Douka et al. 2019) left us a treasure trove of jewellery unequalled in its technical sophistication during the much later Upper Palaeolithic (Mode 4) of both Eurasia and Africa (see Figure 4.1). Still, there is no admission from the African Eve apostles that they misjudged the empirical evidence rather severely.

Their posturing and self-justifying defences of this programme need to be ignored. It was initially based on Protsch's hoax, has since provided no credible empirical support, has never offered any explanatory potential, and now has lost its keystone elements. The only redeeming feature of the replacement hypothesis is its feel-good message that all of humanity is related. Even that was accomplished at the expense of positing the complete extinction of all humans of the world. Nothing is exhilarating about the knowledge that one is a descendant of a 'tribe' that 'replaced' all humans on Earth, and that is the very embodiment of genocide. In the end, there is nothing to emancipate this doomed hypothesis. However, its most significant scientific deficiency is that it explains none of the changes from robust to gracile humans; it just posits that they occurred.

The domestication hypothesis, in stark contrast, provides explanations for literally all of these changes. Many, if not most, of them are attributable to rapid neotenisation, especially in the females initially. All of these changes are entirely consistent with those generally attributable to domestication, and neoteny is a major 'by-product' of domestication. Although deleterious in the context of natural selection, neoteny can, as noted above, also involve evolutionary benefits. Most obviously, it facilitates the retention of neuroplasticity or 'morphological evolvability' (de Beer 1930: 93). This can mean the extension of youthful inquisitiveness and exploratory behaviour as well as playful behaviour patterns continuing into maturity. The effect is easy to see in the difference between the behaviour of the wolf and its domesticate, the dog: domesticates preserve not only juvenile morphology but also aspects of juvenile responses to particular situations or stimuli. The innovations gradually introduced in the course of the Early Upper Palaeolithic, from 45–28 ka ago, may well be explicable by these behavioural changes. They seem to reflect a certain amount of experimentation with, and exploration of, media and with possible solutions. These changes, reflected in technology and palaeoart, are universal from southern Africa to eastern Asia, but they seem to occur first in Siberia. Hence the pattern of an early development in Africa radiating from there is again contradicted by the empirical data.

In a species whose behaviour is increasingly moderated by cultural imperatives, the corresponding plasticity of cultural behaviour is likely to foster the curiosity, inventiveness and inquisitiveness of youth. To be more specific, the introduction of figurative (iconic) palaeoart is probably attributable to children and adolescents (Bednarik 2007; 2008c). Nearly all Pleistocene palaeoart of Europe that can be assigned to particular age cohorts on empirical evidence is the work of non-adults, and while this does not imply that young people made all palaeoart of the time, it does suggest their major participation. The change from the relatively regimented palaeoart forms of the Lower and Middle Palaeolithic traditions to the more permutable figurative system also likely reflects the trend towards neotenous attributes that marks the final Pleistocene. Vestiges of it can be observed in the attitudes of some extant societies such as the Australian Aborigines or the Andaman Jarawas, who tend to regard non-iconic palaeoart as the more profound and mature form (Bednarik and Sreenathan 2012).

Not only does the domestication hypothesis explain the numerous changes from robust to gracile humans, but it also does so with a universal and generic rationalisation that seems to clarify all aspects of what it means to be a present-day human. The theory states that all of what makes us the way we are

(Bednarik 2011c) derives from a single factor. Between 45 ka and 30 ka, robust humans in four continents developed concepts ('memes') of individual and entirely arbitrary properties one can loosely summarise as 'female attractiveness', and these cultural constructs spread as a kind of 'memetic introgression'. They led to a particular interest in female attributes well reflected in palaeoart, but also in preferences in the male selection of mating partners. We have no choice in accepting this proposition because we know from extensive research that such selection is very strong and universal in contemporary humans, and since it cannot have existed in ancestral populations at some distant point in time, it must have been introduced at some stage. This is inescapable. So we need to ask, what would have been the effects of such an unprecedented selection factor? Clearly, it would have impeded purely natural selection and would have set in motion a development favouring specific genes lacking traditional selective values. The name for a process of selecting genes by non-evolutionary changes is 'domestication': phylogenetic progress is not determined by the environment or other dysteleological factors, but by teleological selection.

Next, we need to ask: what would have been the consequences of affecting the genome to meet cultural imperatives? The study of domestication as a universal phenomenon leaves no doubt: beginning with neoteny, all of the characteristics distinguishing us from our ancestors, the Robusts, are explicable as the outcomes of domestication. We can also ask: is there an alternative explanation for the changes from robust to gracile humans? Leaving aside the replacement by a new species, there is not. It would be an incredible coincidence if all of these many changes just happened without a cause, and just accidentally resembled effects of domestication. Does this prove the theory to be true? Certainly not: the theory of evolution has also not been demonstrated to be true. However, until a better explanation is offered, it is the best we have for the phenomena of nature. The same applies to the auto-domestication theory: one day, a superior interpretation may arise, but until then it remains the only logical explanation so far offered for the relevant observations.

Testing the domestication hypothesis

This leads to the question: does it mean that the auto-domestication of humans has been explained in every detail? No, it has not – not by a long shot. It merely provides a credible and workable framework within which questions about the origins of human modernity can be raised, considered and tested. Nevertheless, many specifics remain unclarified. To illustrate this point, let us reflect upon just one issue that emerged in the previous sub-section: why do present-day human males experience a preference for youth in their mate choices? It seems such a simple question, and yet its detailed consideration provides an inkling of the true complexities of the factors defining the domestication case. The elements to be included in this contemplation are not just that modern humans are the only animal species that prefers youth in a female. This is itself a very peculiar feature, but then others might be related to it. Modern humans are almost the only species that experiences menopause. Surely a pre-determined limitation to reproductive ability is not a condition advantageous to natural selection. In the same context, the anomalous human preference of neoteny, especially in females, needs to be explained, because it, too, does not occur in the animal world. Infantile features, in other species, are signals of many kinds that tend to elicit nurturing responses in the adults of the species, and it simply does not seem likely that human male responses developed from these. It has sometimes been suggested that the prolonged post-reproductive lifespan in modern women provides certain fitness benefits (Chapman et al. 2019). Such higher fitness would benefit the daughters of grandmothers by caring for grandchildren, thus ultimately increasing the breeding success of the offspring. There appears to be some merit in this argument, but it does not suffice to clarify how the potential for menopause arose in the first place: for any evolutionary variable to be available for selection it must first exist and be available for adoption. It is difficult to see how that might have happened in this case.

What seems much easier to grasp is how youth got to be chosen, being of genuine value to evolution as noted above, due to potentially longer fecundity. However, that seems to point to menopause having been introduced first. Another possibility is that, as youth became established as a selective factor for females, mutations promoting neotenous features could have been favoured in sexual selection. That may sound like an appealing proposition, but it must be remembered that there are so many other potential variables involved in sexual selection that actual youth should not be expected to have been a significant player. Indeed, since it would appear to be selection neutral (it would not have affected phylogenic direction), it is unlikely to have had an impact. Nevertheless, if females had 'feigned' youth through neoteny, that would be quite different; it would have certainly had genetic consequences. The abolition of oestrus is also closely related to the matter at hand and would have been strongly selected. Particularly noteworthy is the role reversal between males and females, which is another fundamental difference between us and all other animals: present-day humans are the only species in which males exercise sexual selection (directly or indirectly), rather than vice versa. We can appreciate from this example of the constellation of youth, neoteny, oestrus and menopause the full complexity of the countless factors involved in human self-domestication.

There are numerous other details of the spectrum of relevant variables defining the auto-domestication hypothesis that remain unexplored and unclarified for the time being. The almost complete paucity of critical reviews of this theory is, therefore, regrettable. More than ten years after its first publication, we are aware of only one detailed review of it, presented in an obscure doctoral thesis. Most disappointingly there has been no reaction from the high priesthood – to use Thompson's (2014) term – of the replacement (African Eve) hypothesis. Apart from choosing to reject its publication in fashion journals, the response from that quarter has been a deafening silence. We shall, therefore, have to limit our response to critiques to the observations in the thesis of Thomas (2013). He expresses his traditional perception of the phenomenon of domestication:

After all, doesn't domestication definitionally require a domesticat*or*? Isn't it, in other words, something done *by* one species *to* another species? If so, who is 'doing it' to humans? The question here, then, is whether the concept of self-domestication is one that can be talked about with any coherence at all.

(ibid.: 9, original emphases)

He thus echoes Darwin (1871: 29); "civilized men, who in one sense are highly domesticated") and also extends the twentieth-century bias of linking domestication with domesticity (i.e. the 'Neolithic' revolution), thereby following the tradition already established by Boas, Wilson, Hodder and Leach, among others. Thomas is even well aware that the bonobo (*Pan paniscus*) has been suggested to have undergone a self-domestication process (Hare et al. 2012), and that much the same has been said about the dog (Groves 1999; Hare and Tomasello 2005; Miklósi 2007). As noted above, there is no sharp separation of domestication from symbiosis; there is a continuum. Domestication is certainly not defined by who does what to whom; in a scientific sense, it merely refers to mutations that display traits of the domestication syndrome. In many cases, e.g. among invertebrates, domesticator and domesticate are interchangeable entities (e.g. Mueller et al. 2005; Kooij et al. 2011), and the belief that humans are always the domesticators is biologically naive: most domestications occurred without human involvement.

Thomas's ambivalent understanding of domestication is also manifest in the overall tenor of his thesis, by how he links human domestication with language origins. His great emphasis on the latter is understandable: his thesis is a salute to the work of his thesis supervisor, Simon Kirby. However, in considering the beginnings of language, he ignores most of the critical contributions to this great debate (for instance, Everett 2005; 2008; Falk 2009; while Bickerton is misinterpreted). The problem, of course, is that domestication cannot be linked to language formation in any way, as Thomas (2013: 176) does. We have finally overcome the irrational tendency of archaeologists to posit a 'late' introduction of language ability, yet Thomas connects it to an event that has occurred rather late in human history. After that kerfuffle about the hyoid bone by authors ignorant of its role (consider the australopithecine hyoid find; Alemseged et al. 2006), most credible authors on the subject nowadays support the introduction of some form of human language at least one million years ago, probably much earlier. There is also the little matter of the FOXP2 language gene on chromosome 7 from 'Neanderthal' remains (Krause et al. 2007; cf. Enard et al. 2002; Zhang et al. 2002; Sanjuan et al. 2006). The history of maritime colonisation (Bednarik 2003a; 2014d), extending back to the Early Pleistocene, suffices to purge the notion of language appearing as an outcome of domestication. Moreover, no domesticates other than humans have language, yet Thomas links self-domestication, a concept he seems to eschew, with language origins.

This leaves us with the task of finding in his critique elements that might usefully help in testing the domestication theory. Of interest is Thomas's objection that the significant reduction of brain volume towards the end of the Pleistocene (Henneberg 1998; Bednarik 2014e) can be explained with a concomitant reduction in body size, citing Ruff et al. (1997). However, according to these authors, the decline in body mass during the entire Holocene was less than 5 kg, which would translate into a corresponding decline in brain volume of little more than 20 ml. This is only a fraction of the observed atrophy during that time. Henneberg (1988; 1990) reports a 'Mesolithic' mean volume for males of 1,567 ml ($n = 35$); this declined to a mean of 1,391 ml in recent times, a total Holocene reduction of at least eight times greater than 20 ml. Similar decreases over the same period have been recorded in other continents (Carlson 1976; Van Gerven et al. 1977; Brown 1987; 1992; Henneberg and Steyn 1993; Brown and Maeda 2004). Moreover, there is no such correlation between body size and the continuous expansion in endocranial volume marking almost the entire Pleistocene (Rightmire 2004). Hominins of the early part of the Pleistocene were of much the same height as extant humans, as indicated for instance by KNM-WT 15000 (Turkana boy), KNM-ER 736 and KNM-ER 1808.

More importantly, Thomas misunderstands our argument, which rejects the old mantra of encephalisation having been necessary to drive cultural or cognitive complexity. That dogma seems to fall apart when we discover that brain atrophy begins at a time of growing demands on brainpower. Also, while we can measure cranial volume fairly precisely, we cannot measure former body volume, we can at best estimate it (based on assumptions about soft tissue bulk). Nevertheless, the main issue remains that the brain size relative to body size is irrelevant here; the *absolute brain size* is what counts. It determines how early in its development the human embryo is expelled relative to all other species, rendering the infant entirely dependent upon parental support for many years. The evolutionary and social costs of this were enormous, and entirely unnecessary if a smaller brain had been just as good, as noted above.

Another belief of Thomas we need to disagree with is that "sexual selection is actually a subset of natural selection" (Thomas 2013: 214). This does not extend to sexual selection driven by cultural imperatives: if a population decided to limit breeding to docile individuals by sexual selection, it would not be natural selection by any stretch of the imagination. However, it would very likely lead to the introduction of a wide range of other changes by pleiotropy, all deriving from the domestication syndrome (Hammer 1984; Wilkins et al. 2014). Some pages later Thomas argues against himself that "the need to compete for mates [is] not [an] evolutionary" process in itself, but an aspect "of the adaptive landscape against which process[es] such as selection" occurs (Thomas 2013: 237). There are various other perplexing statements, such as "humans and other domesticates have long shared this environment [and] it should really be no surprise that similar changes might have occurred both in humans and domesticates" (ibid.: 238). It is obvious that to assume two species sharing an environment would therefore also share domestication would be absurd.

Overall, Thomas confuses cause and effect in trying to separate the hundreds of changes from robust to gracile into what he calls condition-dependent and condition-independent models. All of these changes are effects, yet he mistakenly thinks we present them as causes. According to the auto-domestication theory, these mostly deleterious changes all result from one single factor: the selection of traits that offer no evolutionary benefits in any environment, traits that may have included attractiveness (a purely cultural construct), neoteny, body ratio, cephalofacial features, or whatever else was considered sexually attractive at the time. Thomas fails to appreciate that it is not even important what the factors were that caused human development to veer off its natural evolutionary course suddenly. Whatever they were, we have extensively documented the effects, and we have sought to determine the cause. The most consequential error he makes is to claim that we "assume ... the favoured trait is detrimental to fitness" (ibid.: 265). The favoured trait, be it attractiveness or whatever, is not in itself detrimental to fitness. Sustained selection for it, however, *can be* detrimental to it, because like selection for docility (or any other persistent selection of trait), it likely activates the domestication syndrome and the very many detrimental traits associated with it. The problem, very simply is that Thomas does not appreciate the role of pleiotropy in all this.

It is, therefore, necessary to elaborate on this again. Pleiotropy occurs when one gene influences two or more seemingly unrelated phenotypic traits. Therefore, mutation in a pleiotropic gene may have effects on several traits simultaneously, and these tend to be deleterious. For instance, deletion in the 22q11.2 region of chromosome 22 has been associated with both autism and schizophrenia (Vorstman et al. 2013), even though these diseases manifest very differently from each other. Numerous further pleiotropic effects have been identified, one of the best known being sickle cell anaemia: a mutated HBB gene produces numerous consequences throughout the body. Albinism is attributable to a pleiotropic mutation of the TYR gene, while Marfan syndrome is caused by one of over 1,000 different mutations of the FBN1 gene. The pleiotropic effects in domestication have been mentioned before: selection for one trait, for instance, tameness, affects a host of other traits collectively defined as the domestication syndrome. This aetiological nature of domestication needs to be fully appreciated in considering the auto-domestication theory.

The surprising aspect of the domestication syndrome is that the circumstances must have differed in every domestication event caused by humans; yet the expressions of the syndrome can be seen to share many commonalities across all these species. It is therefore justified to ask what could have led from very diverse variables to such a uniform outcome. A prime trait in humanly directed domestication was probably docility, but it is also apparent that in many cases other priorities would have been involved. Until the genetics of domestication are better understood, we can only assume that significant distortions of breeding patterns may result in an aetiological shift towards the domestication syndrome. Some aspects of it are not limited to mammals; non-mammalian species can present broadly similar changes.

One hypothesis that could be offered as an alternative to auto-domestication theory as presented here would propose domestication by systematic selection against aggression among humans, rather similar to the purported mechanism of many of the domestication events created by humans. If that had occurred, it could certainly account for the many domestication traits observed in present-day humans. The most plausible rationalisation of this kind presented so far is called the cooking hypothesis (Wrangham et al. 1999; Wrangham and Conklin-Brittain 2003; Wrangham 2009). The digestive system of humans is small relative to other primates and appears to be adapted to eating mainly cooked food. The oldest currently known hearth is a 2 m long, about 8 cm thick, ash layer some 30 m into Wonderwerk Cave, South Africa (Figure 4.2), at the base of one of the several large excavation trenches in that cave. The hearth is of the Lower Pleistocene and in the order of 1.7 million years old (Beaumont 2011). We have studied this feature carefully and confirm that the thick ash layer contains numerous calcined animal bones, many of them fractured. There is no reasonable doubt that this is a fireplace used by early *Homo erectus* to cook meat shortly after the site's Oldowan occupation ended. It may not prove that fire was also used in other parts of the world at that time, but it stands to reason that the occupation of the northern regions of the Northern Hemisphere by later hominins would have involved the use of fire.

Certainly, the rather early use of fire in cooking meat would help explain the human digestive system's preference for such food, and why humans have a relatively small gut, caecum and colon. Less convincing is the second part of the cooking hypothesis: that this development involved a change in behaviour as more 'co-operative' individuals were favoured over more aggressive conspecifics. It is difficult to envisage a scenario in which such more social behaviour would have been selected. The bridging argument offered by Wrangham (2009: 158) is that females and low-status males would have found it "absurdly difficult" to cook their meals, which would have been stolen by aggressive males. Therefore, females had to form bonds with males to protect them (Wrangham et al. 1999: 575), which selected in favour of such supportive males.

This narrative seems not only contrived, but it also clashes with the male provisioning hypothesis, which we have considered above: that the only way the frequently pregnant or lactating females could acquire the level and type of protein and fat-rich nourishment they needed was to be reliably supplied by males. Whether these necessities were raw or cooked seems irrelevant, and if females were provisioned before the introduction of cooking, then that would have continued, and there is no justification for the cooking hypothesis. Moreover, we are here considering the proposition that domestication might be the result of selection against aggression, and if cooking was introduced in the earliest Pleistocene, we need to clarify that it led to no evidence of domestication whatsoever. *Homo erectus* continued almost unchanged, evolving very slowly, for hundreds of millennia, to the best of our present knowledge.

Figure 4.2 Calcined bone fragment in the earliest-known fireplace, 1.7 million years old, Wonderwerk Cave, South Africa

This discussion testing the possibility of reduced human aggression also leads to another issue: the lack of empirical evidence that aggression in late humans is distinctly lower than in, say, chimpanzees. Even the most cursory look at known human history makes it amply clear: humans remain highly aggressive and violent to this day. They kill for thrills and sport; the alpha males wage incredibly destructive wars to get their way and humans are easily conditioned to inflict great pain on their conspecifics. Just three regimes of the twentieth century, those of Mao, Hitler and Stalin, have claimed more than 100 million human lives. Looking at history more generally, it is a story of suffering on an astronomical scale. Defining us as a species that became domesticated through a process of genetically reducing aggression denies the inherent viciousness of human nature and diminishes the sacrifice of those who suffered from it.

More productive than these preliminary considerations of testing the auto-domestication hypothesis is to review the relevant empirical evidence. The veracity of any testable hypothesis is tested by subjecting it to falsification attempts. In the case of the auto-domestication theory, testing is readily possible by various means. The most obvious is by investigating how many genetic markers signalling domestication can be detected in the genome of present-day humans. In 2007, the time of the theory's inception, the genetic markers of domestication were unknown, but in recent years a great deal of such data has become available. The new genetic evidence corroborates the auto-domestication hypothesis, and our proposal has found a good number of supporters. There have been various developments since 2007 that either falsify the African Eve theory or verify our alternative. For instance, the formerly 'brutish' Neanderthals, which in the heyday of the replacement hypothesis were suggested to belong to the apes rather than the hominins (Davidson and Noble 1990), were reported to have produced petroglyphs (Rodríguez-Vidal et al. 2014; actually, that had long been known: Peyrony 1934) and to have used parts of raptor birds for body decoration (Finlayson et al. 2012; Morin and Laroulandie 2012). As recently as 2003, most archaeologists rejected the idea attributing seafaring skills to Robusts and also sought to explain away the Indonesian evidence concerning the island colonisations by *Homo erectus* (Davidson 2003; Rowland 2003; Spriggs 2003). Since then, two American teams have demonstrated the Lower Palaeolithic presence of hominins in the Mediterranean islands of Crete and Gavdos (Mortensen 2008; Strasser et al. 2010; 2011; Simmons 2014) and a Russian team has claimed the presence of people with an Oldowan technology (Mode 1) on the island of Socotra off the Horn of Africa (Amirkhanov et al. 2009).

Relevant new developments have weakened the replacement hypothesis: the purported speciation event supposedly responsible for the appearance of a new gracile species has never occurred. Genetics have shown, indirectly, that robust and gracile *Homo sapiens* types are of the same species (Krause et al. 2007; Green et al. 2010; Reich et al. 2010; Rogers et al. 2017). Moreover, all other relevant new information, e.g. from genetics, supports the auto-domestication theory as we will see below. Incredibly, the African Eve advocates are unmoved (d'Errico and Stringer 2011; Stringer 2014) and continue muddying the waters, for instance, with their ruse of confining the term 'humans' to their 'anatomically modern humans'. In other words, they seem unaware that the word '*homo*' means 'human', and that therefore all hominins since *Homo habilis* are considered human. They also need to take note that, based on current understanding, all hominins of the last half-million years or so appear to have been interfertile; they are therefore all subspecies of *Homo sapiens*. That had not only been known many decades ago when the name *Homo sapiens neanderthalensis* was applied to those 'apes'; it also means that this subspecies, as well as the Denisovans and various other robust hominins of the Middle and Late Pleistocene, are all likely to belong to the *H. sapiens* species. Quite possibly, this includes *H. antecessor* and *H. heidelbergensis*. Who knows, perhaps even late *H. erectus* will find himself in this exclusive club, as Wolpoff (1999) has long suggested.

Recent work in genetics research has suggested significant overlaps between selective sweeps in anatomically modern humans and several domesticated species. This allows one way of testing the domestication hypothesis (Bednarik 2008b), which would predict that, irrespective of the course of that development, it should be reflected in the genome. The domestication syndrome has recently been attributed to a large number of genetic expressions. Ever since Darwin (1868), it had been noted that mammalian domestication yielded particular sets of behavioural and morphological traits: reductions in general brain size and specific parts of the brain; neoteny; significant changes to oestrus; increased tameness and docility; changes to cranial and facial morphology and reduction of tooth sizes; changes in adrenocorticotropic hormone levels; and alterations to the concentrations of neurotransmitters are the main factors recognised today. Not all these factor apply equally to all domesticates, but there are adequate consistencies to justify the application of Hammer's (1984) term "domestication syndrome"; cf. Brown et al. (2008) and Wilkins et al. (2014).

Neither of Darwin's two potential explanations of domestication is accepted today: the traits are not caused by improved living conditions (cf. Leach 2003), nor are they attributable to hybridisation. In the first case, it is well established that feral domesticates preserve their traits for many generations, as noted above (Bednarik 2011c); in the second case, we know from domestication experiments with rats and foxes that there is no hybridisation involved (King and Donaldson 1929; Castle 1947; Belyaev 1969; Trut et al. 2009). The potential involvement of neural crest cells (Wilkins et al. 2014) to explain how selection for tameness might influence embryological developments of tissues is an important consideration. The properties of the domestication syndrome can possibly be traced to mild neural crest cell deficits (but see Sánchez-Villagra et al. 2016). Of relevance is that one of the domestication genes shared by modern humans, dogs and cattle, FAM172A, neighbours NR2FI on chromosome 5, which regulates neural crest specifier genes (Simões-Costa and Bronner 2015). A hypothesis alternative to the implication of neural crest cells, involving the thyroid gland instead, has been proposed by Crockford (2000; 2002; 2009).

The most prominent behavioural and neural aspect of the domestication syndrome (Sapolsky 1992; Künzl and Sachser 1999) is the increase in docility relative to progenitor populations, as noted above perhaps the principal characteristic sought by the human domesticator. It is explained as reduced adrenal size, as noted in domesticated foxes (Osadschuk 1997; Oskina 1997). The reduction of androgen levels and rise in oestrogen levels (Cieri et al. 2014), often associated with lower reactivity of the hypothalamus-pituitary-adrenal system (Trut et al. 2009), is thought to be a key trait in domestication. Similar changes apply in domesticated rats (Albert et al. 2008) and birds (Schütz et al. 2001; Suzuki et al. 2012). As tameness, which is essentially suppression of flight reactions, is the initial selection agent, it is assumed that it is the primary cause of domestication. While this may be valid in the case of mammals and birds domesticated by humans, it seems much less likely to be the driving force in human auto-domestication or other domestications. This is despite the proposal that the digit ratios (D2:D4; Nelson et al. 2006) attributed to Neanderthals imply that these (and many other hominoids) had a higher prenatal androgen exposure than present humans (Nelson et al. 2011). The selection for tameness, sometimes ill-defined as 'pro-sociality', refers to tameness relative to humans (interspecific), not intraspecific tameness. Nor, for that matter, is tameness the only reason for domestication: many domesticated species cannot even be 'tamed' (silkworms, for instance).

Neanderthals appear to have been free of such mental illnesses as schizophrenia: the NRG3 gene, associated with it, seems to be absent in them. As noted in Chapter 3, schizophrenia may well be a very recent condition that may have appeared only a few centuries ago (Jeste et al. 1985; Hare 1988; Bednarik and Helvenson 2012). Selective sweeps in regions associated in genome scans with psychosis, such as 1q21 (Voight et al. 2006), tend to yield relatively recent aetiologies, of less than 20 ka, as predicted by the domestication hypothesis. Much the same applies to many other deleterious alleles in 'modern' humans, such as RUNX2 and CBRA1 (causing cleidocranial dysplasia or delayed closure of cranial sutures, malformed clavicles and dental abnormalities), THADA (associated with type 2 diabetes), the microcephalin D allele (perhaps 14 ka old; Evans et al. 2005) and another contributor to microcephaly, the ASPM allele (5.8 ka; Mekel-Bobrov et al. 2005).

The significant advances made in genomics in recent years have facilitated some preliminary understanding of the phenotypic traits underwriting animal domestication as a generic phenomenon (Morey 1994; O'Connor 1997; Zeder 2006; Zeder et al. 2006). Indeed, the precise meaning of 'domestication' has recently changed: various species are now thought to have 'self-domesticated' rather than having been 'passive parties' in the process (Groves 1999; Marshall-Pescini et al. 2017). Such species have benefitted from the changes just as has the human domesticator (Budiansky 1992; Morey 1994; Driscoll et al. 2009; Zeder 2012b).

Selective sweeps in the genomes of modern humans and several domesticated species have recently revealed dozens of overlapping genes (Prüfer et al. 2014; Racimo 2016; Peyrégne et al. 2017). The following domestication genes overlap in the horse and human: AMBRA1, BRAF, CACNA1D, DLGAP1, NT5DC2, NTM and STAB1. In cattle and humans, they are ERBB4, FAM172A, GRIK3, LRP1B, PLAC8L1, PVRL3, SNRPD1, TAS2R16 and ZNF521. The dog has yielded these 15 domestication genes overlapping with present humans: COA5, COL11A1, COQ10B, FAM172A, GGT7, GRIK3, HSPD1, HSPE1, LYST, MOB4, NCOA6, RFTN2, RNPC3, SF3B1 and SKA2. Finally, the cat shares with us BRAF, GRIA1, HSD3B7, ITGA9, MYLK3, NEK4, PLAC8L1, PPAP2A, PPAPDC1B, PRR11, RNPC3, SYTL1, TEX14, TP53BP1 and ZMYND10. These 41 genes associated with loci under positive selection, both in present humans and in one or more of the four domesticates considered, do not prove that domestication in these five species necessarily proceeded analogously. The circumstances would have differed in the domestication of various species. However, these shared genes do suggest that humans were, in their relatively recent genetic history, subjected to changes that resemble those of domestication in other mammalian domesticates. Nevertheless, it must be cautioned, first, that few of the above genes are shared across the five domesticates; and, second, that numerous genes are under selection in various domesticates, but not in humans.

Another area being clarified by recent insights into genetics is that none of the 17,367 protein-coding genes in two Neanderthals from Spain and Croatia (Castellano et al. 2014) are listed among the 15 genes known to overlap between at least two domesticated species (ADAMTS13, ATXN7L1, BRAF, CLEC5A, DCC, FAM172A, GRIK3, NRG2, PLAC8L1, RNPC3, SEC24A, SMG6, STK10, TMEM132D and VEZT). While not providing finite proof, this factor does seem to confirm the 'pre-domestication' status of robust *Homo sapiens*.

These brief observations about the still severely limited genetic information of domestication processes also provide part of a blueprint of testing the auto-domestication hypothesis. Having been reviewed and corroborated by others in recent years (e.g. Benítez-Burraco et al. 2016a; 2016b; Theofanopoulou et al. 2017), it has now advanced to the status of a theory, but its testing still needs to be continued as new knowledge becomes available. What is clear is that a build-up of deleterious alleles can be observed across numerous domestic subspecies, relative to their progenitor subspecies, that is attributable to pleiotropy. Examples are not limited to mammals but have been reported even from plants. The accumulation of detrimental mutations in rice genomes has been defined as a "cost of domestication" by Lu et al. (2006). Liu et al. (2017) arrived at very similar findings, using the same expression. A mammalian example is the higher frequency of non-synonymous substitutions in domesticated yaks compared to wild yaks (Wang et al. 2011).

However, the most comprehensive comparisons have been made of dogs and grey wolves, perhaps because dogs exhibit more phenotypic variation than any other mammal and are affected by a wide variety of genetic diseases. Björnerfeldt et al. (2006) sequenced the complete mitochondrial DNA genome in 14 dogs, 6 wolves and 3 coyotes. They found that dogs have accumulated non-synonymous changes in mitochondrial genes at a faster rate than wolves, leading to elevated levels of variation in their proteins. The extreme phenotypic diversity in dogs is attributable to this. Similarly, Cruz et al. (2008), examining the nuclear DNA of dogs, report a higher proportion of non-synonymous alleles compared with non-functional genetic variation. Like population bottlenecks and inbreeding, artificial selection influences levels of deleterious genetic variation (Marsden et al. 2016). Marsden and colleagues provide quantitative estimates of the increased burden of deleterious variants directly associated with domestication, finding that dogs have 2–3 per cent higher genetic load than grey wolves.

A clearer picture is beginning to emerge with further developments in genetics in recent years. Makino et al. (2018) compare whole-genome resequencing data from five domestic animal species (dog, pig, rabbit, chicken and silkworm) and two domestic plant species (rice and soybean) with their wild ancestors. They note significantly reduced genetic variation across a range of allele frequencies and increased proportion of non-synonymous amino acid changes in all but one of the domestic species. The exception is the European domestic pig, which is assumed to have continued to breed with wild populations after domestication and thus experienced reduced population bottleneck effects. The most likely 'culprits' in encoding deleterious amino acid variants are perhaps single-nucleotide polymorphisms (SNPs). They are variations in a single nucleotide that occur at specific positions in the genome, where each variation is present to some appreciable degree within a population. SNPs underlie differences in the susceptibility to a wide range of diseases as well as the way the body responds to them and their treatment. They provide insights into the genetic bases of phenotypes, selection dynamics and population histories. That applies especially in domesticated species, where SNPs burden the genome with an increase in the frequency of deleterious SNPs (dSNPs). This enrichment of dSNPs compared with synonymous SNPs (sSNPs) could well account for the detrimental aspects of domestication. Deleterious variants enriched within low recombination regions of the genome experienced frequency increases similar to sSNPs within regions of putative selective sweeps. Future research across the broad spectrum of domestication is expected to provide better resolution of this issue.

The harmful mutations in the genome of each individual can lower its fitness, dependent on their dominance and selection coefficient. Demographic conditions, selection and admixture all affect the occurrence of such mutations. In the case of domesticated species, these factors are rendered even more complicated by artificially induced bottleneck events and introgression, and especially by systematic artificial selection. The precise roles and interplay of these factors, however, remain inadequately explored. In particular, relatively little attention has been paid to how gene flow affects and is affected by the dynamics of deleterious mutations. Especially if these are recessive, the frequency of introgressed ancestry can be increased, and outbreeding enhancement (hybrid vigour or heterosis) is created in hybrid individuals. When recombination rates are low, these factors lead to an increase in introgressed ancestry even without beneficial mutations.

The implications

At this point, we can summarise that, since the auto-domestication theory was first proposed in 2008, new developments in such fields as genetics have generally been in its support. For instance, the separation between robust and gracile members of *Homo sapiens* has faded with the genetic evidence that the two were interfertile. Indeed, we have every justification for questioning the species identifications of many palaeoanthropologists generally. Only a few years ago, we were urged by them to believe in an African speciation event that did not occur. Just as from 1912 to 1953 we had a species called Piltdown man, we now have one called *Homo floresiensis* and another named *Homo luzonensis* that will both probably be declared non-existent one day. We have dozens of hominin species currently, practically all of them identified purely based on skeletal and dental characteristics. There has been a proliferation of career-enhancing discoveries of 'new species', many if not most of them probably interfertile with other 'species' and therefore subspecies at best. Nevertheless, the same boffins who are so careless with the taxonomy of hominins throw a tantrum when a science journalist misuses genetic findings to propose a taxonomy of races (Wade 2014). Admittedly, some of Wade's propositions are beyond the pale, and his racial nomenclature of the genetic variability of humans is as misguided as any other such attempt. But let us be quite clear: so are numerous claims of the replacement hypothesis.

For instance, we often hear about the supposedly decreasing genetic variation (expected heterozygosity) as one moves further from Africa (Wang et al. 2007), claimed to record the migration of modern Africans into other continents. Not only is this an unsubstantiated proposition (the evidence for it seems rather unconvincing; Hellenthal et al. 2008; Campbell and Tishkoff 2010; Wang et al. 2007: Figure 2A), but the empirical evidence contradicts it. The genetic diversity coefficients for Africans and Asians are identical, 0.0046, and at 0.0044, that factor is quite similar for Europeans, as noted in Chapter 2. Even the premise of genetic variation revealing genetic history is false: genetic variation is greater in African farming people than in African hunters-foragers (Watson et al. 1996), yet nobody assumes that the former are ancestral to the latter (cf. Ward et al. 1991). Alternatively, consider that the genetic divergence between the Seri people of Sonora, north-western Mexico, and the Lacandon people in Chiapas, southern Mexico, is more significant than that between Europeans and Asians (Moreno-Estrada et al. 2014). However, they live in the same country and about the same distance from Africa. These examples show the effects of precipitate claims deriving from misunderstandings of genetic assertions just as much as Wade's, and also reflect illusions of dogma. Palaeoanthropology, in common with most archaeology, operates like alchemy and is much in need of a theoretical framework, as Eckhardt (2007) so rightly observes. The incommensurabilities between that field of interest and premature interpretations of genetics do need to be resolved.

Until these matters are settled, it is judicious to be wary of most of the foundation 'truths' the African Eve hoax was based on, as reported in Chapter 2. The ever so slow unravelling of that hypothesis we are currently witnessing speaks for itself. It shows how easy it still is for the entire discipline to be captured by an unlikely fad for decades, as has happened so many times before. It also means that the replacement hypothesis is in dire need of replacement. Only one alternative has so far been proposed, and as we have seen in the previous sub-section, it has remained entirely unchallenged by all new findings of the last decade. On the contrary, they have done nothing but confirmed the domestication hypothesis. So what does that mean in practical terms?

In essence, the theory endeavours to explain the changes that occurred between 40 ka and 30 ka ago that turned natural selection on its head. Most of these changes from robust to gracile humans run counter to evolutionary trends: they offer no selective benefits, but they occurred over a relatively short period – instantly, in geological or evolutionary terms. They were not caused by changing climate, language introduction, new technology, better brains, significant dietary changes, mass migrations, gradual or instantaneous genetic changes, population replacement, cognitive revolutions, sea-level adjustments, or any of the other usual suspects. They were caused by a succession of developments that began with relatively minor behavioural modifications. Towards the last quarter of the Late Pleistocene, mating behaviour commenced changing ever so subtly. Possibly about the same time as menopause became effective, human males developed something unheard of in the animal world: a culturally determined preference for females of specific characteristics. In this, they overturned the principle that it should be the females who, directly or indirectly, determine mating partners. That culturally introduced revolution gradually, and in the end, decisively replaced natural selection as the principal agent of genetic change in humans.

We have a fair idea of what the males sought in their females, simply by reviewing the results – the entire gamut of gracile features – and checking which of them can first be observed in the skeletal remains of the time. Neoteny becomes first evident in the females (Bednarik 2008b) and then develops rapidly. How it relates to the uniquely human preference for youth, to 'male provisioning' or the introduction of menopause remains unexplained, but it is likely that these factors are somehow interrelated. In the end, it is not even particularly important what precisely set off a chain reaction in which specific traits were systematically selected outside of natural selection. Until a better contender is proposed, it is simply the case that female neoteny seems to have been the principal agent. What we can be sure of is that, at that time, non-evolutionary sexual selection began to have significant effects that introduced the aetiology of the domestication syndrome. That much is certain because all important changes from robust to gracile humans are features of domestication. So we can invert the question and ask, which selective process could have set in train the domestication of hominins? Unless we credit this process to an outside entity, such as a deity or a black monolith of unidentifiable material from Jupiter, we have no choice but to accept that there was 'artificial' selection of some kind.

Bayesian reasoning may be of help here. What would be the probability that the numerous robust → gracile subspecies changes all resemble domestication by pure chance? We could apply Bayes's Theorem:

> credence in proposition X given observation D is *proportional to* the likelihood of observation D given proposition X *multiplied* by prior credence in proposition X.

With as many variables involved as those of the robust → gracile conversion, the probability that they all resemble domestication effects by pure accident becomes simply exorbitantly small. Bearing in mind that the domestication theory is perfectly falsifiable by an alternative demanding a lesser number of assumptions to be made, it stands unopposed until a more parsimonious explanation for human auto-domestication is proposed.

Ultimately the crux of this theory as it is currently formulated revolves around the need to identify the circumstances causing the human genome to adopt a raft of domestication syndrome traits. The most parsimonious explanation seeks the immediate causation in the introduction of culturally communicated *sentient* constructs of attractiveness (consciously selected, in contrast to bird song or tail feathers or body size), which are utterly unique in the animal world. The strongly developed preferences existing today cannot have existed in the distant past, so we have no option but to accept that these constructs must have been introduced at some point in human development. This does not imply that standards of 'beauty' are completely 'hard-wired'; in humans, as in other animals, the development of standards of attractiveness is likely to involve a range of mechanisms, from innate templates to imprinting to imitation and other forms of social learning. It does imply, however, that human beings, like other animals, are likely to have genetic adaptations for assessing the 'mate value' of potential mates. Studying attractiveness without considering these adaptations would be like trying to understand the eye without treating it as an organ of vision.

Ultimately it matters little to the domestication theory whether the inversion of natural selection is due to deliberate selection of physical traits (as occurs in all other domesticates, albeit not 'deliberately') or some other, as yet not considered, factor. The theory as such could be adjusted to alternative scenarios. Its unassailable nature is not dependent upon these details but is attributable to the lack of an alternative credible explanation for the many parallels between the robust → gracile changes and the expressions of domestication syndrome factors. To deny these parallels would be no different from denying that these factors account for the differences between cattle, dogs or chickens and their respective progenitor species. It would be just as futile as arguing that the domestication syndrome is itself a figment of imagination and that similarities among domesticates are purely coincidental.

The issue of sentience, however, needs to be clarified, as does its connection with constructs of attractiveness. This involves consideration of self-awareness and consciousness. *Self-awareness* defines the sentience of one's knowledge, attitudes, opinions and existence. The mirror test is often applied to animals to detect this faculty (Gallup 1970; Mitchell 1993; 1997; 2002; Heyes 1998; De Veer and Van Den Bos 1999; Gallup et al. 2002; Keenan et al. 2003). However, there is no universal agreement on its efficacy (Swartz 1997; Morin 2003). Some of the great apes, elephants and bottlenose dolphin are among the species that have passed the mirror test. These are much the same species that have been demonstrated to possess von Economo neurons (Seeley et al. 2006; Butti et al. 2009; Hakeem et al. 2009). Von Economo neurons seem to occur in relatively large species with large brains and extensive social networks (Bednarik 2011c). Constructs of individuality could well have evolved from these complex networks, because the advent of self-awareness is difficult to explain without social complexity well above that of, say, social insects.

Consciousness can be seen as a transparent representation of the world from a privileged egocentric perspective (Trehub 2009). It is self-referential awareness, the self's sense of its existence. It focuses attention on the organism's environment, processing incoming external stimuli (Dennett 1991; Farthing 1992). Self-awareness, by contrast, focuses on the self, processing both private and public information about selfhood. It is defined by the capacity of being the object of one's attention. The individual is a reflective observer of its internal milieu and experiences its mental events (Gallup 1998; Gallup and Platek 2002; Carver 2002). The self is a social construct (Seyfarth and Cheney 2000) shaped by the individual's culture and immediate conspecifics (Leary and Buttermore 2003). Nevertheless, the self is not equivalent to consciousness (Natsoulas 1998), as shown by the observation that many attributes seen as inherent in the self are not available to conscious scrutiny. The neurological computation of the boundaries of personhood derives from people's behaviour and from the narratives they form.

Thus, it needs to be ascertained how the chain of events from sensory input is established and how behaviour is initiated, controlled and produced (Carruthers 2002; Koch 2004; Nelson 2005; Clowes 2007). It appears that subcortical white matter, brainstem and thalamus are implicated in consciousness (Fernández-Espejo et al. 2011), although it is assumed that unconsciousness mainly involves the cortical brain (Velly et al. 2007) and the thalamus is not believed to drive consciousness. Internally directed aspects of cognition, such as the Theory of Mind (ToM), episodic memory, self-evaluation and self-awareness derive from the default mode network, which is considered to be a functionally homogeneous system (Sestieri et al. 2011). Relative to ToM, conscious self-awareness is even less understood and accounted for ontologically. Like consciousness, self-referential awareness is as difficult to explain as it is to find a self-consistent set of axioms to deduce all of mathematics, which we know is impossible.

Self-awareness is believed to be located primarily in a neural network of the prefrontal, posterior temporal and inferior parietal of the right hemisphere (Stuss et al. 2001; Decety and Sommerville 2003; Gusnard 2005; but see critiques in Morin 2002; 2004; Morin and Michaud 2007). It must be the result of an interplay of many variables, starting from the input of the proprioceptors to the engagement of several brain regions. It includes the operation of distal-type bimodal neurons (moderating anticipation and execution; Maravita et al. 2003). Self-awareness can safely be attributed to all hominoids and hominins, and there is a reasonable expectation that it became progressively more established with time. An exciting aspect of it in the present context is that some of its effects can be recognised archaeologically. Body decoration, especially of the self, would be impossible without self-awareness, while in turn, its practice is likely to have contributed to reinforcing the concept of the self. As noted in Chapter 2, the Makapansgat cobble, deposited between 2.9 and 2.4 Ma ago, implies self-awareness and developed Theory of Mind (Bednarik 2015c). More recently, pigment use has been suggested by between 1.3 and 0.8 Ma ago (at Kathu Pan 1) and subsequently becomes very common. The classical archaeological demonstration of self-awareness derives from beads and pendants. Their identification as such is usually secure (Bednarik 2001; 2005) and on present data, they first appear in the Lower Palaeolithic traditions such as the Acheulian in northern France, England, Libya and Israel. Although such items tend to have various roles or meaning besides being 'decorative', all of these are culturally determined and transferred.

Between 40 ka and 30 ka ago, the time when we detect rapid gracilisation in female humans, we also note the appearance of figurines, especially female figurines, as well as the depiction of vulvae or pubic triangles. Irrespective of which of the various 'explanations' of this emerging interest in female sexuality we may or may not subscribe to (Bednarik 1990a; 1996), one thing is certain: nothing indicates a similar interest before that time. Despite the fabulous artistic skills of the producers of these depictions of the earliest Upper Palaeolithic, we seem to lack any figurines of great hunters, of heroic male role models, of muscle-packed idols of the day. Instead, we have an assortment of females ranging from slim 'dancing girl' to obese women with utterly over-emphasised female attributes. They include figurines that seem to have been mass-produced according to some rigorous specification, to examples expressing high levels of individuality. The most interesting are the earliest ones, those pre-dating 30 ka. They include the fragile Galgenberg figurine from Austria (Bednarik 1989); the lion-headed lady from Hohlenstein, Germany; the 'bison-woman' in the Salle in Chauvet Cave, France; the possibly cicatrised woman from the Hohle Fels cave in Germany (Franklin and Habgood 2015); and a gamut of apparent vulva depictions or symbols of the early Aurignacian or perhaps even the final Mousterian (Figure 4.3). What renders them particularly interesting is that they express a definite interest in female attributes, to the point of over-emphasising them very deliberately; and they coincide with the rising incidence of neoteny, especially among females.

Let us consider the most recently discovered of these finds, the mammoth-ivory figurine from the basal Aurignacian deposit at Hohle Fels (Conard 2009). As it is one of the earliest of these objects and has been suggested to be close to 40 ka old, it is of particular relevance here, being the earliest female representations known (Figure 4.3c). It has

> many carefully depicted anatomical features, but also many other features that are not anatomically correct or in proportion, including the exaggeration of sexual characteristics such as large, forward projecting breasts, a greatly enlarged and explicit vulva and bloated belly and large thighs.
>
> (Franklin and Habgood 2015)

132 *Human self-domestication*

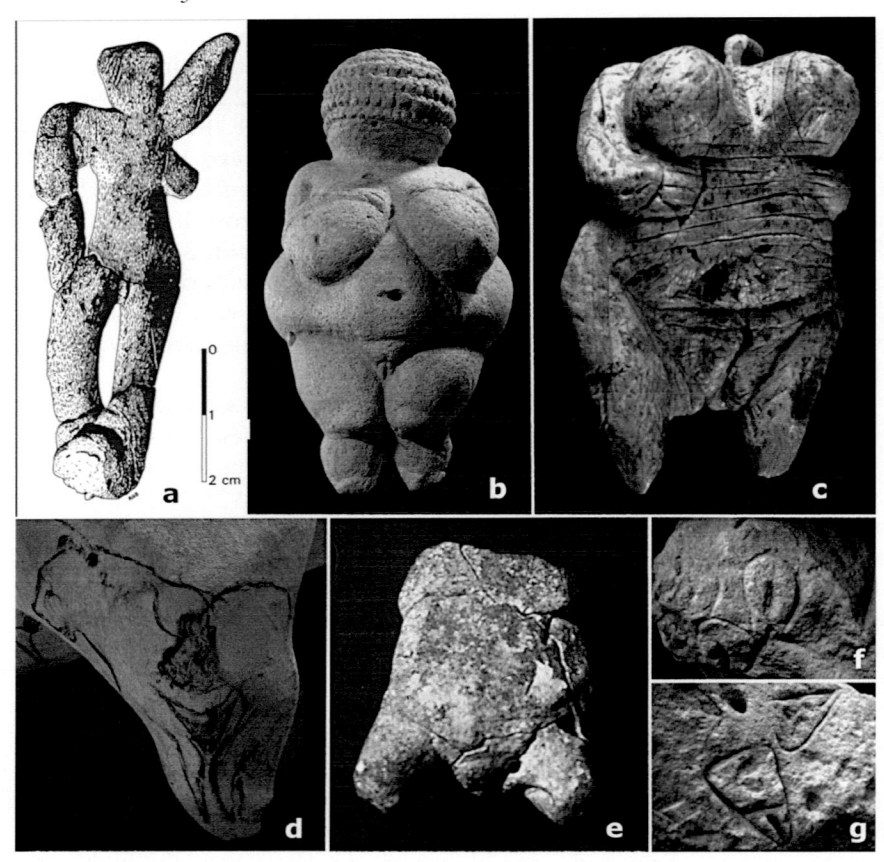

Figure 4.3 Depictions of female genitalia of the Aurignacian and Gravettian, from (a) Galgenberg, (b) Willendorf, (c) Hohle Fels, (d) Chauvet Cave, (e) Avdeevo, (f) Abri Cellier, and (g) La Ferrassie

The arms, as in many other and more recent of these female figurines, rest on the belly as if to emphasise it, tucked under the truly massive breasts. The concern with detail, noted on many of these figurines, implies that the maker attempted to convey a specific message, and whatever the purpose of this figurine was, it involved very deliberate and quite extreme emphasising of female features. The carefully made grooves found on the breasts, shoulders, abdomen, back, arms and legs have been plausibly interpreted as scarification or cicatrices. Such markings can also be found on other figurines, especially the large Hohlenstein Stadel therianthrope, from the same region and also of the lowest Aurignacian (see Figure 2.2). This figure may no longer be a symbol of *Kraft und Aggression* (Hahn 1986), having been reassembled and suggested to depict a female (Schmid et al. 1989), although this cannot be decided conclusively

(Kind et al. 2014). It, too, bears parallel cuts on the upper arm, similar to those on the most certainly very female Hohle Fels figurine.

Or consider the perhaps even more evocative figurine No. 14 from Avdeevo in Russia, made like many objects from that site of limestone marl (Figure 4.3e). Although the figurine is only preserved as a fragment, the tightly folded-back legs and wide-open thighs suggest that the intent was to show the vulva to greatest effect prominently. Just as the Hohle Fels ivory figurine with its grotesque overemphasis of female sexual attributes implies a preoccupation with these aspects, so does the Avdeevo marl specimen with its "apparently upwards turned perineum" and the prominence of its "realistically shown genitalia" (Gvozdover 1995: 27). Its pose is somewhat similar to that of a figurine from Kostenki 13 (Rogachev et al. 1982: 144, Figure 45). There is nothing unique about the emphasis of female traits, which is found in many of these statuettes. That is why they have been called 'Venus figures', although the more recent specimens tend to lack the more overt aspects of this practice (Bednarik 1990a; 1996).

Of course, the advent of traditions of frequent female depictions and symbols coinciding with the onset of female neoteny could be sheer coincidence, but it does seem more likely that there is a connection. The fundamental shift between the sexes, of males taking from the females the initiative in mate selection, would have also involved a significant change in awareness of personal traits. Before its introduction, females (as in all other species) are likely to have selected males in accordance with established behavioural patterns, whatever these were. If these resembled those of other species, they would have lacked the element of deliberate choice. Whatever the traits selected were, they were not conscious choices of awareness. Males may compete for access to females by a wide range of strategies, including through dazzling plumage display, song or prized territory; or by male-male competition through duelling. After a harem of does or lionesses have watched their respective males spar, they accept the outcome of the contest as determining who would father their offspring. In Chapter 3, we have visited the 'sexy son hypothesis' and Fisher's runaway effect deriving from it, explaining the mechanism. Ignoring the precise form that male-male competition or any other sexual selection may take, the females are not assumed to have conscious control over their reactions. This is decidedly different when males determine the traits they desire, traits that have very little if anything to do with objective genetic or any other fitness. First, these traits were purely cultural constructs: there is nothing obvious or 'objective' about them. There are thousands of variables in the human physiology and psychology that could be designated as 'attractive' in a potential partner, and we cannot know how these constructs of attractiveness were selected initially. However, their formulation would have involved a heightened level of awareness of individual characteristics, and the hypothesis here is that this awareness is reflected in the earliest known depictions of the human form. This was no longer consciousness; it was an elevated state of self-awareness.

134 *Human self-domestication*

Another archaeological observation reinforces this point: at the same time as we observe rapid neotenisation and the frequent depiction of females, sometimes in the most explicit poses, we also observe a marked increase in objects that would have been used in body decoration. Beads and pendants, as we have noted, had been in use for hundreds of millennia, but in the last quarter of the Late Pleistocene, they become very prominent, having been recovered in their tens of thousands from that period. They are not even the only indication of body decoration. We assume, as has been noted above that even the final 'Neanderthals' used body parts of large raptors to decorate themselves, but with the appearance of human figurines, we can catch a glimpse of practices about which we could otherwise have no idea. The figurines are generally depicted nude, with apparent evidence of apparel mostly limited to the small, genderless figurines from Siberia (Bednarik 1996). However, their nudity is frequently emphasised by the depiction of various types of 'girdles', 'bandeaux', bracelets and necklaces. These hardly qualify as clothing; they would not have kept anyone warm or covered them. They appear to be decorative and more probably designate specific meanings or information about the wearer (Figure 4.4). Their importance to the artisan seems apparent from the often considerable detail of their depiction. The attention given to these items has even encouraged speculations about the textile materials used in their production (Soffer et al. 2000). These items are lacking in the Siberian and west European samples, but they occur in central Europe and are found on most specimens of the large east European corpus. Again this is evidence of the attention lavished on consciously personalised awareness, in a world pregnant with exogrammically stored meanings.

Finally, we have already visited the belief that these hominins of the last 30 millennia of the Pleistocene probably wore body decorations in the form of cicatrices and, possibly, tattoos and body paint. It is almost self-evident that

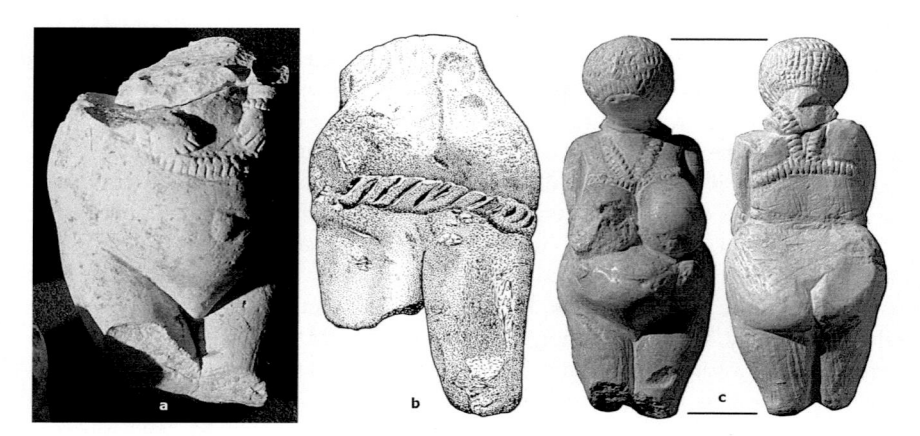

Figure 4.4 Female Early Upper Palaeolithic depictions with typical items of body decoration, of marl and ivory from Kostenki (a, c) and fired clay from Pavlov (b)

collectively these features would have expressed complex cultural meanings, which helps explain their detailed depiction on the figurines. These objects were not just likenesses of females; they were also elaborate exograms: [The Willendorf 1 figurine] "wore the fabric of her culture. She was, in fact, a referential library and a multivalent, multipurpose symbol" (Marshack 1991). As one who delved most deeply into the subject has stated: "We believe that each type of figurine had its own symbolic meaning, conveyed by the pose and accentuation of the female body parts" (Gvozdover 1989). The finds themselves seem to confirm that individuality, which could have only developed when reflective contemplation of individuals' traits had begun to govern social discourse to a significant extent – and, likely, governed mate choice.

All of this could, of course, be pure coincidence. But not according to Thomas Bayes. Ultimately we do agree with the replacement advocates on one point: something extraordinary happened in Europe as well as elsewhere between 40 ka and 30 ka ago. We only disagree on what it was that happened. Our opponents believe that a genocidal 'race' of ultra-smart humans from Sub-Saharan Africa marched into Europe en masse and, unable to breed with the dumb resident Robusts or to colonise or enslave (and thus domesticate) them, they 'replaced' them. We, by contrast, believe that all the factors we have considered cannot be just a series of extraordinary coincidences, devoid of any collective meaning. We believe that the probability of that is so minuscule that we need to consider the possibility that the domestication theory does provide a realistic explanation.

5 The unstoppable advance of exograms

About exograms

We have considered many facets of the human auto-domestication theory, and yet we are far from having exhausted the wealth of corroborating elements. However, rather than pursuing a strategy of piling on confirming evidence, we will now change tack and test some of the contiguous issues arising from Chapter 4. First, there is the glaring problem posed by the atrophy of the human brain towards the end of the Pleistocene, and its accelerating rate during the subsequent Holocene. We have been conditioned by a century of indoctrination proclaiming that encephalisation, the relentless increase of human brain size, was driven by gradually increasing cognitive or intellectual demands. Although it can be formulated in Darwinian terms with a little tweaking, this essentially Lamarckian rationalisation fails to explain the precise correlation between brain volume and demands placed on the brain. Given the high cost of encephalisation, particularly in social and evolutionary terms, it seems unlikely that more brainpower became available for selection. Instead, the theory is that greater intellect was such a decisive trait that random mutations promoting larger brain size were actively selected. Whatever the precise interplay of size and function may have been, the theory seems eminently sensible.

Similarly, there is universal agreement that with the advent of the Upper Palaeolithic, the demands made upon the human brain increased significantly, and these demands intensified relentlessly throughout that era and perhaps even more so in the course of the Holocene. This is reasonably assumed based on archaeological evidence indicating that the complexity of cultural evidence, of technological sophistication and social intricacy all escalated for the last 40 ka.

This leads to a cataclysmic mismatch. During the very same period, from the beginning of the Upper Palaeolithic in Europe to the present day, the size of the human brain declined on average by about 13 per cent. We have noted that the rate of this decrease is 37 times greater than the long-term previous encephalisation rate of 7 ml per 10 ka. This leaves us with the need to account not only for a continued, if not increased, ability of the brain to cope with demands made on it but to manage to do that with an ever-shrinking brain.

This suggests that volume is not necessarily the variable that determines processing ability in the brain, which confirms anecdotal evidence to that effect. However, the remaining problem is how to explain the lengthy process of human encephalisation: if a larger brain is not required, why on earth was the hominin lineage saddled with the enormous handicap of an ever-increasing brain? It was the large cranium that led to prolonged infant dependency in humans, demanding the early ejection of the foetus, in a state of reliance on the mother increasing proportionally to the helplessness of the infant. That prolonged total dependence and lactation period also reduced overall fertility rates and thus evolutionary fitness, just as the enlarged cranium increased the incidence of obstetric fatalities in birth (Joffe 1997; O'Connell et al. 1999; Bednarik 2011c). The most acute effect of the foetal state of new-borns, however, was the enforced dependence of the mother on the support of the band and, as noted above, on being provisioned with the dietary needs of pregnancy (Biesele 1993; Aiello and Wheeler 1995; Deacon 1997). This, one can reasonably assume, involved changed patterns of social behaviour and renegotiated roles of the sexes. It is therefore not an exaggeration to say that our historical patterns of gender roles were to a significant extent determined by the enlargement of the foetal brain.

Therefore, it seems inconceivable that evolution would have selected in favour of encephalisation when a much smaller brain would have been perfectly able not only to meet demands but even to sustain the increased need for processing power made on it during the Upper Palaeolithic (Figure 5.1). This is a significant aspect of human evolution that needs to be addressed, having so far remained largely ignored. On the face of it, there appears to be no easy explanation, but the domestication theory offers a way. It begins with rehearsing an entirely uncontended insight: the period beginning around 40 ka ago features not only the introduction, as far as we know, of human figurines, but it marks a veritable explosion in the making and use of palaeoart. This includes all material finds that remind us of art in the modern sense (for discussion of the term 'palaeoart', see Chapter 1). Palaeoart has a long history reaching back well over one million years, but early examples of it remain few and far between for most of that time (Bednarik 1992a). They become a little more visible archaeologically towards the late part of the Middle Palaeolithic traditions, i.e. with the Late Pleistocene, beginning 130 ka ago. It is with the dawn of the Upper Palaeolithic, around 40 ka ago, that palaeoart finds begin to increase exponentially in numbers (Bednarik 2017a). They become so common that by around 25 ka ago, the examples we are aware of occur in their many thousands, only to become more numerous again with time. By the end of the Pleistocene, known examples survive in their tens of thousands, and a mere millennium ago, they number in their millions. It is essential to appreciate that much of that pattern is attributable to taphonomy (selective survival; Efremov 1940; Bednarik 1994a). Nevertheless, the distinctive increment evident about 40 ka ago cannot be taphonomically derived, because we know of no particular environmental upheavals at that

138 *The unstoppable advance of exograms*

time that could have uniformly truncated the record of many different classes of material evidence (such as rock paintings, petroglyphs, ivory, bone).

It, therefore, appears justified to ask, what could have prompted this upsurge in palaeoart production and use? The manifest proliferation includes the introduction of several new components, especially the rise of figurative or iconic depiction, the introduction of new materials and technologies, and the presumed expression of increasingly complex worldviews. In accounting for these changes, we have to concede that we agree partially with the explanation offered by the replacement advocates: yes, the changes may signal significant transformations in self-awareness, new constructs of reality and a host of other conversions. The replacement scholars, of course, interpret this as a revolution attributable to the wholesale replacement of all robust human populations in the world by superior invaders, but this is not a viable interpretation of the evidence, as we have seen. These cultural and no doubt behavioural modifications were not introduced abruptly but appear progressively over many millennia. More importantly, their early phases seem to be attributable to still fairly robust, 'Neanderthaloid' people, and as we noted, there is no clear evidence of entirely anatomically modern people between 40 ka and 28 ka years ago. Even the Crô-Magnon specimens, we have seen in Chapter 2, are only 27,760 carbon-years old (Henry-Gambier 2002), and especially cranium 3 is quite Neanderthaloid. We have noted that similar 'intermediate' specimens are common right across the width of Eurasia and that presumed 'Neanderthal' remains have been found in several Upper Palaeolithic and even later contexts. There is no hard and fast separation in either human remains or technological features; they grade into one another in both Asia and Europe. Moreover, the distinctive changes in palaeoart were not introduced from outside; in fact, a figurative production is entirely lacking in Africa before the Late Stone Age tablets from Apollo 11 Site (Wendt 1974; 1976; contrary to earlier views, they are not attributable to the Middle Stone Age). Similarly, sophisticated technologies appear in Eurasia well before they make their debut in Africa, so it is simply false to assume that any of these 'Upper Palaeolithic' innovations necessarily derived from Africa.

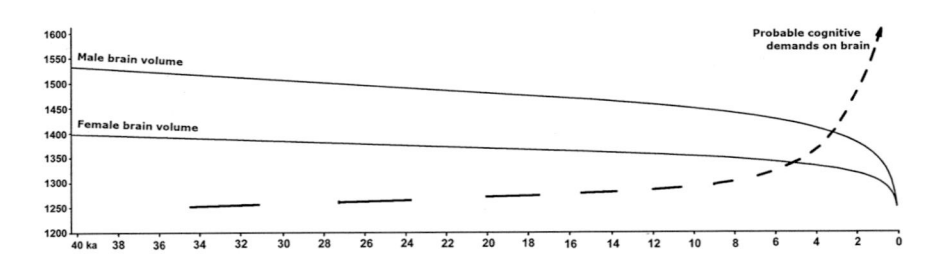

Figure 5.1 The atrophy of the human brain during the last 40,000 years compared with the presumed rise in the cognitive demands on the brain during the same period

So how did they arise? The auto-domestication theory proposes a realistic solution, involving the increasing demand for brainpower, the shrinking human brain and the rapidly expanding diversity of palaeoart expressions. There is one way of compensating for the brain atrophy. It resembles the way the use of external storage can augment a computer's operation. One can imagine a computer's loss of functionality whose memory is reduced by 13 per cent. By connecting it to an external hard drive of immense capacity, the role of that computer would change significantly. Rather than relying on its severely limited (and diminishing!) digitised memory, it would become a hub or portal for processing a great deal of potential storage. It would thus relieve part of its memory banks and gain them for its pivotal role as the centre of an information system. If we transfer this analogy to the brain, we can see that augmenting the working memory capacity with 'external storage' would remove the limitations of biological working memory and thus free up processing power. As the physical medium of memory storage is changed from brain tissue to an external medium, the neural memory system can be retrained towards memory retrieval.

In their book, *The Meaning of Meaning*, Ogden and Richards (1923: 53) state:

> A sign is always a stimulus similar to some part of an original stimulus and sufficient to call up the engram formed by that stimulus. An engram is the residual trace of an adaptation made by the organism to the stimulus. The mental process due to the calling up of the engram is a similar adaptation: so far as it is cognitive, what it is adapted to is its referent, and is what the sign which excites it stands for or signifies.

The concept of the engram and also the word to define it derive from Semon (1904; 1921: 24). The term 'engram' delineates a hypothesised memory trace, a persistent protoplasmic alteration of neural tissue thought to occur upon stimulation of the brain, and it is assumed that memory in animal brains is stored in such structures. For much of the twentieth century, it was thought that external sensory stimuli resulted in discrete biophysical or biochemical changes in neural tissue, and some researchers dedicated their work to discovering these neural tissues. Semon called this the "mnemonic trace", and Richard Dawkins appears to have re-discovered the concept independently 72 years later, calling the phenomenon a "meme". He seems to have been unaware of the considerable efforts to discover the physical evidence of it. Karl Lashley (1890–1958), in particular, spent several decades trying to locate engrams in rodent brains, succeeding instead only in demonstrating that there is no single biological locus of memory, but that there are many (Lashley 1923a; 1923b; 1924; 1930; 1932; 1935; 1943; 1950). Penfield (1952; 1954) and others reported being able to reactivate memory traces by stimulating the temporal lobes, but the reported episodes of recall occurred in less than 5 per cent of patients and could not be replicated by later neurosurgeons (e.g. Jensen 2005).

Subsequent work by others, such as Thompson (1967; 1986; 1990; Thompson et al. 1976; Steinmetz et al. 1987; 1991; 1992; Christian and Thompson 2005), confirmed the finding that the phenomena accounting for memory are widely distributed throughout the cerebral cortex. Memory is stored in the brain in the form of very active neurons that fire persistently, forming or recalling memories (Churchland 1986; Squire and Zola-Morgan 1991; Hooper and Teresi 1992; Kandel and Pittenger 1999; Squire and Kandel 1999). Lashley had focused on the cerebral cortex, but it has emerged that the hippocampus, amygdala, visual cortex and auditory cortex are also involved in memory storage, while associations are stored in the associative cortices. This 'distributed' storage of memory in large groups of neurons probably ensures that they are not lost too easily. A 'complete memory' may contain aspects, such as colour, sound, emotion and others that are stored in vaguely defined brain modules in different parts of the brain (Christos 2003).

The very idea that memory can be stored external to the brain had already been expressed by Plato (in *Phaedrus*, 274e–275a) around 370 BCE. He observed in his dialogue between Socrates and Phaedrus that the use of writing fosters forgetfulness because people were "calling things to remembrance no longer from within themselves but from external marks". Thus the underlying issue had long been identified. It was Penfield's (1958) investigation into the engram that generated the notion of storage of memory traces external to the brain and led to the proposal by Gregory (1970: 148). He realised that this would be a relatively stable and permanent expression. Goody (1977) and Carruthers (1990; 1998) developed the idea of such a 'surrogate cortex'. The first proposal of actually identifying specific phenomena as engram-like, externalised and permanent forms that served as repositories of memory is in Bednarik (1987). We interpreted very early rock art as patterns of cognitive reference frameworks externalised in properties accessible to sensory perception (ibid.: 223), as projections of neural structures (ibid.: 226) and as projections of neural systems perceptible to the senses (ibid.: 225). Most importantly, we argued for a significant communication potential of such externalised, engram-like phenomena, because conspecifics would have possessed "resonating" cerebral systems capable of response; other hominins would have understood that 'he made this meaningful mark'. In this work we referred to these permanently externalised markings as 'psychograms', using the term introduced by Anati (1981: 206) rather than coining a neologism. Our subsequent assessments of the cognitive development of hominins were guided by the insights gained from this model (Bednarik 1990b; 1992b).

Some years later, Donald created the name 'exograms' for the externalised memory traces, but he seemed unaware of both our effort and Semon's much earlier work (Donald 1991; 2001). Indeed, he assumed the engram to be a real entity, rather than a hypothetical phenomenon whose existence has been widely precluded from consideration for some time. Donald considers that due to brain plasticity, the 'interanimation' of exograms and engrams leads to a continuous reformatting of distributed hybrid memory networks that triggers the neural updating and re-wiring. A seminal work in 1998 led to the field of 'extended mind studies' which has been lively ever since. Clark and Chalmers (1998) present the controversial idea that even non-exogrammic objects within the environment function as a part of the mind. It does stand to reason, however, that exograms create and support cognitive processes not accessible to creatures restricted to purely biological memories (Rowland 1999). The storage system they form allows the individual adept in its use to visit memory locations at will, extracting and manipulating their information contents. The processing power needed to recall both the locations and the specific items to be remembered is both economical and flexible (Sutton 2008). Once the virtual architecture was securely internalised, it could be used highly effectively, and the adept's mind had become the equivalent of random access memory (Carruthers 1990; 1998).

However, Donald's account has been considered problematic by many. Terminologically, he uses the adjective 'exographic' instead of the more correct 'exogrammic' (e.g. Sutton 2008; Bednarik 2014f). His assertions that exograms are semi-permanent, unconstrained and reformattable, can be of any medium, have virtually unlimited capacity and size, and can be subjected to unlimited iterative refinement need to be qualified somewhat. Not all external cognitive artefacts are as static and permanent as Donald suggests, and such external representational systems need not necessarily be of unlimited capacity, translatable across media, or endlessly reformattable as Donald's typology suggests (Sutton 2008; 2009). Various of Donald's views have been widely criticised. His belief that "Neanderthals underwent a drastic, rapid extinction" has rightly been defined as "unsupported assertion based on a kind of current 'folk-wisdom' that has to be relegated to the realm of pop-science", "comparable to phrenology" (Brace 1993). This "extinction" is highly debatable, if for no other reason than the presence of Neanderthal and Denisovan genes in modern non-Africans. Donald's contention that the introduction of language would speed up the rate of cognitive evolution is analogous to claiming that the rate of mutation determines the rate of genetic evolution (Cynx and Clark 1993). His ideas of very late language origins had been refuted before he presented them (Falk 1975; 1987; Arensburg et al. 1989, 1990; Bickerton 1990) and they are indefensible today (Enard et al. 2002; Zhang et al. 2002; Sanjuan et al. 2006; Krause et al. 2007; Falk 2009; Bickerton 2010). Other critiques include his "cavalier misuse of information available from anatomy, anthropology, and archaeology" (Brace 1993, cf. 1996; 2005); his inadequate presentation of the relevant neurology and his neglect of cognitive ethology (Cynx and Clark 1993). In response to Donald's (1991) pronouncement that "unlike the constantly-moving contents of biological working memory, the products of thinking, when reformatted exogrammatically, could be frozen in time, held up to scrutiny at some future date, altered and re-entered into storage", Adams and Aizawa (2001: 58) state that "there can be no cognitive science of transcorporeal processes" (cf. Rupert 2004; Malafouris 2004; Aizawa and Adams 2005; Block 2005; Prinz 2006; Adams and Aizawa 2008). Adams and Aizawa define cognition by non-derived, intrinsic or original content; and by cognitive processes of a particular kind, the mechanisms by which organisms remember, perceive, attend and learn. However, of the three 'theories of content' they cite (Dretske 1981; Fodor 1990; Cummins 1996), explaining how original content arises naturally, none is universally accepted today. A generally acknowledged theory of cognition, or of which tasks are cognitive, remains elusive (Menary 2007: 15), although an increasing number of neuroscientific studies are devoted to the subject.

The major flaws in Donald's model are, however, in his take on the archaeological evidence. His three stages of cultural evolution, foreshadowed by Fairservice (1975), are not supported by the relevant evidence. In particular, his reliance on the now-discredited replacement hypothesis and his lack of familiarity with the archaeological information about presumed early 'symbolling abilities' are crucial. Donald is oblivious to pre-Upper Palaeolithic exograms, which renders his chronology of the introduction of exograms and how it affected the cultural development of hominins severely flawed. The most instructive forms of evidence for this process, crucial in understanding the origins of human modernity (Bednarik 2011b; 2011c; 2013d), are the finds from the earliest periods, whose antiquity exceeds the advent of the 'Upper Palaeolithic' by hundreds of millennia. Also, neuroscience and the aetiology of hominin behaviour (Bednarik 2012c) need to be central to any comprehensive consideration of the introduction and roles of exograms.

In appreciating their roles in hominin evolution, it may be useful to begin with our observations (Bednarik 1987) that even the most basic anthropogenic rock markings carry numerous inherent messages, for both the maker and any conspecific. If the latter happens to be equipped with the same 'cortical software' (i.e. cultural understandings) as the maker, those messages would be far more comprehensive than for other humans; but even for culturally unconnected individuals, such markings convey meanings. In examining the roles of exograms in recalling memory as well as in the formulation of reality constructs, we could begin by exploring the female figurines we considered in the last section of Chapter 4. We noted the frequent occurrence of the depiction of body decoration on these 'Venuses', as they are widely called. It can occur as presumed scarification (cicatrices), tattoos or body painting; as items of jewellery; and in the form of presumed girdles or belts that are presumably not intended as clothing but are assumed to have had specific meanings attached to them (see Figure 4.4). The latter point is emphasised by the great artistic care afforded to these features: if they had no definite meanings, they would not have been depicted in such exquisite detail.

Body decoration, like beads and pendants, can only exist in the context of cultural meanings. These can be of great diversity. For instance, they might indicate availability or non-availability for marriage, or they might protect the wearer from real or imagined dangers, or indicate rank or status. They could signify wealth, ethnic group affiliation, membership or alliance with any group or faction, and any of a countless number of other possibilities. All of these connotations are only known to cultural insiders, i.e. individuals familiar with the emic meanings of such symbolic conventions. However, we do not need to know these meanings to understand that they are likely to be intricate and multifaceted. They would have been the outcome of convoluted historical developments of symbol systems, which also are unlikely to become available for valid scholarly contemplation. In the case considered here, the female figurines of the Final Pleistocene, we are dealing with cultural systems that are so remote from our modes of comprehending that the endeavours of archaeologists to identify intention, meaning and similar modern concepts of relevance are misguided. Just as thousands of tomes have been filled with humanist waffle about the meaning of cave art, all of it of untestable propositions, we lack any dependable interpretation of the figurines or their roles in society. Nevertheless, we may credibly speculate that, in combination with the apparent evidence of body decoration on many of them, we are dealing with small surviving vestiges of a once extensive and complex information system. The figurines themselves are imbued with externalised communication potential that would have prompted decoding in any brain equipped with the required software. They no doubt conveyed much relevant information about the producer's skills and the owner's status, perhaps also about the circumstances of their display and handling, or of their context. The decorative elements may have revealed to the adept a great deal about the meaning of the figurine itself. All of these many details are exograms – externalised memory traces: they reside in the material properties of the object. In any non-adept hominin – one not equipped with the internalised random access memory, the required cognitive scaffolding – it would be no more instructive than it is for the modern beholder: that this is a figurine of a female. Nearly all of the information accessible to the adept would be inaccessible to anyone else, be it an archaeologist or a *H. s. heidelbergensis* individual. Such is the power of the exogram if the brain is 'tuned' to it.

So what precisely can be an exogram? Any object, mark or feature can become an exogram. Some applications of the concept are much more apparent or self-explanatory than others, however. For instance, a picture of a dear relative acts as an external memory trace, setting off a variety of emotive reactions in the neural system. These will differ subtly if, for example, the person depicted had died, as the image will then elicit different reactions in the brain. A film, a piece of music or a theatre performance will similarly provoke reactions. Somewhat less obvious are mnemonic devices of countless types, including, for instance, theatre architecture (memory prompters for Shakespearean actors), moral codes, maps, all forms of record-keeping, monuments, crosses on a calendar, timetables, rituals, memory sticks, rhythms and rhymes. Pre-literate peoples depended on the latter two substantially in their impeccably precise recital of lengthy origin myths and ballads extending over many hours. Writing and language both consist of exograms, as do all symbols and signs, and memes are exograms transferred from one brain to others by writing or communication. Billboards, advertisements, instruction manuals, slide rulers, software programs, knots, tracks and many other forms are all exogrammic devices. Then there is the vast number of exograms that are perhaps much more innocuous and personal, and therefore also much less noticeable. They include that unconvincingly stereotypical actor in a specific television commercial, the familiar chip in one's favourite coffee mug, the way something is folded, the colour of ripe vs unripe fruit. All of these factors and millions of others can carry referential information stored in them to which the brain's internalised virtual architecture can connect effortlessly and recall meanings previously deposited in them.

It thus becomes evident that a great deal of the way we process information and even the way we negotiate exograms heavily govern reality. We only need to consider where the latter ability would be, were it not for our skill in employing exograms. Would we be able to experience reality consciously at all? If the sky is heavy with clouds, we tend to expect rain, because this condition has in the past resulted in rain. In all probability, other animals can also sense the approaching downpour, but it is rather doubtful that they do this in the same conscious way as humans. Indeed, it seems entirely possible that the observations of natural patterns of many types helped greatly in establishing routine connections between such external phenomena and neural processes. After all, exograms must have begun and evolved somehow, and it seems perfectly reasonable to consider that routine observations of changes in the environment and their inevitable short-term consequences could have formed the basis of such cause and effect perception that proved to be of selective potential. This raises the question of the origins and the earliest appearance of exograms, which we will explore below. For the moment we are only concerned with observing that there are possible scenarios for their initial development, but that it is inescapable that some form of enhancement of neural abilities and development of observer-relative perception must have taken place. Since it is clear that this has been explored inadequately, and since exograms determine such a massive part of our world today – even define our construct of reality – we have every reason to take a more sustained interest in the issue.

It is of importance to note that we have no evidence suggesting that other animals use or create exograms, and if this were correct, it would mark a most decisive factor separating us from them. The great apes are said to show some aptitude for sign language, and there is no doubt that they and a few other species have shown some ability to communicate with humans. That does not in itself demonstrate that their neural equipment has developed the ability to connect to exograms, nor would a rudimentary ability to do so solve the issue. Based on the evidence to hand, we have two alternative scenarios for the advent of exogram use in humans: a late or an early establishment. However, what would have affected our ancestors much more than the introduction of exograms is the beginning of their heavy use, of relying on them entirely for communication and conscious awareness.

Even that is not central to the present volume, which is concerned primarily with human auto-domestication. The relevance to that subject is that large-scale proliferation of exograms could easily account for the human ability to cope with a reduction of brain volume. That is the sole reason for visiting this topic here. It is amply clear from the above that if the brain connects to the exogrammic world out there, most of its prolific knowledge and information no longer needs to be stored within brain tissue or imagined engrams. Indeed, the brain then becomes somewhat like a random access memory in the sense of Mary Carruthers, of a system now stored primarily external to it. Frankly, this seems to be the only logical explanation for how the human lineage survived the atrophy of its brain, resulting as it did from an 'incidental' domestication process. In a sense, it even seems possible that, in an ironic twist, evolution could have had the last word, and selected in favour of the ability to use and process exograms. In this scenario — and we are not advocating it — exograms might have saved humanity from extinction caused by its inadvertent self-domestication.

This scenario is far removed from the previous models of how 'modern' humans came to be. It is about as far from the replacement hypothesis as that is from the Biblical explanation. It may not be correct, but it is raised here to show how far we have come, and what is possible if one is prepared to challenge the dogma of an inherently unscientific discipline. It may open up entirely new vistas. Even though this particular explanation may be invalid, secure new knowledge has emerged in the process of examining the topic. Just as Tooby and Cosmides (1992: 42) have mocked mainstream sociocultural anthropology, "where scientists are condemned by their unexamined assumptions to study the nature of mirrors only by cataloguing and investigating everything that mirrors can reflect", we have discovered that mainstream Pleistocene archaeology and palaeoanthropology continue as the same progression of misinterpretations they have historically always been (Bednarik 2013a).

We have even touched upon the lofty heights of deeply philosophical questions: what if exograms are involved in the formulation of our "imagined world made real" (Plotkin 2002)? On the face of it, that does sound entirely plausible. After all, archaeology has never embraced the idea of exograms in any meaningful way, just as it has never realised its inherent error of regarding evolution as a teleological process. With its oxymoron of 'cultural evolution' referring to lithocentric "observer-relative, institutional facts" (Searle 1995), which are the basis of unfalsifiable cultural and even ethnic but always etic constructs, it is far removed from science (Bednarik 2011a; 2013a). Not only has archaeology ignored exograms, but it has also severely misinterpreted all surviving exograms of the Pleistocene by explaining them away as entities archaeologists believe they are capable of dealing with: art and signs. Palaeoart is probably neither, and while it has the potential to identify cultures (in contrast to tools), that potential was wholly disregarded; instead, palaeoart was forced into the fictional cultural chronologies archaeology has invented.

Another impediment is the inadequate consideration of taphonomic effects in the interpretation of such remains, whose quantitative and qualitative properties have been severely distorted by a great variety of processes (Efremov 1940;

Solomon 1990; Bednarik 1994a). Taphonomy (which determines the selective survival of material evidence, and thus the qualitative and quantitative composition of all samples in archaeology) has seen to it that most exograms could have never survived from the Pleistocene, and many of those that could endure would probably not be recognisable as having functioned as exograms. Nevertheless, there are several classes of such materials that can, under fortunate circumstances, remain not only recoverable but also identifiable as exograms. Such examples from the Middle Pleistocene have been classified into beads and pendants, petroglyphs, portable engravings and notches, proto-sculptures, pigments and other manuports (Bednarik 1992a; 2003b; 2017a). All African Eve believers universally reject them as being in any way meaningful products of human activity. Their position on this has become increasingly untenable over the years as more and more specimens become available. Some recent finds, such as the headbands, tiaras, pendants and bracelets made and worn by the robust Denisovans are incredible and show spectacularly how wrong the beliefs about robust humans were (see Figure 4.1). In some ways, technologies seem to have been more advanced 50 ka ago than they were 25 ka ago; therefore, it is not justified to assume a linear, purely progressive development. We saw in Chapter 4 that the greenstone ornament from Denisova Cave was made with a complex technique apparently not rediscovered until Neolithic times (Figure 5.2). Therefore, the processes of cultural advance were far more complex than many archaeologists imagine. The most pernicious rejection of early sophistication is, we have seen, by defining evidence of uncomfortably high early development as an indication of a "running ahead of time" (Vishnyatsky 1994). Not only does this amount to a deliberate distortion in archaeological interpretation, which is at the best of times unreliable. It amounts to the argument that some genius hominin invented something in complete isolation, but his conspecifics lacked the nous to grasp the significance. To appreciate the absurdity of this thinking, we may visualise a brilliant inventor of beads wearing his creation around his neck as his associates around him scratched their heads in bewilderment. It is evident that beads, for instance, can only exist if they have meanings, and these must be shared meanings. Neither exograms nor memes of this type could have ever existed, even less attained archaeological visibility, unless many people shared them. Therefore, a running ahead of time explanation is an absurdity, particularly in materials presumed to be symbolic.

Language and other exograms

The recognition of an exogram in the 'archaeological record' is perhaps best illustrated by citing examples. The oldest known or suspected exograms tend to be manuports, which are unmodified objects transported and deposited by hominins. To qualify for that status, they must be of outstanding characteristics that would have attracted attention (usually visual); they must be of a material that does not naturally occur in the site where it was found and could not have been transported there by any natural means; and they must have been found in a sediment layer containing sound hominin occupation evidence. The highest standards of proof

Figure 5.2 Greenstone bracelet from Denisova Cave, Middle Palaeolithic, but of Neolithic technology

need to be applied to the oldest examples because their potential explanatory power is the greatest. The oldest specimen so far presented is the red jaspilite cobble deposited between 2.4 and 2.9 million years ago in the dolomite cave of Makapansgat, northern South Africa (Eitzman 1958; Dart 1974; Bednarik 1998). This stone object is highly prominent because of its red colour, its extreme hardness and 'special' material (seen from the perspective of a stone tool user), but most of all by its shape and the symmetrical deep pit markings that insinuate facial features. It needs to be remembered that facial detection is the most pervasive of all forms of pareidolia: present-day humans detect faces on burnt toast, tree bark, rock cliffs, house facades, even in Martian mountains (Bednarik 2016b).

Numerous animal species perceive pareidolic eye markings (e.g. on the wings of moths), especially those combined with a 'mouth'. In the case of the Makapansgat stone, the markings are perfectly placed on the head-shaped cobble (see Figure 2.9). There can be no doubt that a hominin of, say, 3 million years ago would have been visually fascinated by this object's strong pareidolic effect (cf. Sugita 2009; Taubert et al. 2017). Moreover, it was excavated from cave breccia rich in australopithecine remains that contained no fluvial elements, and the cobble comes from many kilometres away. It was found together with very archaic stone tools, and although there is no proof that it was deposited by australopithecine hand, there can be no doubt that it was carried into the cave by a hominin before the breccia formed. That may well have been by an early version of *Homo*, such as *Kenyanthropus* (Bednarik 1998). We can only think of two ways the stone could have been brought into the cave: as a gastrolith (gizzard stone) in the stomach of a bird many times the size of an ostrich, or by having been carried in by a hominin. In the latter case, the reason for the transport is that the primate concerned detected its pareidolic properties, which presupposes apperceptive capability. This refers to mental perception, especially the process of understanding something perceived in terms of previous experience. Therefore, it indicates that the primate concerned had the capacity of transferring some of its knowledge into a natural piece of stone and later retrieving that information. The stone both resembled and represented a head, and this understanding could be shared with a conspecific, even without language. The rounded cobble had become a symbol and an exogram. Not only did it permanently externalise a memory trace, but it also made that exogram available to any conspecific detecting its pareidolia.

We can safely assume that all very early exograms were of pareidolic potential because their meanings would have been the easiest to communicate, even in the absence of language. Indeed, onomatopoeic words can be suggested to be the phonetic equivalent of visual pareidolia, and in the context of ontogenetic language acquisition, sound symbolism plays an important role. The transfer of meaning by pareidolia or visual similarity (iconicity) is a neurally much simpler process than the transfer of symbolic meaning by any other process. Other primates have no difficulty determining what an image depicts, so we can be confident that this facility was always available to hominins. Even when iconicity has been reduced to a two-dimensional likeness, detection of meaning is simply a matter of disambiguation by pareidolia (Bednarik 2003c; 2014f). However, the meaning of non-iconic entities, such as beads or any aniconic exogram is much more difficult to convey to conspecifics. That applies especially in the absence of reflective or recursive language. Therefore, two factors appear likely: first, that the use of pareidolia-based iconicity facilitated the beginnings of verbal communication, initially perhaps by onomatopoeia; and, second, that the initiation of aniconic exograms would have been a consequence of such language having reached a commensurate level of communicative modulation allowing such arbitrary meanings to be shared. From our perspective, these seem to define the necessary conditions allowing language to take root, together with Bickerton's (2010) dictum that language ability must have been of immediate value to be selected.

From these principles, we can develop an essential line of argumentation. If it is correct that non-pareidolic exograms demand a reasonably sophisticated level of verbal communication (speech), we only need to look for the appearance of such empirical evidence on the archaeological record. There are just two caveats. One concerns the distorted nature of that record, inevitably garbled in favour of the replacement hypothesis. This, of course, is a universal problem we always need to compensate for, because the replacement advocates have long campaigned against the acceptance of pre-modern sophistication of any kind, and long denied the Robusts language use. The second caveat also concerns a systematic distortion: the taphonomy of the surviving record. It stands to reason that very many early exograms would have been of a nature that prevented their preservation. For instance, no Late Pleistocene exograms made in sand surfaces had ever been reported before 2019, but such unlikely 'ammoglyphs' have just been discovered in South Africa (Helm et al. 2020). Moreover, many other exograms may have survived, but we cannot simply detect former exogrammic function. If, for instance, the unmodified tooth of an animal had functioned as an exogram, standing perhaps as a symbol for the animal itself, we may well find the tooth in an excavation. Nevertheless, there is no way we could ever detect its former meaning. In short, the only exograms we can hope to identify with reasonable certainty are the most blatant examples, primarily those that have been suspected of having had symbolic meanings by those archaeologists who had an open mind about the minds of the ancients.

The Makapansgat stone is by far the oldest example of an object that must have had an exogrammic role, and this isolated placement renders it susceptible to rejection. On the other hand, it is perfectly reasonable to expect hominins or hominids of up to 3 million years ago to detect the outstanding properties of such a stone and to be sufficiently intrigued by them to carry the cobble for a considerable distance. The alternative would be to assume that they carried the stone into the cave without knowing why, a rather fatuous argument, especially if it is only made to sustain a failed hypothesis. However, from the next two million years, there are very few, if any, finds that seem to have been used as exograms. They become more numerous in the Middle and Late Acheulian, and especially in the subsequent Middle Palaeolithic or Mode 3 traditions. Many of these exograms feature one of the following characteristics:

1   The *referent* (the object represented, the signified) and *referrer* (the exogrammic element) are cognitively relatable by being 'of the same kind': the obvious examples are fossil casts that were apparently 'curated' or deliberately included in artefacts, such as Acheulian bifaces. The best-known examples are the hand-axes from West Tufts and Swanscombe in England (Oakley 1981: 14–16). The former is fashioned around a *Spondylus spinosus* cast, and the latter comprises a Cretaceous sand dollar (*Conulus* sp.; Figure 5.3). These are among the manuported fossils collected by people of Mode 2 and Mode 3 tool industries, and we can rightly assume that they were collected because their 'special' properties were recognised.

2   Other manuports were collected because of some other unusual properties, e.g. crystal facets, colour, translucency, shape. Most of them allow no speculation about the referent and offer no iconic aspects. The most frequently reported have been rock crystals, found in the Lower Palaeolithic of China, India, Israel, Austria and South Africa. Quartz crystals played significant roles in traditional societies right up the present times, in many cases, exceedingly complex exogrammic roles. It is not suggested that similar complexities also applied to very ancient crystal finds, but the collection of such 'special' stones implies that they conveyed meaning.

3   In considering linear engravings of Modes 2 and 3 assemblages (late Lower and Middle Palaeolithic), the occurrence of phosphene motifs is particularly noteworthy. Phosphene motifs are patterns perceived by the visual system without visual input but can be induced by mechanical stimulation from a great variety of sources (pressure on the eyes, blow on the head, sneezing, laughter, among others), electrical stimulation of the brain, transcranial magnetic stimulation, some medications, meditating, psychedelic drugs or extended visual deprivation, among others (Eichmeier and Höfer 1974). Phosphenes are autogenous and involuntary phenomena of the mammalian visual system whose form constants cannot be influenced by

cultural conditioning and which seem to be ontogenically stable. They reflect the inherent structures of the visual system rather than any external factor or information. Their patterns are of a limited number of geometric form constants, about 15, and we established that all known earliest linear markings made by humans, both in the ontogenetic and phylogenetic sense, matched such phosphene motifs (Bednarik 1987). This might suggest that in the former recapping the latter, we can catch a glimpse of the early development of linear markings.

The earliest engraved markings found on portable objects and rock surfaces consist of parallel lines, convergent lines motifs, dot patterns (e.g. groups of cupules), lattices, zigzags and multiple arcs or wave lines. Later during the Mode 3 conventions, radial figures, circles, circuitous mazes and spirals were eventually added. These patterns all resemble phosphene motifs, and our phosphene theory proposes that the universality of phosphene experiences rendered it possible to interpret these exogrammic externalisations as meaningful. This hypothesis (Bednarik 1984b; 1987; 1990c) has remained unfalsified to this day, despite being easy to test (by reporting pre-iconic graphic marks that are not phosphene motifs), but

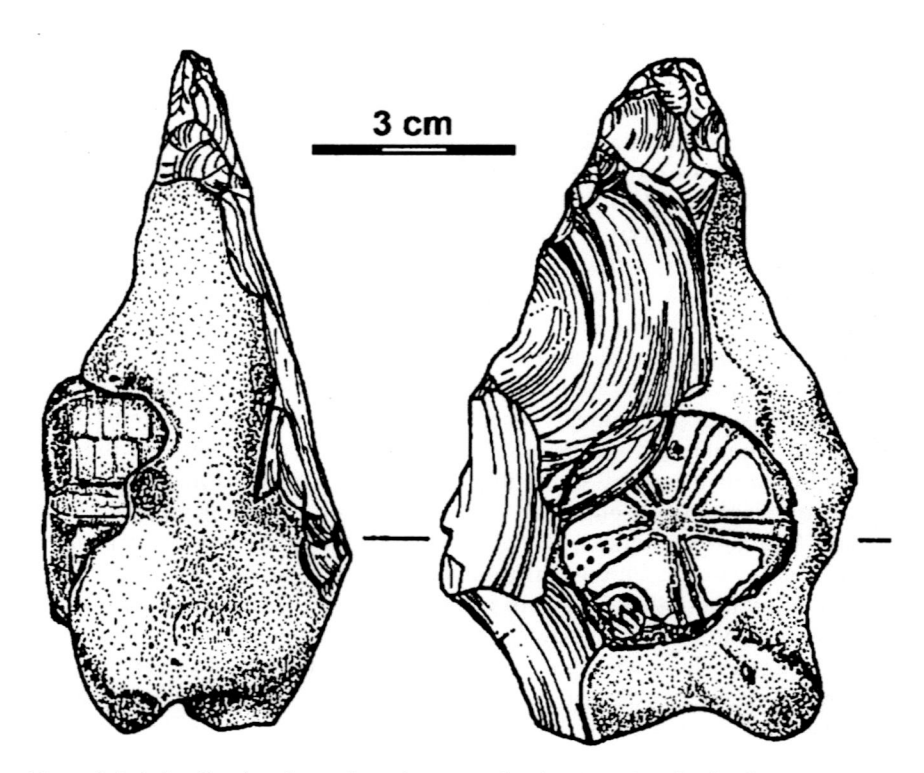

Figure 5.3 Acheulian hand-axe from Swanscombe, bearing *Conulus* fossil cast

has been overshadowed by the controversial and unsupported entoptic phenomena hypothesis which attributes rock art to shamans (Lewis-Williams and Dowson 1988). The relevance of the phosphene hypothesis in the present context is that the close correspondence of practically all Lower and Middle Palaeolithic anthropogenic markings with phosphene pattern types is too compelling not to engage. If there is a connection, it would strongly favour an explanation of early mark-making as an appeal to the visual system of conspecifics. Clearly, this would explain not only the markings but also their role in developing communication systems, especially language. This rationalisation would clarify the increasing involvement of exograms throughout the late Mode 2 and the Mode 3 traditions. The issue is therefore fundamental to understanding the development of language, other forms of symbol use and exogram proliferation. It is simply too important to squander on blind alley explanations such as the simplistic entoptic premise: in the whole wide world there is not a single report of a shaman having made rock art, but there are thousands of ethnographic accounts of non-shamans producing rock art (especially in Australia, where there are no shamans). Moreover, the phosphene motifs mark not only the phylogenetic origins of graphic production but also its ontogenic beginnings: all early markings by infants constitute phosphene patterns (Kellogg et al. 1965; van Sommers 1984).

We shall, therefore, pursue the much more productive line of reasoning involving exograms. Very early linear markings can be problematic. They may be made with stone tools, but they could still be just incidental cut marks or they may be taphonomic marks of many types. For instance, among taphonomic marks on bone, some can be the result of gnawing by animals (rodents, carnivores), transport by gravitation or water, kinetic action within a sediment (e.g. solifluction, cryoturbation), mycorrhizal action (i.e. by aqueous solution of carbon dioxide from micro-organisms active on the surfaces of plant rootlets), by gastric acids of carnivores such as hyenas, kinetic effects of burrowing animals, effects of drying or heat, and by trampling within uppermost sediment layers. Much the same applies to ivory and antler objects, and some of these effects can also be found on mollusc shells. Scanning electron microscopy (d'Errico 1994) and microtopography with laser-scan microscopy (Steguweit 1999) can certainly help in identifying surface markings, but ultimately such tasks are not for the novice: they require high levels of experience (Marshack 1985; 1989; Bednarik 1988; 1991; 1992c; 1994c; 2006b; d'Errico 1991; 1994). Less experienced commentators can easily err in their pronouncements.

Anthropogenic and apparently deliberate engravings and notches constitute one of the most commonly found form of exograms that have been recovered from Lower and Middle Palaeolithic (or, in southern Africa, Early and Middle Stone Age) deposits. Possibly the best known are the several items from the large hand-axe-free Lower Palaeolithic site at Bilzingsleben in Germany. The extensive occupation site called Steinrinne was located on the shore of a long-disappeared lake, on relatively muddy ground. It features worn pavements of stones and animal bones pressed into the mud (Bednarik 1997a: Figure 14), activity zones and several dwelling traces. Dated to the Holstein Interglacial

(c. 400 ka ago), Bilzingsleben yielded several very archaic hominin remains and a series of bone (see Figure 2.5), ivory and sandstone objects bearing a variety of linear markings (Mania and Mania 1988; Bednarik 1995a; Brühl 2018). Other Lower Palaeolithic engravings include those from the classical *Homo erectus* site of Trinil in Java, a zigzag pattern on a freshwater mussel shell dated to between 540 and 430 ka ago (Joordens et al. 2014); an apparently engraved stegodon tusk from Xinglongdong Cave, China (Gao et al. 2004); the fragment of a presumed horse bone from Sainte Anne I, Haute-Loire in France, bearing ten regularly spaced cuts along one edge (Raynal and Séguy 1986); and an engraved forest elephant vertebra from Stránská Skála, the Czech Republic (Valoch 1987). An engraved ivory fragment from Whylen, south-western Germany, appears to be from Rissian deposits but was lost during or immediately after the Second World War (Moog 1939). There is also a report of two engraved bone fragments from Kozarnika Cave, north-western Bulgaria, reported to be between 1.4 and 1.1 million years old, but both their dating and their artefact status need to be confirmed (Bednarik 2017a). One is a bovid bone fragment with ten grooves, the other a cervid bone fragment bearing 27 notches along an edge. South Africa has contributed a stone plaque with seven subparallel, clearly engraved long lines from Wonderwerk Cave (Figure 5.4). The stone was deposited before 276 ka ago (Bednarik and Beaumont 2012). In short, the number of such exograms from the earliest period remains severely limited, but with the transition to the Middle Palaeolithic/Middle Stone Age modes of production, engravings on various portable objects increase markedly in number (Bednarik 2017a).

The rejection of such finds by replacement hypothesis promoters has diverted attention away from these early exograms and thus distorted the study of cognitive evolution. Most of these objections come from commentators who have examined neither the finds nor the find sites. Fortunately, there are also forms of surviving early exograms whose identification is almost impossible to challenge. One of the most distinctive representatives is the beads and pendants. Small objects that have been drilled through with stone tools could be either beads or pendants, or they could be small utilitarian objects such as buckles, pulling handles or quangings (these are pulling handles the Inuit used in sealing; Boas [1888] 1938: Figures 15, 17, 121d; Nelson 1899: Plate 17; Kroeber 1900: Figure 8). Utilitarian objects of this kind need to be very robust and are therefore of distinctive shape, use-wear and material. Small objects that were drilled through either in the centre or close to one end (e.g. teeth perforated near the root); that are too small or too fragile to be utilitarian objects; and that lack the typical wear patterns of such artefacts can safely be assumed to be beads or pendants. The evidence that they were drilled with a stone tool is often indicated by a distinctive bi-conical and 'machined' section and sometimes by rotation striae. The wear traces of beads and pendants have been studied in detail on numerous archaeological as well as replicated specimens (Bednarik 1997d; 2005) and often indicate their mode of use. For instance, if beads were worn on strings for a long time, their flat faces rubbed against those

Figure 5.4 Stone plaque with seven incised grooves, c. 300 ka old, from Wonderwerk Cave

of adjacent beads and became very smooth. The perforations of pendants of any materials worn singly often show the traces of the supporting strings. These always appear on the inside of the perforation, opposite the object's point of gravity (Bednarik 1997e).

These and other tell-tale signs are unmistakeable, and yet the replacement model's advocates have on some occasions even rejected beads in their endeavours to prove the hominins concerned to be too primitive to use beads. The most incongruous objections of this kind are perhaps those relating to the numerous stone beads of the Acheulian from a series of sites in France and England. These were the first Palaeolithic beads identified when Boucher de Perthes (1846) presented his seminal evidence from the gravels of the Somme basin in northern France. He proposed that, based on the stratigraphy of the stone tools he had excavated, humans coexisted with Diluvial (Pleistocene) fauna; and that these hand-axe makers used beads. His first proposition was universally rejected by archaeology, culminating in the famous collective gaffe at the Paris Archaeology Congress of 1858. By that time two British geologists, Joseph Prestwich and Hugh Falconer were quietly testing Boucher de Perthes' claims by excavating next to one of his trenches. Their findings (Prestwich 1859) supported both claims, and Prestwich observed that on some bead specimens, the perforation had been "enlarged and completed".

Nevertheless, while the discipline conceded its miscarriage concerning the stone tools, the issue of the beads remained largely ignored to this day. Prestwich identified the finds incorrectly as "a small fossil sponge, the *Coscinopora globularis*, D'Orb". Boucher de Perthes, Falconer and Prestwich were not the only ones to find these beads in Acheulian deposits, Rigollot (1854) and Wyatt (1862) did so too. Late in the nineteenth century, Smith (1894: 272–276) collected about 200 identical items from an Acheulian site at Bedford, England. He described these as being of the same species and showing identical artificial enlargement of the natural orifice. Smith was confident that his specimens were used as beads, but he made no mention of the earlier finds.

In all, thousands of these stone beads have been collected from the Acheulian of St Acheul, Amiens, Soissons, Le Pecqu and Paris in France; and from Biddenham, Limbury and Leagrave in England. In 2003, we examined a collection of 325 of these French and English stone beads held at the Pitt Rivers Museum, Oxford (Bednarik 2005). They had been labelled *Coscinopora globularis* and collected before the early twentieth century. We subjected them all to detailed microscopic study but focused on the best-provenanced specimens, those from the Acheulian deposit of the Biddenham quarry at Bedford, England, and acquired by the Pitt Rivers Museum in 1910 (see Figure 2.6). One of the first things to note is that all of these objects had been incorrectly identified since the 1850s. We demonstrated that they are in fact of the species *Porosphaera globularis*PHILLIPS 1829, which is a Cretaceous sponge. The genus *Coscinopora* is a lychnisc hexactinellid sponge, for instance, *Coscinopora infundibuliformis*GOLDFUSS 1833 is funnel or cup-shaped, with a distinctive stem. It belongs to the order Lychniskida of the class Hyalospongea, whereas *Porosphaera* is of the Pharetronida, one of the two orders of the Calcispongea. Therefore, the species are not even closely related.

Despite its name, even *Porosphaera globularis* is only rarely of truly globular shape, which accounts for only about a quarter of all specimens. Their sizes vary from <1 mm to about 50 mm, the average of collected specimens (among which smallest sizes would be under-represented) being roughly 10 mm. Some specimens possess cylindrical tunnels of about 2 mm diameter that enter to various depths, ranging from mere indentations to nearly complete penetration. The cause of these tunnels has not been conclusively established, but most likely they were bored by parasites. The tunnels are usually but not always reasonably central, and there are occasional specimens with more than one such tunnel. After examining 2,734 randomly collected specimens, Nestler (1961) reported that only 390 of them (14 per cent) show any degree of tunnel development, and this confirms similar findings by other palaeontologists. These observations are significant because all the Acheulian specimens examined consist entirely of spherical specimens that are all *fully perforated* through their centre, i.e. the tunnel has two entries. Since it is estimated that only 14 per cent (maximal) of the natural specimens have any degree of tunnelling, and only a small proportion of these, say, less than one fifth, have tunnels penetrating to within 1–3 mm of the surface; and bearing further in mind that only about a quarter of the naturally occurring specimens are of reasonably spherical shape, it becomes evident that less than 0.7 per cent of a natural *random* sample of *Porosphaera globularis* can be expected to have *both the shape and the nearly full tunnel development* which are evident on *all the Acheulian specimens* known from England and France. When it is further considered that the Acheulian finds are mostly between 10 mm and 18 mm in diameter, whereas a natural sample would include sizes from 50 mm down to under 1 mm, with the smaller sizes probably much dominating, it becomes evident that the Acheulian sample is representative of perhaps 0.1 per cent or 0.2 per cent of a random sample.

Moreover, such specimens that are fully perforated by *natural* agency alone are incredibly scarce, accounting for certainly far fewer than 1 per thousand. Breaking through the barrier at the far end of the tunnel would defeat the purpose of the parasite that used the stone for protection, so the only specimens that could have two naturally formed openings would be the result of natural fluke processes. It is therefore statistically absurd to suggest that some form of natural selection could account for the Acheulian accumulations of numerous specimens of one size range, complete perforation and roundness. Nevertheless, some archaeologists have tried to explain this evidence away, simply because they find it inconceivable that Lower Palaeolithic hominins could have used beads (e.g. Rigaud 2006–2007; Rigaud et al. 2009). This is a classical demonstration of the lengths replacement advocates will go to in order to preserve the dogma that all pre-modern hominins were too primitive to have used beads. This is despite the clear evidence that these hominins had possessed developed self-awareness for millions of years and had used red pigment for well over a million years.

In addition to the careful selection process of the French and English beads made from fossil sponges, a close examination of 325 of them even provides extensive tribological evidence of their use (Bednarik 2019). This comes in two forms. Many of the specimens bear traces of deliberate action to break through the barrier preventing the tunnel from having a second opening. Evidence of flaking and percussion or pressure damage occur at the formerly closed end of the fossil's tunnel on many specimens. Indication of reaming out of this aperture is also evident in some cases (Figure 5.5). These traces are always entirely limited to the area of the formerly closed end of the tunnel. Besides reaming or pressure damage, some specimens display flaking scars around the opened aperture, showing that they were struck with some force in order to gain access to the tunnel in preparation for threading cordage through the bead. Some of these flake scars exhibit even the typical ripple patterns of silica stone fractured by spalling (Figure 5.6). The silicified fossils are as hard and brittle as chert or chalcedony. The form in which this damage occurs is distinctly anthropogenic and deliberate, it occurs not randomly, and it cannot reasonably be attributed to any natural process. In some cases, as many as six or seven impact flake scars can be discerned, indicating the difficulties in removing the remaining wall at tunnels that stopped some millimetres from the surface opposite the original tunnel entry.

156 *The unstoppable advance of exograms*

Figure 5.5 Microphotograph of the artificially enlarged orifice of one of the Bedford Acheulian beads

Second, many of these beads possess distinctive wear facets where they rubbed against other beads while worn on strings. These facets formed around the tunnel openings, i.e. on the surfaces that would have been in contact with other beads had they been threaded on a string and subjected to long-term abrasive wear from neighbouring beads (Figure 5.7). The facets range from small patches (Figure 5.7 a, c) of abrasion damage to very extensive depressions (Figure 5.7 d), in extreme cases covering much of the entire side surface of specimens (Figure 5.7 b). The worn areas range from flat-angled to quite steep recesses of hemispherical shape. Their extent is always distinctly delineated. Some of these deep wear facets are almost perfectly circular and central, so that the resulting concave ring of worn surface is evenly wide around the tunnel entry, while others are distinctly asymmetrical. Of particular interest are those specimens, usually rather small, that are distinctly wedge-shaped when viewed

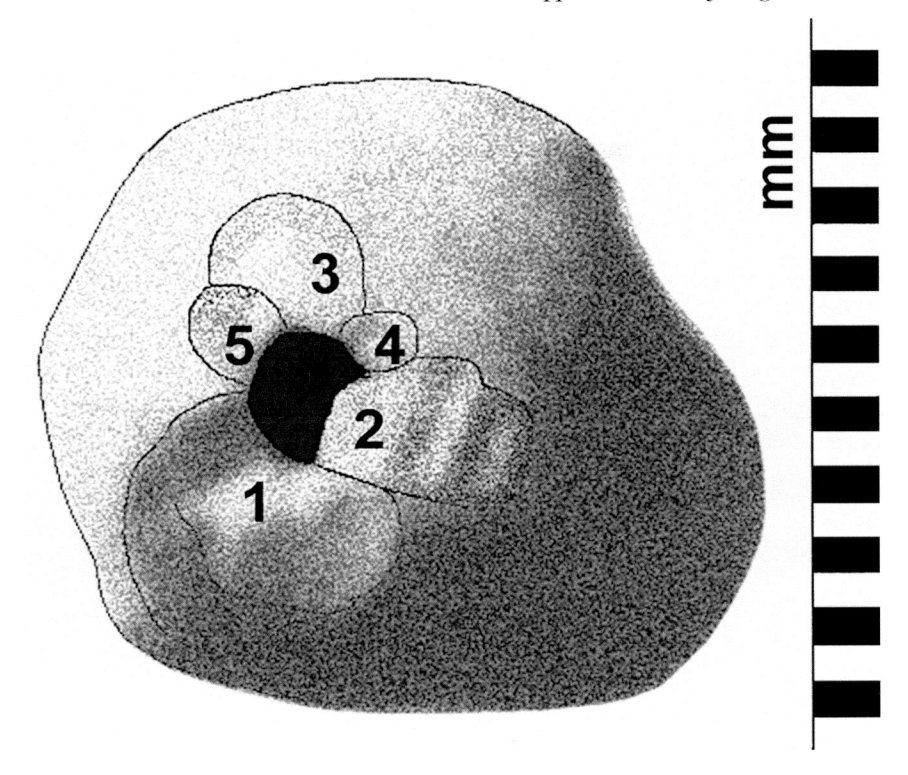

Figure 5.6 Flake scars at the closed end of the tunnel of one of the Bedford Acheulian beads and their sequence. Five scars are clearly visible, No. 2 displays rippling typical of impact fractures on silica stone

perpendicular to the direction of the central tunnel (Figure 5.7 b). They show the most non-symmetrical wear facets, evidently because if beads were worn as a necklace, i.e. forming a circular arrangement, there was inevitably more wear on the inside of a string's loop (Figure 5.8 d). Smaller beads were more affected by this and may have taken on a 'keystone function': the two wear facets are then distinctly non-parallel so that the two tunnel orifices can both be seen from one perspective (from the centre of the circular arrangement of the beads). This was necessary to accommodate the bulk of the fully spherical and larger beads, such as those perhaps added to a necklace at a later time (see Figure 5.8). Unless discoloured by the sediment, the *P. glob.* specimens are of the same buff colour as the weathering rind or cortex on sedimentary silica (which is indeed what they consist of). The wear facets, however, are always of a notably lighter colour, and significantly they never bear any taphonomic markings as found on the rest of the surfaces of these fossil casts.

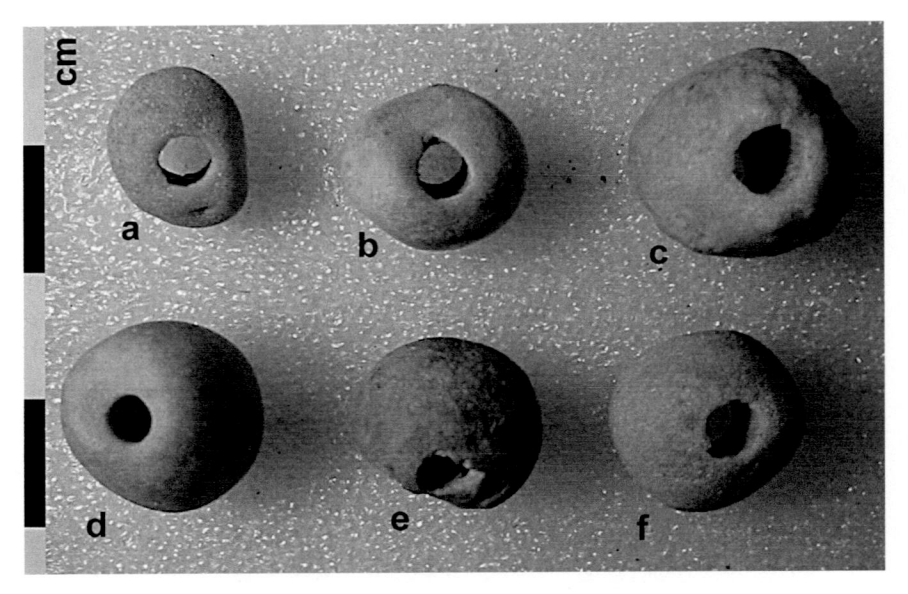

Figure 5.7 Six Acheulian *Porosphaera globularis* beads showing different degrees of wear at tunnel opening, including significant asymmetrical concave wear (b, d)

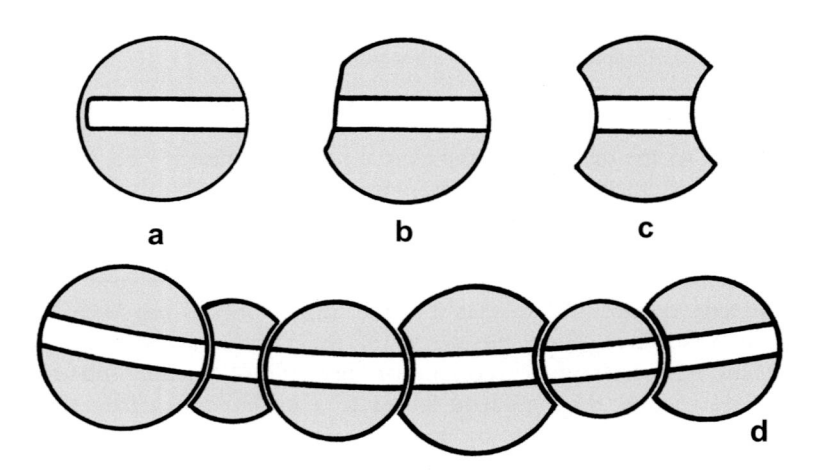

Figure 5.8 Schematic depiction of (a) the initial beads before anthropogenic action; (b) flaking to open the second tunnel entrance; (c) heavy wear from rubbing against other, fresher beads for many years; and (d) the outcome of beads of different ages on a string having been worn, some for very long periods of time

Only one type of tribological wear can account for such consistent wear patterning: the stones must have been arranged with their tunnels permanently aligned to be worn in this way, and for a very long time (Bednarik 2019). Such consistent wear patterns cannot possibly be explained as natural phenomena; the beads can only have been subjected to this wear through hominin intervention.

The bizarre thing about the insistence of African Eve advocates that there can be no beads older than 40 ka is that they seem unaware of other early examples, such as the 43 Acheulian ostrich eggshell beads from El Greifa Site E in Libya (Bednarik 1997d; Ziegert 2007); the Acheulian beads from Gesher Benot Ya'aqov in Israel (Goren-Inbar et al. 1991); or the pendants from Repolust Cave in Austria (Mottl 1951). The latter have also been claimed to be Lower Palaeolithic, although their dating is somewhat tenuous. However, it was not for that reason that the latter were rejected: they were precluded because the penetrations were defined as evidence of animal chewing (d'Errico and Villa 1997). Why an animal would chew a perfectly executed hole into the end of a wolf's canine, or how it would even manage to do this remains a mystery. It is also unknown why an identical perforated tooth of the Upper Palaeolithic is perceived as a pendant, while the African Eve's adherents reject an older one. Moreover, the perforation in question is perfectly shaped and shows extensive wear from suspending cordage (Figure 5.9). Here we have an example of an irrational dismissal of evidence when a more rational reaction would have been to reject the dating claim. It illustrates that the exponents were not familiar with the object, its circumstances of recovery or any other relevant details. Their only objective was to discredit the find. This is bizarre because elsewhere, d'Errico advocates the acceptance of an engraving by *Homo erectus* (Joordens et al. 2014) although it could easily be explained away as a taphonomic marking, or the acceptance of a group of quartz crystals as manuports (d'Errico et al. 1989) when they could reasonably be interpreted as a chance find.

So far, only two Lower Palaeolithic proto-sculptures, another form of very early exograms, have been reported. The first is the specimen from the Late Acheulian at Berekhat Ram, Israel (Goren-Inbar 1986), which even d'Errico has accepted as authentic upon examination (d'Errico and Nowell 2000) – after first rejecting it. The second, from Tan-Tan in southern Morocco, is a modified quartzite manuport bearing several pecked grooves intended to emphasise its human form (Bednarik 2003c). It is from a Middle Acheulian occupation site, and microscopic flakes of red, iron-rich material suggest that it was coated with ochre or haematite (see Figure 2.8). Although undated, it is assumed to be in the order of 400 ka old on the basis of the accompanying rich lithic assemblage, which makes it the second-oldest known pareidolic exogram after the Makapansgat cobble.

Another form of exogram that has managed to survive from the Lower Palaeolithic is the petroglyph, no doubt because of its high durability. Petroglyphs are anthropogenic rock markings made by a reductive process,

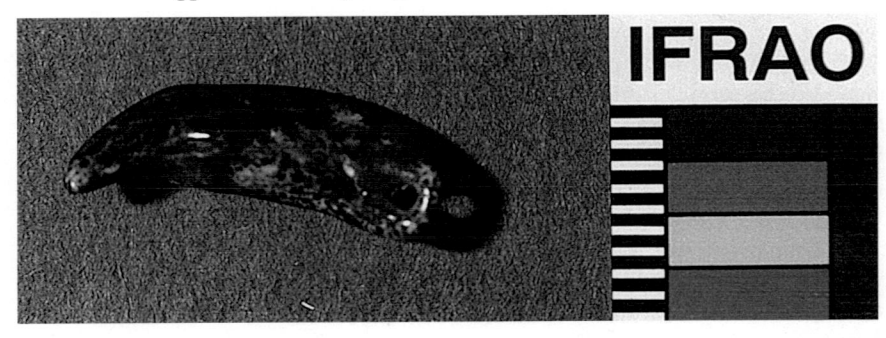

Figure 5.9 Carefully perforated wolf canine, probably c. 300 ka old, Repolusthöhle, Austria

in this case, percussion with a hammer-stone. All known examples are very basic designs: cupules and simple linear grooves. Cupules are spherical cap-shaped (rather than hemispherical) depressions made by repeated, carefully targeted impact that constitute the most common rock art motif, occur throughout much of the world and in most periods of human culture, and have provided few clues to their significance (Bednarik 2008d). The apparently oldest examples known are those of Auditorium Cave at Bhimbetka (Bednarik 1993a) and Daraki-Chattan Cave near Bhanpura (Bednarik et al. 2005), both in central India. They appear to have been created by people with a pre-Acheulian technology, with Oldowan-like stone implements. At both sites, they co-occur with linear groove markings. In both cases the petroglyphs are found on very hard quartzite inside caves and replication has shown that to produce one cupule on this rock requires many tens of thousands of blows with hammer-stones (Kumar and Krishna 2014). There are more than 530 cupules just at Daraki-Chattan, the creation of which involved immense efforts. Their purpose is entirely mysterious (Figure 5.10).

Cupules are also the earliest known petroglyphs in southern Africa. Some of those at Nchwaneng (see Figure 2.7) and Pothole Hoek in the southern Kalahari Desert have been attributed to the Fauresmith technocomplex and are thought to be about 410–400 ka old (Beaumont and Bednarik 2013). A smaller group of cupules at Sai Island, Sudan, are about 200 ka old (Van Peer et al. 2003). The site at East London in South Africa yielded a lattice pattern thought to be in the order of 400 ka old (Laidler 1933; 1934). The prominence of cupules among the world's earliest rock art is almost certainly a taphonomic phenomenon because they are among the most deterioration-resistant humanly made rock markings. Apart from the observation that the makers sought to constrain their diameter to the smallest possible size while achieving the greatest depth, little can be said about them. Given the great effort of their production, we can be sure that they had distinctive meanings to their makers and observers, one of which appears to be a message of perfection. These cupules are the

Figure 5.10 Some of the hundreds of cupules on the walls of Daraki-Chattan Cave, central India, of the Lower Palaeolithic

most common surviving form of exograms from the earliest phase of exogram use, the Lower Palaeolithic period. Providing a glimpse of conceptual sophistication that is entirely incompatible with the purported primitiveness of the hominins concerned, cupules have been made ever since and were still produced in the twentieth century, at least in Bolivia and Australia.

The last category of evidence suggesting the use of exograms in this very earliest period of human history relates to mineral pigments, most notably iron oxides and hydroxides, including haematite and ochre. Extensive evidence of pigment use begins at least 1.1 million years ago in southern Africa, e.g. at Wonderwerk Cave (Chazan et al. 2008; Beaumont 2004a; 2011; Beaumont and Bednarik 2013) and Kathu Pan 1 in South Africa (Beaumont 2004b; Beaumont and Bednarik 2013), and Kabwe in Zambia (McBrearty and Brooks 2000; Beaumont and Bednarik 2013). In Europe and India, pigment use has also been demonstrated from many Acheulian deposits (Bednarik 1992a; 2003b). Pigment as such is not an exogram, but its use is likely to have involved exogrammic information. Mineral pigments are likely to have been applied to the human body, to artefacts of various types, to clothing, tents, and to rock surfaces as dry pigment drawings or wet paintings, stencils or prints. Most of these uses do involve exogrammic roles, by externalising information that can be recovered by conspecifics.

The vast majority of very early exograms have no chance of surviving from the Lower Palaeolithic. Particularly important in this context is language, i.e. forms of exograms externalised as sounds that represent agreed meanings. One of the critical areas where catastrophist explanations of human modernity, such as the replacement or African Eve model, clash most severely with the gradualist scenario is on the question of human language beginnings. The replacement model places them as late as possible in the course of hominin history, seeking to preserve its standard narrative of the primitiveness of all 'pre-Moderns'. Its adherents have recently even adopted the extreme language of reserving the term 'human' for fully 'modern' man, or *Homo sapiens sapiens*. This perverse practice ignores the simple fact that the term 'homo' means 'human', hence all hominins since *Homo habilis* have been human. Their position is untenable because it assumes that pre-modern humans are not human, and yet many of them were the same species as us; some of them bred with the purported Moderns. The preferred explanation of the replacement scholars has long been that spoken language is limited to African Eve's progeny and therefore began only between 40 and 30 ka ago in Europe. It is part of the archaeological agenda of maintaining a separation between humans and other animals that ultimately has its subconscious origins in religion: the need to differentiate between those who have souls and those who lack them.

A gestural origin of language was first proposed in the eighteenth century, and by the middle of the nineteenth century, language hypotheses had become so rampant in Europe that in 1866 the Société de Linguistique de Paris saw itself obliged to ban the topic from its meetings and publications altogether. The theories then in circulation included the 'bow-wow', 'ding-dong' and 'heave-ho' versions, purely speculative, essentially onomatopoeic hypotheses. Nevertheless, the positing of naive and unfounded archaeological hypotheses about language origins has continued right up to the end of the twentieth century, with such examples as the hypothesis of Noble and Davidson (1996). Their idea, first enunciated by them seven years earlier (Davidson and Noble 1989), is that language was only possible after the advent of iconography, i.e. the production of figurative palaeoart. It posits that language began only after drawings of objects were made because these were needed to communicate meaning.

Nevertheless, as there is almost no evidence of figurative Pleistocene palaeoart outside of Europe, the traditions for which we have no evidence of figurative 'art' production have also failed to provide evidence of language, according to these authors. This is obviously absurd and is contradicted by their footnote to their hypothesis, that seafaring does prove the use of language (Davidson and Noble 1992). The Tasmanians, for instance, have never had any tradition of iconic rock art, so according to Davidson and Noble, they could not have had language. Nor, for that matter, should any other people lacking iconic palaeoart, such as the Jarawas (Bednarik and Sreenathan 2012) and various South American groups. Indeed, Davidson and Noble (1990) went as far as to claim that Neanderthals were closer to the apes than they were to humans, which illustrates their ideological encumbrance clearly enough. The entire artifice of the replacement hypothesis is in the final analysis subconsciously predicated on the demand that there must be a fundamental difference between animals and the crown of evolution – us, the likeness of a deity.

Leaving aside such irrational hypotheses as those assuming the existence of souls and minds, and instead viewing human species as part of the natural world, it soon becomes evident that the short-range model of human language is bereft of any evidence in its favour, yet it has been heavily promoted until most recent times. Much of the language origins debate has focused on the Neanderthal hyoid bone from Kebara Cave (Arensburg et al. 1989; Marshall 1989; notwithstanding Lieberman's 2007 speculations) which only indicates its unproductive and even irrelevant nature. The Dikika infant's hyoid bone (Alemseged et al. 2006) renders these discussions entirely superfluous: it shows poignantly how the historical sequence of finds determines the profound transience of our constructs of the past. More substantive discussions of language ability have focused on Broca's and Wernicke's areas, which have both been claimed to be detectable on cranial endocasts of *Homo habilis*. They may not necessarily be reliable indicators of language ability, but their early presence may indicate that some of the required structures were available to habilines. Such structures can only evolve if selective pressures favour them, so selection in favour of speech must have preceded them. The major syntheses of recent decades tend to return to linguistic and archaeological perspectives (Aitchison 1996; Bickerton 1996; 2010; Dunbar 1996; Falk 2009), and their authors arrive at the same basic finding: human language is such a complex phenomenon that its evolution, in every sense, must have been a very lengthy process. It cannot possibly be accommodated in the replacement model whose advocates were consistently unaware of the precariousness of their ideas in that respect.

This is not a new take on the topic:

> As Horne Tooke, one of the founders of the noble science of philology, observes, language is an art, like brewing or baking; but writing would have been a better simile. It certainly is not a true instinct, for every language has to be learnt. It differs, however, widely from all ordinary arts, for man has an instinctive tendency to speak, as we see in the babble of our young children; whilst no child has an instinctive tendency to brew, bake, or write.
>
> (Darwin 1871: 58)

The neural structures underlying language ability have been established in the human brain for eons; they are not an add-on feature dating from the final Pleistocene. The short-term model of human language origins is highly implausible, even based on simple biological and linguistic considerations (Bradshaw and Rogers 1993). For instance, the human system of producing verbal sounds differs profoundly from that of all other terrestrial mammals in one striking way. Darwin ([1859] 1959) had already observed "the strange fact that every particle of food and drink which we swallow has to pass over the orifice of the trachea, with some risk of falling into the lungs". Every year, thousands of humans choke to death on their food, whereas other mammals have separate pathways for breathing and feeding or drinking. Moreover, the problem is limited to human *adults* and is caused by the relatively low position of the adult human larynx. This appears to be the result of an evolutionary trade-off, indicating a significant advantage in having complex, finely modulated verbal communication. The relatively short palate and lower jaw are less efficient for chewing than those of non-human primates and hominoids, and they provide less space for teeth. However, the design of the human mouth and throat provides optimum conditions for differentiated sound production. The large size of the supra-laryngeal tract allows us to modulate and filter the frequencies of the sounds we make, in combination with the tongue and lips. These features could not have evolved in the space of a few tens of millennia, and they had to come into existence first before they could be selected.

Then there is the need for the brain structures responsible for the 'voluntary', 'intentional' control of speech (Bradshaw and Rogers 1993). In this respect, humans again differ significantly from other primates. Even chimpanzees have great difficulty controlling their verbal expression (e.g. concealing pleasure). They do, however, possess a rudimentary ability to deceive (Byrne and Whiten 1988). Nevertheless, in the area of deceptive behaviour, humans are the undisputed masters. This, indeed, involves self-reflection and great neural control over the speech production centres. Lieberman (1991) attributes our control over language to specific changes in the brain. These include the evolution of what is referred to as Broca's area, as well as the enlargement of the prefrontal cortex and rewiring of concentrations of neurons, the basal ganglia. Moreover, the biological implausibility of the short-term model of language evolution is implied by its inability to account for several simple observations. For instance, children are born with a genetic predisposition towards language acquisition, with an innate syntactic mechanism that appears to be biologically determined by neural structures. As Bickerton (1990) observed, there are 3,628,800 ways in which ten words can be arranged in a sentence. Consider 'Try to arrange any ordinary sentence consisting of ten words': only one sequence provides a correct and meaningful message; 3,628,799 variations are ungrammatical. Nevertheless, humans develop the correct understanding of syntax and grammar rapidly within the first years of their life.

A recent attempt of correlating language origins and self-domestication has been made by linking them to the Williams syndrome (Niego and Benítez-Burraco 2019). There have been numerous previous endeavours of relating changes from robust to gracile humans (e.g. the purported introduction of palaeoart or blade stone tools, none of which coincide with the advent of gracility) to a variety of syndromes and illnesses (see their discussion and rebuttal in Bednarik 2013f). Williams syndrome refers to a deficiency of between 25 and 34 genes from the region q11.23 of one member of the pair of chromosome 7, rendering the affected person hemizygous (only one copy of a chromosome is present) for those genes (Tassabehji et al. 1999). This genetic disorder involves specific facial features, moderate intellectual disability, heart and teeth problems, high blood calcium level episodes, short stature and an overly friendly personality. Although Niego and Benítez-Burraco (2019) present an enthusiastic and very detailed case for a connection with domestication – significantly better supported than previous attempts to explain human evolution via disorders – the few apparent connections they offer present no adequate case. They assume, a priori, that "people with the [Williams] syndrome seem to exhibit more exaggerated domesticated features than typically developing people", but this is not substantiated in a critical reading of their arguments. Their extensive discussion of genetics yields minimal support, as exemplified by their attempt to recruit support from gene GTF21. Benítez-Burraco has previously attempted to involve human disorders in explaining our most recent evolution and language origins: autistic spectrum disorder (Benítez-Burraco et al. 2016a) and schizophrenia (Benítez-Burraco et al. 2016c). In both cases, he and his co-authors failed to cite others who had previously offered similar explanations (see Bednarik 2013f for a fuller bibliography of these topics), just as they have written extensively about the subject of human self-domestication in the last few years without ever citing the auto-domestication theory (Bednarik 2008b).

The roles of exograms

As we have seen, for much of the twentieth century it was thought that memory traces are protoplasmic alterations of neural tissue thought to occur upon stimulation of the brain, and these were called engrams. With significant advances in neuroscience in the second part of that century, this hypothesis gave way to the model that animal memory operates not through specific localised physical changes but through processes more widely distributed in the cortices. At the time of writing the favoured explanation involves short-lived, high-frequency oscillations in the brain called ripples, which have been detected in the medial temporal lobe and temporal association cortex (Figure 5.11). Episodic memory retrieval appears to be through such ripple oscillations that have been detected by intracranial recordings in human subjects (Vaz et al. 2019). These electroencephalographic oscillations occur just before successful memory retrievals. The process seems to involve communication between different parts of the brain, which confirms that memory does not reside at specific loci, but is a feature involving different sites. In that respect, it differs from the memory traces encoded in exograms, which reside in very discrete entities that may take an unlimited number of forms.

An object or surface marking becomes meaningful when it is invested with a content that enables it to stand for something else. Such externalisations of properties of the visual system would resonate in the neural system of a

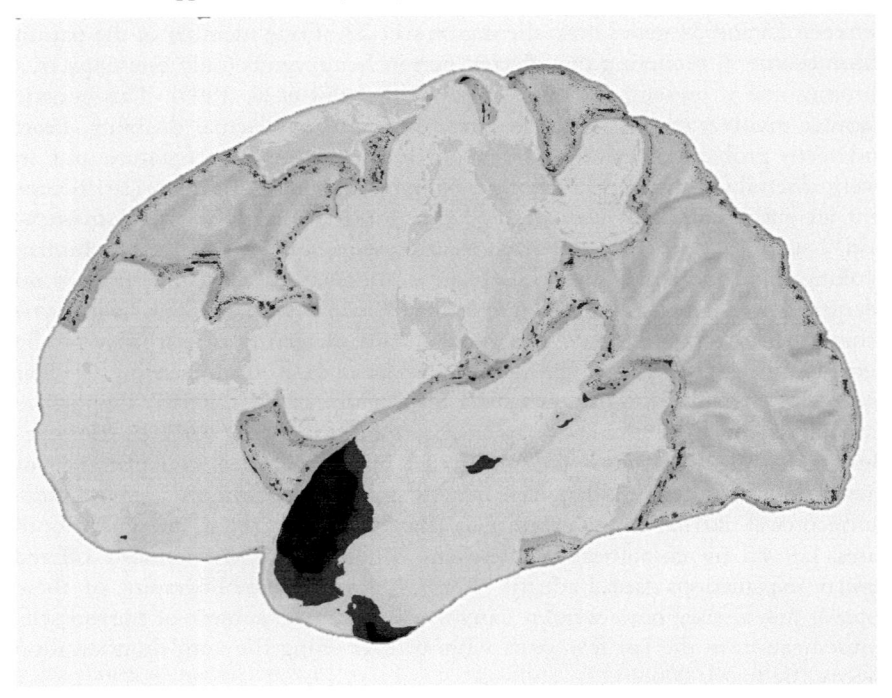

Figure 5.11 Just before a memory is activated, fast ripples of brain activity occur simultaneously, in this example in two areas, the temporal association cortex (very dark areas in lower parts of the image) and the medial temporal lobe (very light areas)

beholding conspecific (the interpretant). As Barrett (2013) contends, in semiotics we need to fundamentally distinguish between the representant (the thing being represented), the representation (the thing doing the representing) and the interpretant (the agent acknowledging or recognising the representation as standing for the representant). This form of communication made the autopoiesis that underlies all human constructs of reality feasible. When the response of one individual to a material quality was recognised empathetically by another, the patterning resonated in the interpretant's neural system. "Thus would the self have become objectified in the enactments of the other" (ibid.: 11).

The idea is not new: Descartes had already recognised the possession of the awareness of the self and the self's place in the world. Constructivism, however, holds the view that the only reality humans can know is that which is represented by human thought (Bednarik 1985; 1990b). Thus the key question is how the brain recalls and interprets cognitive data or experiences and represents those interpretations externally, and this remains unresolved. As Maturana states, "the content of cognition is cognition itself" (Maturana and Varela 1980:

xviii): everything said is said by an observer. His and Varela's theory of autopoiesis implies that "[w]hen we refer to our interactions with a concrete autopoietic system ... we project this system on the space of our manipulations and make a description of this projection" (ibid.: 89). "[A]utopoietic mechanisms operating as self-generating feedback systems ... cannot be separated from those who manipulate and use them" (McGann 1991: 15). Human cognition is a particularly complex autopoietic system, i.e. a system that possesses sufficient processes within it to maintain the whole. It yields precisely what Plotkin (2002) describes as an imagined world made real. "Provided that the internally consistent logical framework is not challenged by it, there is no reason to assume that an entirely false, cultural cosmology or epistemological model could not be formed and maintained indefinitely by an intelligent species" (Bednarik 1990b: 2). Most importantly, the evolution of human sensory facilities and intellect can be assumed to have only equipped us with adequate faculties to make them useful; there is no evolutionary benefit in the ability to define the reality of the cosmos correctly (Bednarik 1984b).

Why the nature and origin of human constructs of reality are so hard to define resembles the impossibility of finding a self-consistent set of axioms to deduce all of mathematics, which Kurt Gödel has shown to be impossible due to the self-referential nature of mathematical statements (Hofstadter 2007). Ultimately consciousness is self-referential awareness, the self's sense of its existence, and this is why its aetiology remains unsolved. Nevertheless, the issue *can* be resolved by involving the role of exograms (Bednarik 1987). The sustained use of every reference system, be it writing, diagrams, imagery, language, numbers, computer language or whatever, changes the structure, chemistry and operation of the human brain (by neuroplasticity). No such system, however, could be assumed to be as effective and all-pervasive in effecting such changes as the continuous use of externalised memory traces. Without it, the human brain as it exists today would be rather like an unconnected computer terminal, rendering the individual's ability to relate to what is experienced as the 'real world' severely impaired. Numerous neuropathological conditions illustrate such a state.

It is proposed, rather uncontroversially, that in the late part of the Pleistocene, competence in employing and exploiting exograms became the primary selecting factor in maximising cognitive fitness, gradually replacing traditional, 'natural' selection criteria. This process is by its very nature autocatalytic, and its effects can be observed throughout present-day societies, being evident virtually everywhere. Exograms generate not only frames of reference through being memory prompters; they also create self-referential realities. The mechanism of establishing these remains unknown, but it may well resemble the better-understood system of body awareness, or of how the individual makes a judgement about a conspecific's body movements (Bednarik 2012a). The former is established in the right hemisphere's superior parietal lobule (Bednarik 2013d: 27); the latter has been suggested to be obtained by running a virtual reality simulation of the corresponding movements in one's own brain

(Ramachandran 2009). Mirror neurons (Stern 1985; Di Pellegrino et al. 1992; Rizzolatti et al. 1996; Bråten 2004; 2007; Ramachandran 2009; Bednarik 2012a; 2013d) are probably involved in this process, as deduced from certain neuropathologies (Bednarik 2011b). Therefore, the most likely explanation of how human constructs of reality are established is that the brain creates a virtual-reality-like model of the external world, perhaps in the parietal lobe, in much the same way as the mental image of the body is shaped (Bednarik 2012a). In this, the exograms are indispensable, forming the most reliable link between brain activity and the external world. This is the mechanism by which humans experience 'reality' 'consciously', and it also seems to be the neural basis of what is termed 'volition'. This human ability of deriving abstract goals from the prefrontal cortex is unique in the animal world, but it would have been rendered possible by the described system.

In the context of human auto-domestication, the role of exograms is particularly important, because of the brain atrophy that is a typical symptom of the domestication syndrome. Indeed, it seems impossible to account for the reduction in brain volume in recent human evolution by any other factor. There is universal agreement that with the advent of the Upper Palaeolithic technology and mode of production, the demands made on the brain increased significantly, and the concurrence of this development with brain reduction demands an extraordinary explanation. It contradicts all canons of evolution: if brain physiology is more critical than brain anatomy or size, why did selection favour encephalisation? Brain atrophy must be occurring outside of evolution. This one factor alone demands that the change from robust to gracile humans was by a development flouting evolution – leaving only one alternative: domestication. Moreover, the self-evident outcome that humans overcame this severe impediment to their future as a species demonstrates that something must have occurred that rendered it possible to compensate for the loss of brain volume. That something can have only been the rapid increase in the use of extended memory via exograms.

The question remaining is this: was the gradual increase in the employment of exograms determined by new cognitive demands and diminishment of brain mass, or did newly acquired memory devices make the Upper Palaeolithic advances possible? We cannot know this, but that process seems most likely to have been evolution-based. Individuals who were more adept at working with and relying upon exograms can be assumed to have had an advantage over those who were slower in taking up that medium. Such natural selection would have furthered and accelerated exogram use. This model derives support from the *rate* of brain volume reduction over time: between 40 ka ago and the end of the Pleistocene, it seems to have been relatively modest (Bednarik 2014e; see also Figure 3.1). However, throughout the Holocene, the rate of reduction increased exponentially (Henneberg 1988; 1990; 2004; Henneberg and Steyn 1993; Bednarik 2014e). The development seems to mirror the increasing complexities of technology, social systems and culture over the same time interval. It would be an unlikely coincidence if the two developments were completely unrelated. If we accept the notion of some form of correlation, the same exponential curve would also be a rough measure of the extent to which humans relied increasingly on exograms.

It needs to be emphasised that there is no solid proof of this correlation, but as noted before, it is essential to find a consistent explanation for the relentless encephalisation of humans until the last part of the Late Pleistocene, in the face of the enormous burden it placed on human society of evolution; and the sudden reversal of encephalisation at a time of sharply increasing demands on the brain. This Gordian knot can be easily untied with the help of exograms. They had been in use for hundreds of millennia, but only on a relatively small scale. It was the domestication-induced brain atrophy that *actively selected in favour of those individuals who were best at working with exograms*. They were better at language use, communication generally, palaeoart use, body decoration, remembering things. They were very probably preferred in sexual selection and brute natural selection. Here, in a nutshell, we have a simple explanation for the emergence of *Homo sapiens sapiens*: the change from robust to gracile *H. sapiens* was in part caused by natural/sexual selection compensating for a condition, brain atrophy, caused by domestication.

The strength of this rationalisation is that it is in full accord with the empirical evidence as it stands, and it accounts for all those vexing inconsistencies that have plagued all previous interpretations. For instance, it resolves most elegantly the Keller and Miller paradox, which we visited in some detail in Chapter 3; it explains why deleterious neuropathologies were not selected against. Moreover, this model clarifies practically everything else about us: what made us the way we are as a sub-species (Bednarik 2011c). For instance, it elucidates the precise reasons for the incredible human proficiency in using external memory traces, developed by Upper Palaeolithic people to a fine art. Competency in exogram use became a highly prized natural selection factor that saved the human species from evolutionary ruin; no wonder we got so good at it. Not only does it now determine our revolutionary ways of using our brain; it even determines the origins of our constructs of reality: how we form a sentient impression of the world.

The profundity of the insights facilitated by the domestication theory presented here therefore extends well beyond merely explaining the change from robust to gracile *Homo sapiens*. They touch, in one way or another, on many, if not most, aspects of human modernity as we perceive it. Theory of Mind, self-awareness, consciousness, technology and culture were all available to certain non-human species, but they were not developed to the integrated system of the self-reflective human brain that observes itself, generating volitional decisions through excitatory/inhibitory neural functions. Homology implies a state of self-awareness in *Homo erectus* resembling that of a present human of about 8–12 years of age (Bednarik 2012a). The faculty of established verbal communication, or speech, can reasonably be attributed to that species (or subspecies), which orthodox Pleistocene archaeology has been denying vehemently. Archaeology tends to view exograms as symbols, yet various forms of them are not symbolic (not involving referent and referrer). Symbols are widely shared with conspecifics, generally via culture, whereas there is a distinct separation of personal exograms (not shared with conspecifics) and shared exograms (culturally determined). The language boards and other communication devices primatologists use in communicating with apes help define the difference between symbol and exogram: they are not native or naturalised systems of external storage, and apes could not create them. Exograms, by contrast, do not necessarily have referents but express abstract concepts. For instance, palaeoart, such as rock art, is undoubtedly comprised of exograms, but there is no more justification for defining it as symbolic as there is for consigning it to 'art' in the Western sense of that term.

Ultimately it is the consistent and skilled use of exograms that most separates humans from other animals and that serves as the clearest indicator of essentially modern behaviour (Bednarik 2011b; 2011c; 2012a; 2013d). If Pleistocene archaeology ignores the early presence of exograms in the archaeological record or is preoccupied with explaining it away, it fails in its professed task, the clarification of hominin history. These phenomena need to be studied by the relevant sciences (e.g. neurosciences and cognitive sciences) rather than be subjected to humanistic word games (Bednarik 2011a). This requires the formulation of empirically-based and testable propositions about the neural processes involved in the establishment, application and transmission of exograms within societies. No precedents exist for such work, but the sciences are better equipped to deal with these subjects than the humanities. Exograms are too important to be left to the attention of archaeology: not only are those that survived the only available physical evidence of the cognitive, cultural and intellectual evolution of humans; they probably saved humanity from abject decline towards the final Pleistocene. That period is marked by the sudden and so far unexplained cessation of several million years of continuous encephalisation when the cranial volume of humans abruptly began to plummet (Henneberg 1988; 1990; 2004; Bednarik 2011c; 2012a). For a few tens of millennia, the human brain shrank rapidly, an atrophy that has not been explained or even considered in any consequential fashion (but see Bednarik 2014e). Caused in all probability by the self-domestication of humans as an incidental effect, as were numerous other sudden and simultaneous changes in the recent 'evolutionary' trajectory of our species, this took place during a period when it is assumed that the demands made on the human brain escalated significantly. Many of the brain's functions, in particular much of its memory, became increasingly encoded in externally stored memory traces. The human brain became the central processing unit of a large system of memory expressions, externalised in numerous forms and media. The ensuing exponential increase in the complexity of human culture became possible by the burgeoning reliance of the brain on exograms – even as brain volume declined precipitously. As Saniotis and Henneberg (2013) noted:

> Extrasomatic mode of information transmission has a potential to be more effective as a mechanism of adaptation. This is due to a multitude of

coding mechanisms (construction of infrastructure-shelters, dwellings, roads etc., modification of plants and animals through domestication, art, writing, electronic storage), to the ability to transmit information laterally within the same generation without the need for biological reproduction, and targeted, non-random generation of new information based on focused collection of data and their logical processing in the context of previous knowledge.

Let us hypothesise. Exograms are not just an extrasomatic mode of information transmission; they are our means of investing the extrasomatic world – about which we are not very well informed – with meanings. All our understanding of the world beyond our individual nervous systems and proprioceptors derives from sensory input, which is severely limited because evolution never 'planned' to enable us to determine reality. If evolution had a purpose, a rationale – which it does not – it could be defined as equipping organisms with the aptitudes of surviving and procreating. This is far removed from gaining an insight into how the world actually works; in fact, it is more likely to result in a distorted construct. The human ability of processing extrasomatic information derives from the attachment of meanings to aspects of the external world. This was accomplished through the deployment of external memory traces that were arbitrarily conferred upon such aspects, linking features of the external world to whatever terminals in the brain, be they ripples of brain activity, synapses or neural tissue. Thus exograms established tangible links to the external world that allowed us to contemplate it in a conscious format.

Perhaps these considerations are not well substantiated here, but they do help to illustrate the complexity of the potential roles and effects of exograms. If it is true that our reality is an imagined world made real (Plotkin 2002), which we emphatically agree with, it seems essential to explore the process this could have involved. If Plotkin (and Plato before him; see his simile of the cave) were right, conscious sentience of the kind humans possess would be rather a little like illusion. The above considerations raise the possibility that it is the use of exograms and its effects that distinguish the introspective ('higher-order') consciousness of modern humans from the consciousness of non-human animals. This proposition would resolve the great difficulties the study of the evolution of consciousness faces, at least to the extent of explaining the difference between animal consciousness and the individual awareness we believe we possess.

Once upon a time, we had many arbiters of humanness, separating us from other animals: upright walk, tool use, culture, language use, forward planning, recursion, Theory of Mind, self-awareness, consciousness, symbolling ability. We were the 'naked ape' (Morris 1967); now, we are the 'neotenous ape' (Bednarik 2011c). Nearly all that separated us from other animals has dissolved under the probing light of science. Being 'introspectively conscious' is one of the last refuges for those striving to accentuate that surprisingly thin separating line. One of the very few characteristics endemic to gracile humans of the most recent times is that we are the only animal species that in mate selection has preferences of youth, specific body ratios (e.g. hips vs waist), facial features and symmetry, skin tone or hair (Bednarik 2012a). This is a very curious condition, endemic to 'modern humans', but ultimately our adeptness of creating and using exograms seems to be the ability most defining humanness. It had not even been considered to be a 'contender' (nor had culturally imposed sexual selection), which only illustrates that Todd Preuss was right to call us the "undiscovered primate" (Preuss 2001): we had not even understood what had made us human. In determining the neural and psychological correlates of consciousness, we had become too absorbed in effects to understand causes. It is reasonably assumed that mental processes, such as consciousness, and the physical processes in the brain are correlated, but the specific nature of that connection remained profoundly unknown. The role of exograms is as central here as it is in considering how hominins coped with the domestication-induced atrophy of the brain of recent millennia.

6 Effects of the domestication hypothesis

Palaeoanthropology and archaeology

This book began by lamenting the states of disciplines such as palaeoanthropology and Pleistocene archaeology: their accident-prone history, their apparent inability to learn from their mistakes and their great reluctance to concede errors. To a significant degree, these characteristics are attributable to the lack of falsifiability of disciplines that in many ways rely instead on the perceived authority of leading practitioners. We have seen how the idea that humans first evolved in England, dominating the field for several decades, was resolved by the sciences. However, the similarly improbable idea that art and thus sophisticated culture first arose in south-western Europe remains as widely held as ever. Indeed, it is reinforced by factors such as UNESCO's World Heritage List, which features dozens of Pleistocene rock art sites from south-western Europe, but not a single one from the rest of the world, where such ancient rock art is far more common. In their eagerness to demonstrate the cultural primacy of Europe, archaeologists have even had numerous late Holocene sites inscribed on the WHL as Pleistocene. This farcical state completely dominates public perceptions of this topic, and even the accepted wisdom of the popular science writers.

The reason why the falsity of the Piltdown finds was accepted but the similar falsity concerning Palaeolithic 'art' remains widely believed is simply that for most of the twentieth century the latter remained virtually unopposed. By now it is so well entrenched that it will require decades to take it down. The vested interests resisting such correction illustrate some of what is wrong with archaeology. Political masters (nearly all archaeologists are in some way answerable to the state) and the fantasies of religious, political and ethnic fancies influence how the human past is viewed. So does the desire to be popular with the public. Thus both palaeoanthropology and Pleistocene archaeology are torn between various demands and expectations, some of which are incommensurable with others, and the result is a bland amalgam lacking in clarity and a sense of purpose. The many facets of archaeology range from Pliocene hominins to industrial archaeology, from numismatics to aerial archaeology, from garbology to space debris archaeology. There are hundreds of such specialisations, and there is little connecting them. Most methodology is opportunistic, and there is no specific archaeological method.

However, most of all, archaeology lacks a unifying theory as well as a unifying purpose: it is simply impossible to subsume Marxist, cognitive, Christian, feminist, maritime, nationalist, Pharaonic, post-processual or all other archaeologies (Bednarik 2013a) under a universal agenda. Many deal with the distant human past, but others do not. Many archaeologies are centred on the method of systematic excavation, but others apply no excavation, and that method is in any case shared with several other fields of inquiry. In scientific disciplines, their inherent purpose is clear-cut and explicit, even if the range of their coverage, i.e. the phenomena they deal with, can seem a little bewildering and decidedly counter-intuitive. To the non-scientist, the movement of continental plates, the design of ball bearings and the friction of rock climbing boots seem entirely incompatible subjects and yet they, together with countless other phenomena are studied by one science: tribology (Bednarik 2019). Compare this approach to the phenomenal world with the common-sense expectations attached to archaeology, and it is easy to see the difference between a science and humanity. Archaeology thus resembles a hobby that always wanted to become a science, but lacked a clear strategy of how it could achieve that.

What would help in this would be to formulate a precise agenda or purpose. Most of the field deals with the 'human past', in one way or another, principally by studying the material remains taphonomy has spared. So, what precisely is the human past? Is it defined by the human responses to environmental challenges posed by climatic changes over the Quaternary period? Or is it human behaviour through the ages (Bednarik 2016a)? Could it be the history of the social systems that were developed over time? What about cultures and their so-called evolution? We come a little closer to the subjects focused on by archaeology by asking if the technologies used might throw some light on the human past. After all, what can be recovered are only objects, and their meanings have to be imposed. For well over 99 per cent of hominin history, most of these objects we recover, apart from food remains, are tools. Implements seem to permit the formulation of snippets of information about technology, although it would be careless to assume that any tool assemblage can fully define the technology in question. Consider, for instance, the polished greenstone bracelet from Denisova Cave in Siberia which is entirely incompatible with its Middle Palaeolithic or Mode 3 technological context (see Figure 5.2). Had it not been found, we would not know that *some* robust humans were capable of using what is typically regarded as a Neolithic innovation. This may be an extreme example, but there are many others to show that the technological pigeonholes of archaeology are far from reliable.

Therefore, even technology is not a dependable variable in learning about the human past. Moreover, tools or the combinations in which they occur in assemblages do not define cultures. However, for over one and a half centuries, archaeology has delineated early human cultures exclusively by technological parameters, notably those it extracted from stone tools. There have to be better, more apposite ways of explaining the human past. One perspective is to focus on the physical appearance of hominins, based necessarily on recovered skeletal remains. Palaeoanthropology seems to provide us with reasonably secure information about individual specimens and has been used extensively in determining past aspects of human nature.

Nevertheless, its interpretations of the available evidence are often problematic and have led to many excesses. Most obviously, the invention of so many hominin species may have enhanced the academic careers of many palaeoanthropologists (Henneberg and Schofield 2008), but several examples suggest that many, if not most, of these species designations are unwarranted. For instance, there is cumulative genetic evidence that most or all robust humans of the Final Pleistocene either interbred with more gracile humans or could have done so had they not been separated by time or space. That means that they can only be subspecies, not separate species: biological species ("with their evolutionary potential"; Mayr 2001) are defined by their members being able to interbreed viably. They do not necessarily coincide with the taxonomies of palaeontologists or palaeoanthropologists any more than the implement types imposed by archaeologists on stone tools should be regarded as emically valid. Based on the genetic evidence, it now appears likely that practically all humans of the last several hundred millennia were of one single species, *Homo sapiens*, but the mainstream in the discipline still has to concede that. Another example is the assignment of the small-bodied humans from Liang Bua Cave in Flores and from Callao Cave in the Philippines to two separate species. Both finds seem to characterise populations presenting island dwarfism, and in both cases, it was ignored that similar populations exist elsewhere (e.g. Sankhyan 1997; Berger et al. 2008). In the first case, it was ignored that the type specimen, named *Homo floresiensis* (Brown et al. 2004; Morwood et al. 2004), is pathological in several respects (Eckhardt 2007; Eckhardt and Henneberg 2010; Eckhardt et al. 2014; Henneberg et al. 2014) and rather than being a new species, is probably the result of genetic isolation. In the second case, we have only teeth, phalanges and a femur fragment in hand of the putative *Homo luzonensis* – not sufficient material to declare a new species (Détroit et al. 2019). After the revision of the age of the Flores 'Hobbit' (Sutikna et al. 2016), from 11–18 ka to 60–100 ka, both populations are now said to be in the order of 50–100 ka old. That places them well within the ambit of *Homo sapiens*, which is the proper species to which both groups can safely be assumed to belong. In both cases, there is a glaring absence of genetic evidence. While it is true that there is even less skeletal material available from the Denisovans, they have been endowed with a proper genome that implies that they are more 'archaic' than Neanderthals. However, even *H. sapiens denisova* and *H. sapiens neanderthalensis* are of our species – and, as mentioned, one of them was capable of applying Neolithic technology.

These examples and others like them show how scholars, in trying to secure maximum impact for their work, present sensationalist results in what Randy Schekman (2013) called "luxury journals" (see Chapter 1). By operating through a narrow base of referees, these establishment outlets are subjected to 'intellectual inbreeding' in the presumed 'upper echelons' of the discipline (the "high priests" of Thompson 2014), which suits the journals because those cliques are likely to rebut any refutation attempts that might arise. Their rejection of challenges to their authority coincides with the reluctance of the journals to admit having made an error of judgement, which they tend to regard as damaging to their status. If we take the example of the Flores 'Hobbit', the dissenting voices were rejected time and again by the luxury journals and had to be presented in journals of lower 'citation indexes'. Although these impact factors are misleading, academic authors are virtually forced to take them seriously. (Another example is the many sensationalist uranium-series dates for rock art reported in certain journals, despite the warnings by many authors that such results are often greatly exaggerating ages.) As we have noted in Chapter 1, numbers of citations are deceptive: the search engines tallying citations cannot distinguish between approving and disapproving literature references. If a paper is rejected in 80 citations and approvingly cited 20 times, its citation index of 100 will be severely misleading, particularly when one considers the number of times it was cited without the citing authors consulting the reference correctly. Furthermore, self-citation distorts the issue even more. Finally, it bears repeating that many journals are much more diligent in dealing with errors than the 'top-ranking', often commercialised, luxury journals, which says more about their editorial integrity than the exalted reputations of these periodicals.

Some of the remedies we advocate are rather far-reaching. For instance, the near-monopolies of the most influential academic journals in providing the popular science industry need to be addressed. They are precisely what a scientific pioneer of the calibre of Huxley railed against when he said nothing does more damage to science than cliques and schools of thought. Many of these journals are answerable to corporate interests, and indeed quite powerful interests, and if they regard having to concede errors as damaging to their prestige, they lack scientific integrity. This approach to science, together with the reward systems often attached to publishing in such journals is very corrosive, as is the counter-scientific rationale of much of academic publishing. However, it would be naive to expect the big science publishers to relinquish any of their territory, and Schekman's call to boycott three such journals does not seem to have had much effect. In the case of Pleistocene archaeology and palaeoanthropology, untested sensationalist claims and downright falsities continue to be disseminated. That is where the problem begins: the narratives of these disciplines churned out by elite outlets determine the disciplines' dogmas – just as Thomas Henry Huxley anticipated. Unfortunately, they are still just as often wrong as they were in the nineteenth and twentieth centuries. Moreover, sciences are not meant to operate just by trial and error.

Palaeoanthropology and Ice Age archaeology remain in the state that Kuhn (1970) defines as 'pre-paradigmatic': theoretically as well as methodologically fragmented, lacking universal theories and subject to sectarianism and faddish routines. Interpretations are universally invented. The available methods are incapable of divining the intangible variables crucial to correct interpretation, such as behaviour, intention, meaning, cognition. Behaviour, for instance, is the outcome of neuronal activity and its interplay with the endocrine system (the hypothalamus, the pituitary gland and the pineal gland). How could it possibly be deduced from the empirical evidence archaeology is capable of providing, and how would such propositions be rendered falsifiable?

As an example, the stone tool types archaeologists have invented are generally Searle's (1995) "observer-relative, institutional facts", lacking emic authenticity. Nevertheless, if we look more closely, we cannot help but notice the inability of stone tool experts even to distinguish conclusively between lithic implements and eoliths or geofacts – natural products that are very similar to stone tools (Figure 6.1). Let us be quite clear about this: the ability to discriminate between lithic artefacts and geofacts was a crucial factor in the rejection of Boucher de Perthes' proposition in the mid-nineteenth century. The basis of the very existence of Pleistocene archaeology was rejected for some decades by all archaeologists because they could not recognise stone implements. We can appreciate that this was mostly due to the pain of a new idea then, but it does come as a bit of a shock that the same impotence still applies today. No specialist is capable of conclusively, irrefutably, determining whether hominins knapped a piece of fractured flint or whether it was merely fractured by natural processes (Nash 1993; Gillespie et al. 2004; Shea 2010; Andrefsky 2013; Ellen and Muthana 2013; Lubinski et al. 2014; Bednarik 2019).

This, surely, illustrates the ineffectualness of archaeology in resolving its inherent limitations. In more than a century and a half since scorning Boucher de Perthes, the discipline has not found a way to determine the artefactuality of a flint confidently. Most sciences were established less than a century ago, and many have soared to dizzying heights in just the last fifty years. Archaeology seems to have stood still by comparison. This is not to suggest that all determinations of stone tools are unreliable; most are entirely credible. If several knapped chert fragments are found in the sediments of a limestone cave, perhaps together with charred animal bones, it would be wrongheaded to demand absolute proof. However, there are thousands of sites around the world that have yielded large quantities of fractured stone pieces over which archaeologists agonise and, in many cases, have heated arguments. We have investigated some of the most famous examples, such as the Calico Site in California (Haynes 1973; Duvall and Venner 1979; Payen 1982) and Pedra Furada (Guidon 1984; Guidon and Delibrias 1986) in Brazil. Similar finds have been reported from Toca da Esperansa in Brazil; Diring Yurlakh and Gorno-Altaisk in Siberia; Riwat and Pabbi Hills in Pakistan and many other places.

The reason for these controversies is that there are no clear and agreed diagnostic characteristics for the discrimination of lithic tools and geofacts. If we

178 *Effects of the domestication hypothesis*

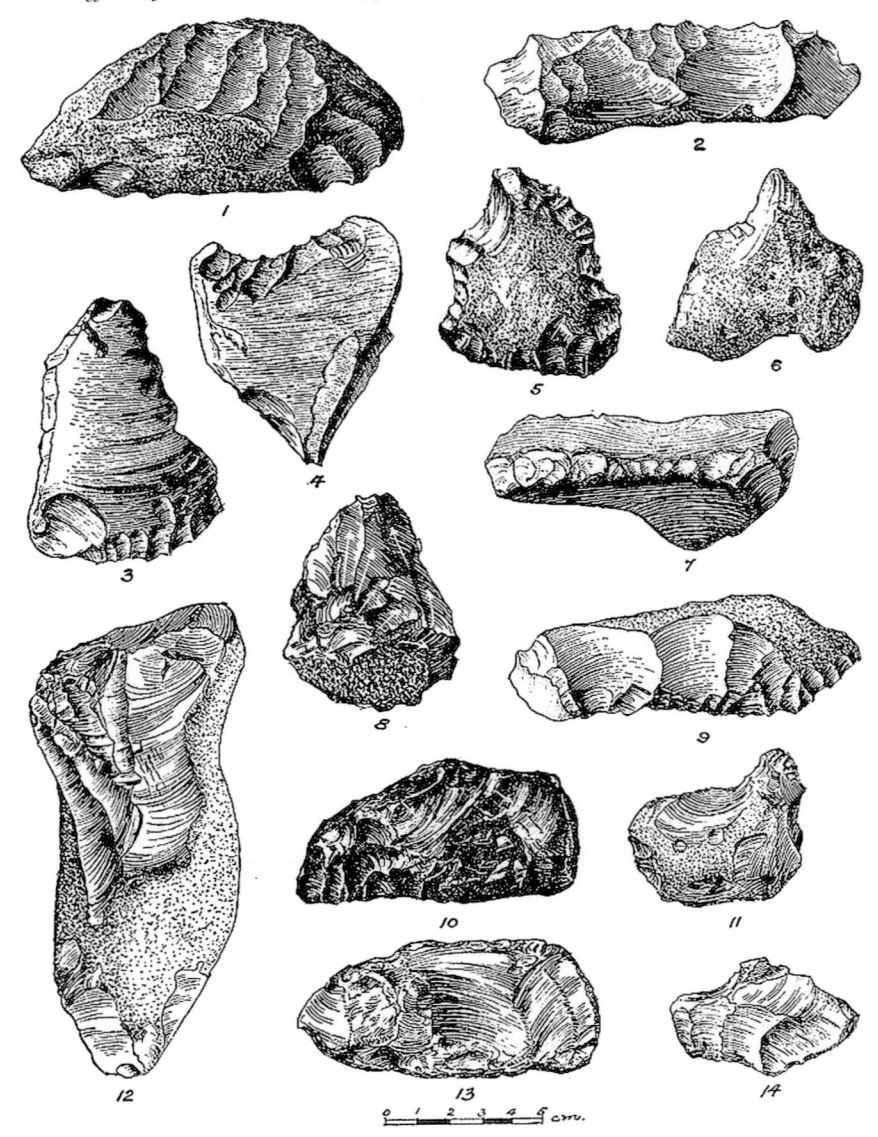

Figure 6.1 Some of the millions of presumed geofacts found in nature

consult any number of lithic typologists, we find little agreement among them and realise that many of the diagnostics they apply lack credibility. Compare, for instance, the diagnostic attributes employed by Barnes (1939), Patterson (1983), Luedtke (1986), Peacock (1991), Nash (1993) and Bradbury (2001):

- Patterson believes the presence of cortex on a flake's striking platform indicates it is a geofact; Peacock does not.
- They both regard a bulb of percussion as typical for artefacts, but Luedtke and Nash do not.
- To Patterson, Peacock and Luedtke, a diffuse bulb of percussion signifies a geofact; to Barnes and Nash, it does not.
- The presence of eraillure scars denotes artefacts, Patterson, Peacock and Nash believe; Luedtke rejects that view.
- Barnes regards an exterior platform angle of <90° as indicative of an artefact; Patterson, Luedtke and Nash do not.

This suffices to illustrate that there is limited consensus on crucial points. In fact, of a total of 13 supposedly diagnostic variables, there is not a single one on which all six experts agree. Many of their other pronouncements appear to be unsupportable. For instance, Bradbury holds the view that the ventral and dorsal surfaces can be identified on artefacts but not on geofacts. Such surfaces occur on any detached flake, irrespective of its causation and are usually identifiable. Luedtke, Peacock and Nash view the presence of ripple lines on the ventral face as indicative of a geofact, which is certainly not the case. Differently weathered dorsal flake scars, although probably more common on geofacts, are also not diagnostic. There are countless instances where tool knappers have recycled ancient stone implements that bore older flake scars. The absence of dorsal flake scars does not, as Patterson, Luedtke and Nash propose, designate a geofact. Nor does a dorsal flake scar orientation parallel to the medial axis indicate the flake is an artefact, as believed by Barnes, Patterson, Luedtke, Peacock and Nash. Three of these experts, Patterson, Luedtke and Peacock, even view the absence of dorsal cortex as indicative of an artefact. Numerous artefacts in fact bear remains of the cortex; and decortication flakes, which are also artefacts, *always* do so. Similarly, Peacock's propositions that platform faceting or radial fissures are diagnostic features seem to have very little merit. In short, not only is there disagreement among the scholars about key features; many of their beliefs are unwarranted or downright absurd. The discipline lacks any general standard of discriminating between stone artefacts and the geofacts resembling them. Therefore, it is fair to say that we have made no significant progress in the secure identification of stone tools since the mid-1800s, or since the times of Boucher de Perthes.

Nevertheless, this is a crucial prerequisite for the credibility of a discipline that makes pronouncements about aspects of the human past that are unfalsifiable at the best of times. We then invent etic categories of these *putative* stone tools, providing them with names and taxonomic pigeon holes. Not content with this level of deduction, we then concoct cultures based on these invented artefact types and the relative combinations in which they occur in assemblages as if tools had ever defined cultures. Once we have extracted lithocentric 'cultures' from the bits and pieces we found, we overextend our narratives even further and people these 'cultures' with ethnic groups, tribes or whatever other social entities. Nothing in this process is falsifiable or scientific; it is based on opinions from the very bottom up.

This example illustrates the difficulties with 'consensus models' of Pleistocene archaeology, and to some degree also those of palaeoanthropology. Proper science, unfortunately, is not a democratic process and the majority view, formed by fashion journals and the popular science industry, tends to be wrong much of the time. Therefore, the vital question, how does one learn about the human past, is in most circumstances dealt with inadequately. Stone tools help in illuminating some aspects of it, but they certainly do not define it. Culture, we have noted, is defined by a set of variables (such as customs, beliefs, ideas, social behaviour, languages, constructs of reality, symbols, shared cognition, rituals, among others) that cannot be recovered by the methods of archaeology as we know it. The only cultural dimensions of long-gone people we could recover are forms of evidence of intellectual activities that can survive in favourable conditions. We call such materials 'palaeoart' because they remind us of art in the modern sense of that term. As we have also noted, archaeology has misused such finds by simplistically regarding them as art and symbols, and second, by treating them as *artefacts*. This is a fundamental flaw because it has transferred the archaeological penchant for creating taxonomies of artefacts, already precarious in the case of lithic or other implements, to a corpus of evidence even less suited to such etic taxonomisation. All taxonomies created in archaeology, like most of those in palaeoanthropology, are observer-relative and institutional devices, offering little if any testability and falsifiability. Archaeologists seem to believe that there is something in their training that enables them to 'assess' or 'analyse' early 'arts' and 'symbols', and even to divine their meanings. Perhaps they feel the reality in which they see themselves existing is 'true' and, therefore, the taxonomies they create are 'objective'. So they view themselves as 'chosen' to interpret the histories of indigenous peoples and of societies that existed tens or hundreds of thousands of years ago. The latter, at least, cannot object, whereas many extant traditional societies have objected to such attention by those who to them represent colonising societies. More fundamentally, as argued cogently by Helvenston (2013), the brains of our distant ancestors differed significantly from ours. Therefore, it should be expected that colonising the cognitive world of the ancients with the simplistic tools of an unfalsifiable and ultimately political discipline would be epistemologically unsound.

The debates about behavioural modernity well convey the relativity of our ideas about the human past. Archaeological notions of this perceived milestone include its introduction with the purported anatomically modern humans in Africa as well as the advent of what is seen as Upper Palaeolithic technology. The modernity of behaviour is not determined by modern explanations of what are purported to be archaeological traces of ancient behaviour, but by the state and operation of the neural structures that are involved in moderating behavioural patterns (Bednarik 2011b; 2013d). The simplistic traditions of the discipline equate perceived patterns of production or economy as revealing modernity, or they link it to skeletal morphologies. However, the former are inadequately understood, and the latter are evidently irrelevant to the modernity of behaviour. One notion derives from the teleological concept of archaeology, the other imposes teleology on the dysteleological nature of evolution. If modernity were favoured by natural selection, as archaeology assumes, it would first have to be expressed to be selected for, just as language or a larger brain can only be selected after it has appeared by mutation. Therefore, the discipline is unable to explain how something of this nature would have developed to the point of being available for selection.

Moreover, nearly all of the variables suggested to indicate modernity can be shown to have been present not just in the Middle Palaeolithic, but even in the Lower Palaeolithic. They include the production of bone tools, microliths, blade tools, long-distance transport and exchanges, rock art and portable palaeoart, maritime colonisation, resin use, pigment use, interment, dwelling construction, use of garments, hunting of large mammals, exploitation of marine resources and others. None of these developments often described as heralding the Upper Palaeolithic technology does this. Consider, for instance, marine resources. Our knowledge about the utilisation of maritime resources is severely limited by the simple taphonomic fact that all coastal regions of the world have been repeatedly submerged under the sea and subjected to coastal erosion by successive sea-level variations. Therefore, such information is scarce, but it does exist (Bednarik 2014d: 83–84) and ignoring it does the discipline a disservice. Second, the very notion that behavioural modernity is needed to engage in a coastal economy is simply absurd. Numerous terrestrial animal species have developed coastal hunting or foraging skills, and the null-hypothesis surely is that all hominins would have managed this too where they lived on coasts. Until archaeologists develop strategies to refute the proposition that exploitation of marine resources is limited to humans of 'modern behaviour', their just-so claims should be ignored.

If Lower Palaeolithic humans possessed the proficiencies some archaeologists (e.g. Spikins 2009) define as Upper Palaeolithic innovations, does this mean that they were behaviourally modern? Some scholars have even considered that 'surprisingly early' advances may be due to a "running ahead of time" (Vishnyatsky 1994), a particularly absurd rationalisation. Just as today some humans manage to reach the Moon (which among other things indicates the present capability of human technology), but most of us do not, capability in the Pleistocene would have been marked by trailblazing – and not by the contents of refuse deposits, which is what archaeology usually excavates. If behavioural modernity is not marked by the variables supposedly indicating Upper Palaeolithic economies, how do we define it archaeologically? The behaviour of humans of medieval times differed considerably from today, so we have in effect "never been modern" (Latour 1993). Archaeology has failed even to explain what it means by 'human modernity': if the Mode 3 Tasmanians of 250 years ago were modern, how would we know that Pleistocene humans with similar Mode 3 technologies were not? These simplistic pronouncements derive from inadequate epistemologies.

The narratives of Pleistocene archaeology and palaeoanthropology may have been honed and perfected for many generations, but does that necessarily mean that they must be the best explanations we can find for the human past? Does the continual elaboration and embellishment of an initial model inevitably warrant a compelling outcome? We have seen in this book that there is an alternative to the dominant model of recent hominin evolution that differs from it as day does from night, even though it is based on virtually the very same empirical evidence. Not only does it present a realistic alternative to evolution; in contrast to the traditional model, which does not explain at all how humans 'became modern', it explains the reasons for all the changes this involved, and it does so in one single, elegant hypothesis. The mainstream model has modern humans suddenly appear out of nowhere and taking over the world in a matter of a few millennia. It derives from the observation that the change from robust to gracile humans occurred in a relatively short time, virtually in a geological instant. Such significant somatic changes tend to take a much longer time, it was reasoned, so there must have been an intrusive species that replaced the superseded type. This replacement model explains virtually nothing: why the neotenisation, the brain atrophy, the proliferation of exograms? It failed even to consider the most important aspects of becoming gracile humans in the Final Pleistocene; it was a compromise solution to a set of circumstances that seemed impossible to explain in a way other than replacement and the limited genetic information available appeared to support it, in a fashion. Nevertheless, it was plagued by glitches and contradictions all along, and since our presentation of the domestication hypothesis in 2008, contrary evidence of various kinds began to contradict the replacement hypothesis. In particular, it became apparent that replacement was based on Protsch's academic hoax, and that the Graciles were far from a separate species: they bred with more robust people. Moreover, new discoveries such as the 'Hobbits' in Flores and the Philippines as well as Chinese fossils did not fit the replacement scenario.

Reassessing human evolution

Instead of drawing the obvious conclusion that the replacement or African Eve hypothesis was slowly disintegrating, its supporters sought to prop it up by tinkering at its edges. This is the kind of reaction we have come to expect from these disciplines over the past couple of centuries: preservation of academic authority takes precedence over academic veracity. The just-so explanation that clashes with most evidence and in the final analysis explains virtually nothing has so far mainly been preserved, with more than a little help from the fashion journals who would otherwise have to acknowledge their lapses of reason over several decades. However, a very few perceptive commentators capable of thinking outside the box have begun to support our auto-domestication theory in recent years. The vast bulk of the disciplines' mainstream practitioners has not responded at all to the 2008 hypothesis, has not sought to refute it or commented on it. Perhaps they are ignorant of it, or perhaps they hope by ignoring it, the idea will go away.

So what will be the effects of the domestication theory, and what would be the effects if it became mainstream?

The immediate effects are already being felt. The disciplines' "high priests", while slightly modifying their declared views, are taking steps to reinforce their positions at the same time. For instance, they have taken to limit the term 'human' to what they regard as 'anatomically modern humans'. This extreme reaction relegates the Robusts, such as Denisovans or Neanderthals, to the status of non-humans. This is in response to the discovery that genetically, they are the same species as us. So genetics, which was cited endlessly to prop up the African Eve hypothesis, is wrong when contradicting that model. Bearing in mind that Graciles universally derived from Robusts in four continents, this is precisely the wrong reaction. It implies a hardening of attitudes, an escalation of opposition. Another response we have seen is the clarification that new genetic data has not turned them into multiregionalists. As if that were of any consequence: the only hypothesis that was not essentially multiregional was full replacement after a speciation event, i.e. the African Eve hoax. However, since we have accepted the single species explanation, because of the intermediate human forms (misleadingly called 'hybrids') we have found between Denisovans and Neanderthals as well as between them and 'moderns', the replacement advocates have taken to the absurd splitting of *Homo sapiens* into humans and non-humans. The strawman of multiregionalism has long become irrelevant: if full replacement did not take place, if 'Moderns' did not evolve in one region, they can only have appeared in multiple regions.

These developments demonstrate that the two disciplines have not learnt from their mistakes in the nineteenth century; these are being repeated today. The advocates of refuted theories will, in most cases, defend them until they die. Therefore, we have to expect that the alternative domestication theory will be ignored, if not rejected, for about three or four decades. That is the usual pattern, so it should not surprise us. What should be more disturbing is that around the middle of the twenty-first century, the domestication theory will be widely accepted, but in a significantly corrupted form. This, too, is the usual pattern. It is the outcome of the enforced delay in acceptance, during which pressure tends to build up to find a solution to the dissonance. The eventual change also leads to an over-reaction, as we have witnessed historically. For instance, when the rejection of Pleistocene rock art suddenly collapsed at the beginning of the twentieth century, the realisation that such palaeoart occurred in south-western Europe led to a preoccupation with the self-delusion that art originated there. This, in turn, caused the complete neglect of all other Pleistocene rock art in the world, even though it is more numerous and older in other continents. Another example is the collapse of the case that hominins first arose in England after the australopithecines had been ignored for decades. Here, the over-reaction led to an obsessive focus on Africa, ignoring the potential of Asia for many decades as a consequence.

It cannot be predicted in what form the domestication theory will eventually be accepted, but we can expect that it will be in some corrupted form. That is the fundamental problem with these two disciplines: lacking in the characteristics of sciences, they are at the mercy of fads and academic cliques, and will presumably remain so for some time. The domestication theory is likely to become a similarly faddish fixation and lead to a series of vogue subsidiaries. It is, therefore, also difficult to predict how it will eventually affect the mainstream. What can be determined, however, is how we would *like* it to affect Pleistocene archaeology and palaeoanthropology, and what would help to improve these fields.

To begin with, this would be an opportunity for them to realise how susceptible they are to short-term fashions. Their epistemological dynamics appear to be as Kuhn (1970) predicts: the proposals of revolutionary concepts punctuate long periods of 'normal' activity. However, whereas in the sciences the latter elicit interest, in archaeology they are typically rejected, scorned or ignored for considerable periods, then either grudgingly accepted or ruefully embraced to the point of over-compensation. An appreciation of these subtleties and undercurrents would help combat the discipline's susceptibility to conformist fads.

More interesting than this generic observation are the reflections on the effects the auto-domestication theory has on our understanding of the human past. After all, this is what archaeology is supposed to be addressing. Archaeology is not worth having if it fails in this, and the replacement hypothesis was an abysmal failure by all those who helped it to gain and maintain currency. Why, for instance, did it take three decades to check its academic pedigree and to discover that it derives from a hoax? Such negligence would be unthinkable in the sciences, particularly when it involves such a fundamental, direction-giving feature of a discipline. Archaeology needs to reflect not only on this but also on the control exercised by its high priests who can sweep challenges to their authority under the carpet seemingly with impunity. This is a serious matter that needs to be faced with integrity: is it worth having a discipline that consists of the invention of narratives about the human past? In what way is such a pursuit superior to the mythologies of past millennia addressing the same subject? How was it possible that a cock-and-bull story about an African Eve and her genocidal progeny garnered so much support that it became the virtual dogma of the discipline for several decades? It does indeed suggest that archaeology is incapable of self-correction. Its previous major controversies have already implied that, and scientists have had to clarify all of them.

The human past implied by the domestication theory involves no wandering tribes, no Exodus, Eden, Eve or Adam. Instead, the emergence of people physically resembling us entails the laws of genetic inheritance, of sexual selection and the domestication syndrome. This implies the consideration of cause and effect, testability and falsifiability. There is no origins hoax involved, and the effects of this change on the paradigm are clearly spelt out. They are virtually all the factors that define the differences between robust and gracile members of the species *Homo sapiens*. These include, but are not limited to, reduction of robusticity and brain size, the proliferation of detrimental genetic mutations and pathologies, universal neotenous changes, the abolition of oestrus, the rise of exclusive homosexuality, of neurodegenerative conditions, the introduction of the domestication syndrome with all it entails, and the incredible proliferation of exograms. These massive changes are the characteristics that make us what we are today, and they have led to many derivative changes, social and cultural, that define the reality in which we see ourselves existing. Nobody can deny that these factors are of immense importance in identifying what archaeologists call 'modern humans', that they are absolutely central in understanding us as a species. However, not a single one of these factors describing the human condition has been considered, let alone explained, by traditional, mainstream Pleistocene archaeology and palaeoanthropology. Instead, humanity has been given a simplistic explanation augmenting the Biblical narrative. In other words, this academic mainstream has wholly failed in its brief of delivering a credible account of how we became what we are.

That is the state of the discipline. Will it mend its ways? Not in the short term, we predict. It has never experienced a change in paradigm without opposing it fervently and without distorting it when finally accepting it. Will it be different this time?

References

Acharyya, S. K. and P. K. Basu 1993. Toba ash on the Indian subcontinent and its implications for the correlation of Late Pleistocene alluvium. *Quaternary Research* 40: 10–19.

Adams, F. and K. Aizawa 2001. The bounds of cognition. *Philosophical Psychology* 14: 43–64.

Adams, F. and K. Aizawa 2008. *The bounds of cognition*. Blackwell Publishing, Malden, MA.

Agrawal, D. P., R. Dodia, B. S. Kotlia, H. Razdan and A. Sahni 1989. The Plio-Pleistocene geologic and climatic record of the Kashmir valley, India: a review and new data. *Palaeogeography, Palaeoclimatology, Palaeoecology* 73: 267–286.

Agrawal, D. P., B. S. Kotlia and S. Kusumgar 1988. Chronology and significance of the Narmada formations. *Proceedings of the Indian National Science Academy* 54A: 418–424.

Ahern, J. C. M., I. Karavanic, M. Paunović, I. Janković and F. H. Smith 2004. New discoveries and interpretations of fossil hominids and artifacts from Vindija Cave, Croatia. *Journal of Human Evolution* 46: 25–65.

Aiello, L. and C. Dean 1990. *An introduction to human evolutionary anatomy*. Academic Press, New York.

Aiello, L. C. and P. Wheeler 1995. The expensive-tissue hypothesis: the brain and the digestive system in human and primate evolution. *Current Anthropology* 36: 199–221.

Aitchison, J. 1996. *The seeds of speech: language origin and evolution*. Cambridge University Press, Cambridge.

Aizawa, K. and F. Adams 2005. Defending non-derived content. *Philosophical Psychology* 18: 661–669.

Albert, F. W., O. Shchepina, C. Winter, H. Römpler, D. Teupser et al. 2008. Phenotypic differences in behavior, physiology and neurochemistry between rats selected for tameness and for defensive aggression towards humans. *Hormones and Behavior* 53: 413–421.

Alemseged, Z., F. Spoor, W. H. Kimbel, R. Bobe, D. Geraads, D. Reed and J. G. Wynn 2006. A juvenile early hominin skeleton from Dikika, Ethiopia. *Nature* 443: 296–301.

Allen, J., C. Gosden, R. Jones and J. P. White 1988. Pleistocene dates for human occupation of New Ireland, northern Melanesia. *Nature* 331: 707–709.

Allen, J. S. and V. M. Sarich 1988. Schizophrenia in an evolutionary perspective. *Perspectives in Biological Medicine* 32: 132–153.

Allman, J., A. Hakeem, J. M. Erwin, E. Nimchinsky and P. Hof 2001. The anterior cingulate cortex: the evolution of an interface between emotion and cognition. *Annals of the New York Academy of Science* 935: 107–117.

Allman, J., A. Hakeem and K. Watson 2002. Two phylogenetic specializations in the human brain. *Neuroscientist* 4: 335–345.

Allman, J. M., N. A. Tetreault, A. Y. Hakeem, K. F. Manaye, K. Semendeferi, J. M. Erwin, S. Park, V. Goubert and P. R. Hof 2010. The von Economo neurons in frontoinsular and anterior cingulate cortex in great apes and humans. *Brain Structure and Function* 214(5–6): 495–517.

Allman, J. M., K. K. Watson, N. A. Tetreault and A. Y. Hakeem 2005. Intuition and autism: a possible role for von Economo neurons. *Trends in Cognitive Science* 9(8): 367–373.

Allsworth-Jones, P. L. 1986. The Szeletian: main trends, recent results, and problems for resolution. In M. Day, R. Foley and Wu Rukang (eds), *The Pleistocene perspective*, pp. 1–25. Allen and Unwin, London.

Ambert, P. and J.-L. Guendon 2005. AMS estimates of the age of parietal art and human footprints in the grotte d'Aldène (southern France). *International Newsletter on Rock Art* 43: 6–7.

Ambert, P., J.-L. Guendon, P. Galant, Y. Quinif, A. Grunesein, A. Colomer, D. Dainat, B. Beaumes and C. Requirand 2005. Attribution des gravures paléolithiques de la grotte d'Aldène (Cesseras, Hérault) à l'Aurignacien par la datation des remplissages géologiques. *Comptes Rendus Palevol* 4: 275–284.

Amirkhanov, K. A., V. A. Zhukov, V. V. Haumkin and A. V. Sedov 2009. Эпёха ёлдёвана ёткрыта на ёстрёве Сёкётра [Oldowan Age discovered at Socotra Island]. *Pripoda* 7(1127): 68–74.

Anati, E. 1981. The origins of art. *Museum* 33: 200–210.

Anderson, E. 1949. *Introgressive hybridization*. Wiley, New York.

Andrefsky, Jr, W. 2013. Fingerprinting flake production and damage processes: toward identifying human artifact characteristics. In K. Graf, C. Ketron and M. Waters (eds), *Paleoamerican odyssey*, pp. 415–428. Texas A&M University Press, College Station, TX.

Anikovich, M. 2005. Early Upper Paleolithic cultures of eastern Europe. *Journal of World Prehistory* 6(2): 205–245.

Aoki, K. and M. W. Feldman 1991. Recessive hereditary deafness, assortative mating, and persistence of a sign language. *Theoretical Population Biology* 39: 358–372.

Arajärvi, R., J. Haukka, T. Varilo, J. Suokas, H. Juvonen, J. Suvisaari *et al*. 2004. Clinical phenotype of schizophrenia in a Finnish isolate. *Schizophrenia Research* 67: 195–205.

Arca, M., F. Martini, G. Pitzalis, C. Tuveri and A. Ulzega 1982. Il deposition Quaternario con industria del Paleolitico Inferiore di sa Padrosa-Pantallinu (Sassari). *Rivista di Scienze Preistoriche* 37: 31–53.

Arensburg, B., L. A. Schepartz, A.-M. Tillier, B. Vandermeersch and Y. Rak 1990. A reappraisal of the anatomical basis for speech in the Middle Pleistocene hominids. *American Journal of Physical Anthropology* 83: 137–146.

Arensburg, B., A.-M. Tillier, B. Vandermeersch, H. Duday, L. A. Schepartz and Y. Rak 1989. A Middle Palaeolithic human hyoid bone. *Nature* 338: 758–760.

Armand, J. 1983. *Archaeological excavations in the Durkadi Nala: an early Palaeolithic pebble-tool workshop in central India*. Munshiram Manoharlal Publishers, Delhi.

Armitage, S. J., S. A. Jasim, A. E. Marks, A. G. Parker, V. I. Usik and H. P. Uerpmann 2011. The southern route 'out of Africa': evidence for an early expansion of modern humans into Arabia. *Science* 331: 453–456.

Armstrong, E. 1980. A quantitative comparison of the hominoid thalamus: II. Limbic nuclei anterior principalis. *American Journal of Physical Anthropology* 52: 43–54.
Armstrong, E. 1991. The limbic system and culture: an allometric analysis of the neocortex and limbic nuclei. *Human Nature* 2: 117–136.
Arnold, B. 1996. *Pirogues monoxyles d'Europe central: construction, typologie, évolution.* Musée cantonal d'archéologie, Neuchâtel.
Arzarello, M., F. Marcolini, G. Pavia, M. Pavia, C. Petronio, M. Petrucci, L. Rook and R. Sardella 2007. Evidence of earliest human occurrence in Europe: the site of Pirro Nord (southern Italy). *Naturwissenschaften* 94: 107–112.
Arzarello, M., C. Peretto and M.-H. Moncel 2015. The Pirro Nord site (Apricena, Fg, southern Italy) in the context of the first European peopling: convergences and divergences. *Quaternary International* 389: 255–263.
Ashley Montagu, M. F. 1960. *An introduction to physical anthropology.* Thomas, Springfield, IL.
Ashley Montagu, M. F. 1989. *Growing young.* Bergin & Garvey, Westport, CT.
Asmus, G. 1964. Kritische Bemerkungen und neue Gesichtspunkte zur jungpaläolithischen Bestattung von Combe-Capelle, Périgord. *Eiszeitalter und Gegenwart* 15: 181–186.
Atkinson, Q. D. 2011. Phonemic diversity supports a serial founder effect model of language expansion from Africa. *Science* 332: 346–349.
Aubert, M., A. Brumm, M. Ramli, T. Sutikna, E. W. Saptomo, B. Hakim, M. J. Morwood, G. D. van den Bergh, L. Kinsley and A. Dosseto 2014. Pleistocene cave art from Sulawesi, Indonesia. *Nature* 514: 223–227.
Awadalla, P., A. Eyre-Walker and J. Maynard Smith 1999. Linkage disequilibrium and recombination in hominid mitochondrial DNA. *Science* 286: 2524–2525.
Ayala, F. J. 1996. Response to Templeton. *Science* 272: 1363–1364.
Ayalew, M., H. Le-Niculescu, D. F. Levey, N. Jain, B. Changala, S. D. Patel *et al.* 2012. Convergent functional genomics of schizophrenia: from comprehensive understanding to genetic risk prediction. *Molecular Psychiatry.* doi:10.1038/mp.2012.37.
Bächler, E. 1940.*Das alpine Paläolithikum der Schweiz im Wildkirchli, Drachenloch und Wildenmannlisloch.* Vol. II, Monographien zur Ur- und Frühgeschichte der Schweiz, Basel.
Bächler, H. 1957. Die Altersgliederung der Höhlenbärenreste im Wildkirchli, Wildenmannlisloch und Drachenloch. *Quartär* 9: 131–146.
Badam, G. L. 1973. Pleistocene fossil studies. *Bulletin of the Deccan College Research Institute* 33(1–4): 21–40.
Badam, G. L. 1979. *Pleistocene fauna of India.* Deccan College, Pune.
Badam, G. L. 1995. Palaeontological research in India: retrospect and prospect. In S. Wadia, R. Korisettar and V. S. Kale (eds), *Quaternary environments and geoarchaeology of India*, pp. 437–495. Geological Society of India, Bangalore.
Badam, G. L. and S. N. Rajaguru 1994. Comment on 'Toba ash on the Indian subcontinent and its implications for the correlation of Late Pleistocene alluvium' by S. K. Acharyya and P. K. Basu. *Quaternary Research* 41: 398–399.
Badcock, C. R. 1980. *The psychoanalysis of culture.* Basil Blackwell, Oxford.
Bader, O. N. 1978. *Sungir': verkhnepaleoliticheskaya stoyanka.* Izdatel'stvo "Nauka", Moscow.
Bagehot, W. 1905. *Physics and politics, or thoughts on the application of the principles of 'natural selection' and 'inheritance' to political society.* Kegan Paul, Trench, Trübner, London.
Bahn, P. G. 1993. The 'dead wood stage' of prehistoric art studies: style is not enough. In M. Lorblanchet and P. G. Bahn (eds), *Rock art studies: the post-stylistic era or where do we go from here?* pp. 52–59. Oxbow Monograph 35, Oxford.

Bailey, J. 2006. A brief overview of chimpanzees and aging research. Written for project R & R: release and restitution for chimpanzees in US Laboratories. Available at: www.releasechimps.org

Balbín Behrmann, R.de, J. J.Alcolea Gonzáles and M. A. González Pereda 2003. El macizo de Ardines, Ribadesella, España. Un lugar mayor del arte paleolítico europeo. In R. de Balbín Behrmann and P. Bueno Ramírez (eds), *Primer Symposium Internacional de Arte Prehistórico de Ribadesella*, pp. 91–152. Asociación Cultural Amigos de Ribadesella, Ribadesella.

Balter, M. 2007. A mind for sociability. *Science Now* 27 July, p. 1.

Balter, M. 2012. Did Neanderthals paint early cave art? *Science Now* 4 June.

Barandiarín, J. M. 1980. Excavaciones en Axlor (Campaña de 1969). *Obras Completas* 17: 129–384.

Barinaga, M. 1992. 'African Eve' backers beat a retreat. *Science* 255: 686–687.

Barkow, J. H. 2001. Universals and evolutionary psychology. In P. M. Hejl (ed.), *Universals and constructivism*, pp. 126–138. Suhrkamp Verlag, Frankfurt.

Barnes, A. S. 1939. The differences between natural and human flaking on prehistoric flint implements. *American Anthropologist* 41: 99–112.

Baron-Cohen, S. 1991. Precursors to a theory of mind: understanding attention in others. In A. Whiten (ed.), *Natural theories of mind: evolution, development and simulation of everyday mindreading*, pp. 233–251. Basil Blackwell, Oxford.

Baron-Cohen, S. 2000. Is Asperger syndrome/high-functioning autism necessarily a disability? *Development and Psychopathology* 12: 489–500.

Baron-Cohen, S. 2002. The extreme male brain theory of autism. *Trends in Cognitive Sciences* 6(6): 248–254.

Baron-Cohen, S. 2006. Two new theories of autism: hyper-systemizing and assortative mating. *Archives of Disease in Childhood* 91: 2–5.

Barrantes-Vidal, N. 2004. Creativity and madness revisited from current psychological perspectives. *Journal of Consciousness Studies* 11: 58–78.

Barrett, J. C. 2013. The archaeology of mind: it's not what you think. *Cambridge Archaeological Journal* 23(1): 1–17.

Bastir, M., A. Rosas, P. Gunz, A. Peña-Melian, G. Manzi, K. Harvati et al. 2011. Evolution of the base of the brain in highly encephalized human species. *Nature Communications* 2: 588.

Bayer, J. 1929. Die Olschewakultur: eine neue Fazies des Schmalklingenkulturkreises in Europa. *Eiszeit und Urgeschichte* 6: 83–100.

Beals, K. L., C. L. Smith and S. M. Dodd 1984. Brain size, cranial morphology, climate and time machines. *Current Anthropology* 25: 301–330.

Beaumont, P. 1990. Wonderwerk Cave. In P. Beaumont and D. Morris (eds.), *Guide to archaeological sites in the Northern Cape*, pp. 101–134. McGregor Museum, Kimberley.

Beaumont, P. 2004a. Wonderwerk Cave. In D. Morris and P. Beaumont (eds), *Archaeology in the Northern Cape: some key sites*, pp. 31–35. McGregor Museum, Kimberley.

Beaumont, P. 2004b. Kathu Pan and Kathu Townlands/Uitkoms. In D. Morris and P. Beaumont (eds), *Archaeology in the Northern Cape: some key sites*, pp. 50–53. McGregor Museum, Kimberley.

Beaumont, P. B. 2011. The edge: more on fire-making by about 1.7 million years ago at Wonderwerk Cave in South Africa. *Current Anthropology* 52: 585–595.

Beaumont, P. B. and R. G. Bednarik 2013. Tracing the emergence of palaeoart in sub-Saharan Africa. *Rock Art Research* 30(1): 33–54.

Beaumont, P. B. and R. G. Bednarik 2015. Concerning a cupule sequence on the edge of the Kalahari Desert in South Africa. *Rock Art Research* 32(2): 163–177.

Bednarik, R. G. 1984a. Die Bedeutung der paläolithischen Fingerlinientradition. *Anthropologie* 23: 73–79.
Bednarik, R. G. 1984b. On the nature of psychograms. *The Artefact* 8: 27–33.
Bednarik, R. G. 1985. Editor's response to letter by B. J. Wright. *Rock Art Research* 2(1): 90–91.
Bednarik, R. G. 1986. Parietal finger markings in Europe and Australia. *Rock Art Research* 3(1): 30–61.
Bednarik, R. G. 1987. Engramme und Phosphene. *Zeitschrift für Ethnologie* 112(2): 223–235.
Bednarik, R. G. 1988. Comment on D. Mania and U. Mania, 'Deliberate engravings on bone artefacts of *Homo erectus*'. *Rock Art Research* 5: 96–100.
Bednarik, R. G. 1989. The Galgenberg figurine from Krems, Austria. *Rock Art Research* 6(2): 118–125.
Bednarik, R. G. 1990a. More to Palaeolithic females than meets the eye. *Rock Art Research* 7(2): 133–137.
Bednarik, R. G. 1990b. On the cognitive development of hominids. *Man and Environment* 15(2): 1–7.
Bednarik, R. G. 1990c. On neuropsychology and shamanism in rock art. *Current Anthropology* 31: 77–80.
Bednarik, R. G. 1991. Comment on F. d'Errico, 'Microscopic and statistical criteria for the identification of prehistoric systems of notation'. *Rock Art Research* 8: 89–91.
Bednarik, R. G. 1992a. Palaeoart and archaeological myths. *Cambridge Archaeological Journal* 2(1): 27–43.
Bednarik, R. G. 1992b. On Lower Paleolithic cognitive development. In S. Goldsmith, S. Garvie, D. Selin and J. Smith (eds), *Ancient images, ancient thought: the archaeology of ideology*, pp. 427–435. Proceedings of the 23rd Annual Chacmool Conference, University of Calgary.
Bednarik, R. G. 1992c. Natural line markings on Palaeolithic objects. *Anthropologie* 30 (3): 233–240.
Bednarik, R. G. 1993a. Palaeolithic art in India. *Man and Environment* 18(2): 33–40.
Bednarik, R. G. 1993b. Wall markings of the cave bear. *Studies in Speleology* 9: 51–70.
Bednarik, R. G. 1994a. A taphonomy of palaeoart. *Antiquity* 68(258): 68–74.
Bednarik, R. G. 1994b. The Pleistocene art of Asia. *Journal of World Prehistory* 8(4): 351–375.
Bednarik, R. G. 1994c. The discrimination of rock markings. *Rock Art Research* 11: 23–44.
Bednarik, R. G. 1995a. Concept-mediated marking in the Lower Palaeolithic. *Current Anthropology* 36(4): 605–634.
Bednarik, R. G. 1995b. Response to Colin Groves. *Bulletin of the Archaeological and Anthropological Society of Victoria* 1995/5: 8–11.
Bednarik, R. G. 1996. Palaeolithic love goddesses of feminism. *Anthropos* 91(1): 183–190.
Bednarik, R. G. 1997a. The origins of navigation and language. *The Artefact* 20: 16–56.
Bednarik, R. G. 1997b. The earliest evidence of ocean navigation. *The International Journal of Nautical Archaeology* 26(3): 183–191.
Bednarik, R. G. 1997c. Direct dating results from rock art: a global review. *AURA Newsletter* 14(2): 9–12.
Bednarik, R. G. 1997d. The role of Pleistocene beads in documenting hominid cognition. *Rock Art Research* 14: 27–41.
Bednarik, R. G. 1997e. Pleistocene stone pendant from Western Australia. *Australian Archaeology* 45: 32–34.
Bednarik, R. G. 1998. The 'austrolopithecine' cobble from Makapansgat, South Africa. *South African Archaeological Bulletin* 53: 4–8.

Bednarik, R. G. 1999. Maritime navigation in the Lower and Middle Palaeolithic. *Comptes Rendus de l'Académie des Sciences Paris, Earth and Planetary Sciences* 328: 559–563.
Bednarik, R. G. 2001. Beads and pendants of the Pleistocene. *Anthropos* 96: 545–555.
Bednarik, R. G. 2003a. Seafaring in the Pleistocene. *Cambridge Archaeological Journal* 13 (1): 41–66.
Bednarik, R. G. 2003b. The earliest evidence of palaeoart. *Rock Art Research* 20(2): 89–135.
Bednarik, R. G. 2003c. A figurine from the African Acheulian. *Current Anthropology* 44 (3): 405–413.
Bednarik, R. G. 2005. Middle Pleistocene beads and symbolism. *Anthropos* 100: 537–552.
Bednarik, R. G. 2006a. The Middle Palaeolithic engravings from Oldisleben, Germany. *Anthropologie* 44: 113–121.
Bednarik, R. G. 2006b. The methodology of examining very early engravings. *Rock Art Research* 23(1): 125–128.
Bednarik, R. G. 2007. Antiquity and authorship of the Chauvet rock art. *Rock Art Research* 24(1): 21–34.
Bednarik, R. G. 2008a. The mythical Moderns. *Journal of World Prehistory* 21(2): 85–102.
Bednarik, R. G. 2008b. The domestication of humans. *Anthropologie* 46(1): 1–17.
Bednarik, R. G. 2008c. Children as Pleistocene artists. *Rock Art Research* 25(2): 173–182.
Bednarik, R. G. 2008d. Cupules. *Rock Art Research* 25: 61–100.
Bednarik, R. G. 2009. Fluvial erosion of inscriptions and petroglyphs at Siega Verde, Spain. *Journal of Archaeological Science* 36(10): 2365–2373.
Bednarik, R. G. 2011a. Rendering humanities sustainable. *Humanities* 1(1): 64–71; doi:10.3390/h1010064. Available at: www.mdpi.com/2076-0787/1/1/64/
Bednarik, R. G. 2011b. The origins of human modernity. *Humanities* 1(1): 1–53; doi:10.3390/h1010001. Available at: www.mdpi.com/2076-0787/1/1/1/
Bednarik, R. G. 2011c. *The human condition*. Springer, New York.
Bednarik, R. G. 2011d. Genetic drift in recent human evolution? In Kevin V. Urbano (ed.), *Advances in genetics research*, vol. 6, pp. 109–160. Nova Press, New York.
Bednarik, R. G. 2012a. An aetiology of hominin behaviour. *HOMO — Journal of Comparative Human Biology* 63: 319–335.
Bednarik, R. G. 2012b. U-Th analysis and rock art: a response to Pike et al. *Rock Art Research* 29(2): 244–246.
Bednarik, R. G. 2012c. The origins of human modernity. *Humanities* 1(1): 1–53. Available at: www.mdpi.com/2076-0787/1/1/1/
Bednarik, R. G. 2013a. *Creating the human past: an epistemology of Pleistocene archaeology*. Archaeopress, Oxford.
Bednarik, R. G. 2013b. Pleistocene palaeoart of Africa. Special issue 'World rock art', ed. R. G. Bednarik, *Arts* 2(1), 6–34. Available at: www.mdpi.com/2076-0752/2/1/6
Bednarik, R. G. 2013c. Pleistocene palaeoart of Asia. Special issue 'World rock art', ed. R. G. Bednarik, *Arts* 2(2): 46–76. Available at: www.mdpi.com/2076-0752/2/2/46
Bednarik, R. G. 2013d. The origins of modern human behaviour. In R. G. Bednarik (ed.), *The psychology of human behaviour*, pp. 1–58. Nova Press, New York.
Bednarik, R. G. 2013e. African Eve: hoax or hypothesis? *Advances in Anthropology* 3(4): 216–228. Available at: www.scirp.org/journal/PaperInformation.aspx?PaperID=39900#.U5JvUnYUqqY
Bednarik, R. G. 2013f. Brain disorder and rock art. *Cambridge Archaeological Journal* 23 (1): 69–81.

Bednarik, R. G. 2014a. Pleistocene paleoart of Australia. Special issue 'World rock art', ed. R. G. Bednarik, *Arts* 3(1): 156–174. Available at: www.mdpi.com/2076-0752/3/1/156.
Bednarik, R. G. 2014b. Pleistocene paleoart of the Americas. Special issue 'World rock art', ed. R. G. Bednarik, *Arts* 3(1): 190–206. Available at: www.mdpi.com/2076-0752/3/2/190.
Bednarik, R. G. 2014c. Pleistocene paleoart of Europe. Special issue 'World rock art', ed. R. G. Bednarik, *Arts* 3(2): 245–278. Available at: www.mdpi.com/2076-0752/3/2/245.
Bednarik, R. G. 2014d. *The first mariners*, 3rd edn. Bentham Science Publishers, Oak Park, IL.
Bednarik, R. G. 2014e. Doing with less: hominin brain atrophy. *HOMO — Journal of Comparative Human Biology* 65: 433–449.
Bednarik, R. G. 2014f. Exograms. *Rock Art Research* 31(1): 47–62.
Bednarik, R. G. 2015a. An etiology of Theory of Mind in deep time. In E. Sherwood (ed.), *Theory of Mind: development in children, brain mechanisms and social implications*, pp. 115–144. Nova Science Publishers, Inc., New York.
Bednarik, R. G. 2015b. The tribology of cupules. *Geological Magazine* 152(4): 758–765.
Bednarik, R. G. 2015c. Hominin mind and creativity. In B. Půtová and V. Soukup (eds), *The genesis of creativity and the origin of the human mind*, pp. 35–44. Karolinum Press, Prague.
Bednarik, R. G. 2016a. An etiology of human behaviour. In R. G. Bednarik (ed.), *Understanding human behavior: theories, patterns and developments*, pp. 63–93. Nova Biomedical, New York.
Bednarik, R. G. 2016b. Rock art and pareidolia. *Rock Art Research* 33(2): 167–181.
Bednarik, R. G. 2017a. *Palaeoart of the Ice Age*. Cambridge Scholars Publishing, Newcastle upon Tyne.
Bednarik, R. G. 2017b. Continuing the wild goose chase: a response to d'Errico and Stringer. *AURA Newsletter* 34(1): 9–12.
Bednarik, R. G. 2019. *Tribology in geology and archaeology*. Nova Science Publishers, New York.
Bednarik, R. G. in press. First Palaeolithic rock art found in central Europe. *Anthropologie*.
Bednarik, R. G. and P. B. Beaumont 2012. Pleistocene engravings from Wonderwerk Cave, South Africa. Special issue, *Préhistoire, Art et Sociétés, Bulletin de la Société Préhistorique Ariège-Pyrénées* LXV–LXVI: 96–97.
Bednarik, R. G. and P. A. Helvenston 2012. The nexus between neurodegeneration and advanced cognitive abilities. *Anthropos* 107(2): 511–527.
Bednarik, R. G., G. Kumar, A. Watchman and R. G. Roberts 2005. Preliminary results of the EIP Project. *Rock Art Research* 22: 147–197.
Bednarik, R. G. and M. Sreenathan 2012. Traces of the ancients: ethnographic vestiges of Pleistocene 'art'. *Rock Art Research* 29(2): 191–217.
Belyaev, D. K. 1969. Domestication of animals. *Science* 5: 47–52.
Benítez-Burraco, A., L. Di Pietro, M. Barba and W. Lattanzi 2016c. Schizophrenia and human self-domestication: an evolutionary linguistics approach. *BioRxiv preprint*, doi:10.1101/072751.
Benítez-Burraco, A., W. Lattanzi and E. Murphy 2016a. Language impairments in ASD resulting from a failed domestication of the human brain. *Frontiers in Neuroscience* 10: 373. doi:10.3389/fnins.2016.00373.

Benítez-Burraco, A., C. Theofanopoulou and C. Boeckx 2016b. Globularization and domestication. *Topoi*. doi:10.1007/s11245-11016-9399-9397.
Berger, L. R., S. E. Churchill, B. De Klerk and R. L. Quinn 2008. Small-bodied humans from Palau, Micronesia. *PLoS ONE* 3(3): e1780. doi:10.1371/journal.pone.0001780.
Berry, D. S. and L. Z. McArthur 1986. Perceiving character in faces: the impact of age-related craniofacial changes on social perception. *Psychological Bulletin* 100: 3–18.
Bhushan, B. 2013. *Principles and applications of tribology*, 2nd edn. John Wiley and Sons, New York.
Bickerton, D. 1990. *Language and species*. University of Chicago Press, Chicago.
Bickerton, D. 1996. *Language and human behaviour*. UCL Press, London.
Bickerton, D. 2010. *Adam's tongue: how humans made language, how language made humans*. Hill and Wang, New York.
Biesele, M. 1993. *Women like meat: The folklore and foraging ideology of the Kalahari Ju/'Hoan*. Witwatersrand University Press, Johannesburg.
Bini, C., F. Martini, G. Pitzalis and A. Ulzega 1993. Sa Coa de Sa Multa e Sa Pedrosa Pantallinu: due 'paleosuperfici' clactoniane in Sardegna. In *Atti della XXX Riunione Scientifica, 'Paleosuperfici del Pleistocene e del primo Olicene in Italia, Processi si Formazione e Interpretazione', Venosa ed Isernia, 26–29 ottobre 1991*, pp. 179–197. Istituto Italiano di Preistoria e Protostoria, Firenze.
Bird, R. and E. A. Smith 2005. Signaling theory, strategic interaction, and symbolic capital. *Current Anthropology* 46: 221–248.
Birks, J. D. S. and A. C. Kitchener 1999. *The distribution and status of the polecat Mustela putorius in Britain in the 1990s*. Vincent Wildlife Trust, London.
Bischoff, J. L., K. R. Ludwig, J. F. Garcia, E. Carbonell, M. Vaquero, T. W. Stafford and A. J. T. Jull 1994. Dating of the basal Aurignacian sandwich at Abric Romaní (Catalunya, Spain) by radiocarbon and uranium series. *Journal of Archaeological Science* 21: 541–551.
Bishop, R. C. 2007. *The philosophy of the social sciences: an introduction*. Continuum International Publishing Group, London.
Björnerfeldt, S., M. T. Webster and C. Vilà 2006. Relaxation of selective constraint on dog mitochondrial DNA following domestication. *Genome Research* 16(8): 990–994.
Block, N. 2005. Review of Alva Noë. *Journal of Philosophy* 102: 259–272.
Boas, F. [1888] 1938. *The mind of primitive man*. Macmillan, New York.
Bogdashina, O. 2010. *Autism and the edges of the known world: sensitivities, language and constructed reality*. Jessica Kingsley Publishers, London.
Boitani, L. and P. Ciucci 1995. Comparative social ecology of feral dogs and wolves. *Ethology, Ecology & Evolution* 7(1): 49–72.
Bonifay, E. 2002. *Les premiers peuplements de l'Europe*. La Maison des Roches, Paris.
Bonifay, E., Y. Bassiakos, M. F. Bonifay, A. Louchart, C. Mourer-Chauviré, E. Periera et al. 1998. La Grotte de la Coscia (Rogliano Macinaggio): étude préliminaire d'un nouveau site do Pléistocène Supérieur de Corse. *Paléo* 10: 17–41.
Bonifay, E. and P. Vandermeersch 1991. Vue d'ensemble sur le très ancient Paléolithique de l'Europe. In E. Bonifay and P. Vandermeersch (eds), *Les premiers Européens*, pp. 309–319. CTHS, Paris.
Bookstein, F., K. Shaefer, H. Prossinger, H. Seidler, M. Fieder, C. Stringer, G. W. Weber, J. L. Arsuaga, D. E. Slice, F. J. Rohlf, W. Rechesis, A. J. Mariam and L. F. Marcus 1999. Comparing frontal cranial profiles in archaic and modern *Homo* by morphometric analysis. *The Anatomical Record (New Anat.)* 157: 217–224.

Bosinski, G. 1992. Die ersten Menschen in Eurasien. *Jahrbuch des Römisch-Germanischen Zentralmuseums Mainz* 39: 138–181.

Boucher de Perthes, J. 1846. *Antiquités celtiques et antédiluviennes*. Treuttel et Wurtz, Paris.

Bouissac, P. 1997. New epistemological perspectives for the archaeology of writing. In R. Blench and M. Spriggs (eds), *Archaeology and language*, vol. I, pp. 53–62. Routledge, London.

Boyd, R. and P. J. Richerson 2005. *The origin and evolution of cultures*. Oxford University Press, New York.

Brace, C. L. 1993. 'Popscience' versus understanding the emergence of the modern mind. *Behavioral and Brain Sciences* 16(4): 750–751.

Brace, C. L. 1996. Racialism and racist agendas: race, evolution, and behavior: a life history perspective. *American Anthropologist* 98(1): 176–177.

Brace, C. L. 1999. An anthropological perspective on 'race' and intelligence: the non-clinal nature of human cognitive capabilities. *Journal of Anthropological Research* 55(2): 245–264.

Brace, C. L. 2005. *'Race' is a four-letter word: the genesis of the concept*. Oxford University Press, New York.

Brace, C. L., K. R. Rosenberg and K. D. Hunt 1987. Gradual change in human tooth size in the Late Pleistocene and post-Pleistocene. *Evolution* 41(4): 705–720.

Bradbury, A. P. 2001. Modern or prehistoric: experiments in distinguishing between culturally and mechanically produced chipped stone artifacts. *North American Archaeologist* 22: 231–258.

Bradshaw, J. and L. Rogers 1993. *The evolution of lateral asymmetries, language, tool use, and intellect*. Academic Press, San Diego, CA.

Bråten, S. 2004. Hominin infant decentration hypothesis: mirror neurons system adapted to subserve mother-centered participation. *Behavioural and Brain Sciences* 27(4): 508–509.

Bråten, S. (ed.) 2007. *On being moved — From mirror neurons to empathy: advances in consciousness research*. John Benjamins Publishing Company, Philadephia, PA.

Bräuer, G. 1980. Die morphologischen Affinitäten des jungpleistozänen Stirnbeins aus dem Elbmündungsgebiet bei Hahnöfersand. *Zeitschrift für Morphologie und Anthropologie* 71: 1–42.

Bräuer, G. 1984a. Präsapiens-Hypothese oder Afro-europäische Sapiens-Hypothese? *Zeitschrift für Morphologie und Anthropologie* 75: 1–25.

Bräuer, G. 1984b. The 'Afro-European sapiens hypothesis' and hominid evolution in East Africa during the late Middle and Upper Pleistocene. In P. Andrews and J. L. Franzen (eds), *The early evolution of man, with special emphasis on Southeast Asia and Africa*, pp. 145–165. Vol. 69, Courier Forschungsinstitut Senckenberg, Seckenberg.

Bräuer, G. 1992. Africa's place in the evolution of *Homo sapiens*. In G. Bräuer and F. H. Smith (eds), *Continuity or replacement? Controversies in Homo sapiens evolution*, pp. 83–98. Balkema, Rotterdam.

Breunig, P. 1996, The 8000-year-old dugout canoe from Dufuna (N.E. Nigeria). In G. Pwiti and R. Soper (eds), *Aspects of African archaeology*, pp. 461–468. University of Zimbabwe Publications, Harare.

Brod, J. H. 1997. Creativity and schizotypy. In G. Claridge (ed.), *Schizotypy: implications for illness and health*, pp. 274–299. Oxford University Press, Oxford.

Brodar, S. 1957. Zur Frage der Höhlenbärenjagd und des Höhlenbärenkults in den paläolithischen Fundstellen Jugoslawiens. *Quartär* 9: 147–159.

Brookfield, J. F. Y. 1997. Importance of ancestral DNA ages. *Nature* 388: 134.
Brown, P. 1987. Pleistocene homogeneity and Holocene size reduction: the Australian human skeletal evidence. *Archaeology and Physical Anthropology in Oceania* 22: 41–67.
Brown, P. 1992. Recent human evolution in East Asia and Australasia. *Philosophical Transactions of the Royal Society of London, Series B* 337: 235–242.
Brown, P. and T. Maeda 2004. Post-Pleistocene diachronic change in east Asian facial skeletons: the size, shape and volume of the orbits. *Anthropological Science* 112: 29–40.
Brown, P., T. Sutikna, M. J. Morwood, R. P. Soejono, P. Jatmiko, E. Wayhu Saptomo and R. Awe Due 2004. A new small-bodied hominin from the Late Pleistocene of Flores, Indonesia. *Nature* 431: 1055–1061.
Brown, T. A., M. K. Jones, W. Powell and R. G. Allaby 2008. The complex origins of domesticated crops in the Fertile Crescent. *Trends in Ecology and Evolution* 24: 103–109.
Brühl, E. 2018. A new intentional pattern on a bone from the Lower Palaeolithic of Bilzingsleben. Paper presented at the NeanderART2018 International Conference, Campus 'Luigi Einaudi', University of Turin, 22 August 2018.
Brumm, A., G. M. Jensen, G. D. van den Bergh, M. J. Morwood, I. Kurniawan, F. Aziz and M. Storey 2010. Hominins on Flores, Indonesia, by one million years ago. *Nature* 464: 748–752.
Brüne, M. 2004. Schizophrenia: an evolutionary enigma? *Neuroscience and Biobehavioral Reviews* 28: 41–53.
Brüne, M. and U. Brüne-Cohrs 2007. The costs of mental time travel. *Behavioral and Brain Sciences* 30(3): 317–318.
Buchen, L. 2011. When geeks meet. *Nature* 479(7371): 25–27.
Budiansky, S. 1992. *The covenant of the wild: why animals chose domestication: with a new preface.* Yale University Press, New Haven, CT.
Burack, J. A., T. Charman, N. Yurmiya and P. R. Zelazo (eds) 2009. *The development of autism: perspectives from theory and research.* Routledge, London.
Burns, J. K. 2004. An evolutionary theory of schizophrenia: cortical connectivity, metarepresentation, and the social brain. *Behavioral and Brain Sciences* 27(6): 831–855.
Burns, J. K. 2006. Psychosis: a costly by-product of social brain evolution in *Homo sapiens*. *Progress in Neuro-Psychopharmacology and Biological Psychiatry* 30(5): 797–814.
Buss, D. M., M. Abbott, A. Angleitner, A. Biaggio, A. Blanco-Villasenor, M. Bruchon-Schweitzer *et al* 1990. International preferences in selecting mates: a study of 37 societies. *Journal of Cross Cultural Psychology* 21: 5–47.
Buss, D. M. and M. L. Barnes 1986. Preferences in human mate selection. *Journal of Personality and Social Psychology* 50: 559–570.
Butti, C., C. C. Sherwood, A. Y. Hakeem and J. M. Allman 2009. Total number and volume of von Economo neurons in the cerebral cortex of cetaceans. *Journal of Comparative Neurology* 515: 243–259.
Buxhoeveden, D., W. Lefkowitz, P. Loats and E. Armstrong 1996. The linear organization of cell columns in human and nonhuman anthropoid Tpt cortex. *Anatomy and Embryology* 194(1): 23–36.
Byrne, R. W. and A. Whiten (eds) 1988. *Machiavellian intelligence: social expertise and the evolution of intelligence in monkeys, apes, and man.* Oxford University Press, Oxford.
Cabrera Valdés, V. and J. Bischoff 1989. Accelerator 14C dates for Early Upper Palaeolithic (Basal Aurignacian) at El Castillo Cave (Spain). *Journal of Archaeological Science* 16: 577–584.
Cabrera Valdés, V., J. M. Maillo Fernández, A. Pike-Tay, M. D. Garralda Benajes and F. Bernaldo de Quirós 2006. A Cantabrian perspective on late Neanderthals. In N. J.

Conard (ed.), *When Neanderthals and modern humans met*, pp. 441–465. Kerns Verlag, Tübingen.

Callaghan, R. and C. Scarre 2009. Simulating the western seaways. *Oxford Journal of Archaeology* 28: 357–372.

Campbell, F. 2006. Molesting the past. *The Weekend Australian*, 25 February: R15.

Campbell, M. C. and S. A. Tishkoff 2010. The evolution of human genetic and phenotypic variation in Africa. *Current Biology* 20: R166–R173.

Camps, M. and P. R. Chauhan (eds) 2009. *Sourcebook of Paleolithic transitions: methods, theories, and interpretations*. Springer, New York.

Cann, R. L., M. Stoneking and A. C. Wilson 1987. Mitochondrial DNA and human evolution. *Nature* 325: 31–36.

Cannon, T. D. 2005. The inheritance of intermediate phenotypes for schizophrenia. *Current Opinion in Psychiatry* 18: 135–140.

Capitan, L. and D. Peyrony 1921. Les origines de l'art à la Aurignacien moyen – nouvelles fouilles à La Ferrassie. *Revue d'Anthropologie* 31: 92–112.

Carbonell, E., J. M. Bermúdez de Castro, J. M. Parés, A. Pérez-González, G. Cuenca-Bescós and A. Ollé 2008. The first hominin of Europe. *Nature* 452: 465–469.

Cardno, A. G. and I. I. Gottesman 2000. Twin studies of schizophrenia: from bow-and-arrow concordances to Star Wars Mx and functional genomics. *American Journal of Medical Genetics* 97: 12–17.

Carlier, L., J. Couprie, A. le Maire, L. Guilhaudis, I. Milazzo, M. Gondry, D. Davoust, B. Gilquin and S. Zinn-Justin 2007. Solution structure of the region 51–160 of human KIN17 reveals an atypical winged helix domain. *Protein Science* 16: 2750–2755.

Carlson, D. S. 1976. Temporal variation in prehistoric Nubian crania. *American Journal of Physical Anthropology* 45: 467–484.

Carruthers, M. 1990. *The book of memory*. Cambridge University Press, Cambridge.

Carruthers, M. 1998. *The craft of thought*. Cambridge University Press, Cambridge.

Carruthers, P. 2002. The cognitive functions of language. *Behavioral and Brain Sciences* 25: 657–674.

Cartailhac, E. 1902. Les cavernes ornées de dessins. La grotte d'Altamira, Espagne. 'Mea culpa d'un sceptique'. *L'Anthropologie* 13: 348–354.

Carver, C. S. 2002. Self-awareness. In M. R. Leary and J. P. Tangney (eds), *Handbook of self and identity*, pp. 179–196. Guilford Press, New York.

Castellano, S., G. Parra, F. A. Sánchez-Quinto, F. Racimo, M. Kuhlwilm, M. Kircher et al. 2014. Patterns of coding variation in the complete exomes of three Neandertals. *Proceedings of the National Academy of Sciences of the U.S.A.* 111(18): 6666–6671.

Castle, W. E. 1947. The domestication of the rat. *Proceedings of the National Academy of the Sciences of the U.S.A.* 33: 109–114.

Cavalli-Sforza, L. L. and M. W. Feldman 1973. Cultural vs. biological inheritance. *American Journal of Human Genetics* 25: 618–637.

Cavalli-Sforza, L. L., A. Piazza, P. Menozzi and J. Mountain 1988. Reconstruction of human evolution: bringing together genetic, archaeological, and linguistic data. *Proceedings of the National Academy of Sciences of the U.S.A.* 85: 6002–6006.

Changeux, J. P. and J. Chavaillon (eds) 1996. *Origins of the human brain*. Clarendon Press, Oxford.

Chapman, S. N., J. E. Pettay, V. Lummaa and M. Lahdenperä 2019. Limits to fitness benefits of prolonged post-reproductive lifespan in women. *Current Biology*; doi:10.1016/j.cub.2018.12.052.

Chase, P. G. and A. Nowell 1998. Taphonomy of a suggested Middle Paleolithic bone flute from Slovenia. *Current Anthropology* 39: 549–553.

Chauhan, P. R. 2007. The Indian subcontinent and 'Out of Africa I'. In J. Fleagle, J. Shea and R. Leakey (eds), *Proceedings of the Life Matters-Stony Brook Human Evolution Workshop Sept. 26–Oct. 1, 2005*. Kluwer-Academic Press, New York.

Chauhan, P. R. 2009. Comment on 'Lower and early Middle Pleistocene Acheulian in the Indian sub-continent' by Gaillard et al. (2009). *Quaternary International* 223–224: 248–259.

Chauhan, P. R. and R. Patnaik 2008. The Narmada Basin Palaeoanthropology Project in central India. *Antiquity* 82(317). Available at: http://antiquity.ac.uk/projgall/chauhan/index.html

Chazan, M., H. Ron, A. Matmon, N. Porat, P. Goldberg, R. Yates, M. Avery, A. Sumner and L. K. Horwitz 2008. Radiometric dating of the Earlier Stone Age sequence in Excavation 1 at Wonderwerk Cave, South Africa: preliminary results. *Journal of Human Evolution* 55: 1–11.

Cherry, J. F. 1992. Palaeolithic Sardinians? Some questions of evidence and method. In R. Tykot and T. Andrews (eds), *Sardinia in the Mediterranean: a footprint in the sea*, pp. 28–39. Sheffield Academic Press, Sheffield.

Chiaroni, J., P. A. Underhill and L. L. Cavalli-Sforza 2009. Y-chromosome diversity, human expansion, drift, and cultural evolution. *Proceedings of the National Academy of Sciences of the U.S.A.* 106: 20174–20179.

Cho H. J., I. Meira-Lima, Q. Cordeiro, L. Michelon, P. C. Sham, H. Vallada and D. A. Collier 2005. Population-based and family-based studies on the serotonin transporter gene polymorphisms and bipolar disorder: a systematic review and meta-analysis. *Molecular Psychiatry* 10: 771–781.

Christian, K. M. and R. F. Thompson 2005. Long-term storage of an associative memory trace in the cerebellum. *Behavioral Neuroscience* 119: 256–537.

Christos, G. 2003. *Memory and dreams: the creative human mind*. Rutgers University Press, New Brunswick, NJ.

Churchill, S. E. and F. H. Smith 2000a. A modern human humerus from the early Aurignacian of Vogelherdhöhle (Stetten, Germany). *American Journal of Physical Anthropology* 112: 251–273.

Churchill, S. E. and F. H. Smith 2000b. Makers of the early Aurignacian of Europe. *American Journal of Physical Anthropology* 113: 61–115.

Churchland, P. S. 1986. *Neurophilosophy: toward a unified science of the mind-brain*. MIT Press, Cambridge, MA.

Cieri, R. L., S. E. Churchill, R. G. Franciscus, J. Tan and B. Hare 2014. Craniofacial feminization, social tolerance, and the origins of behavioral modernity. *Current Anthropology* 55(4): 419–443.

Ciochon, R. L. 2009. The mystery ape of Pleistocene Asia. *Nature* 459: 910–911.

Claidière, N. and J.-B. André 2012. The transmission of genes and culture: a questionable analogy. *Evolutionary Biology* 39(1): 12–24.

Clark, A. and D. Chalmers 1998. The extended mind. *Analysis* 58: 7–19.

Clark, J. D., K. P. Oakley, L. H. Wells and J. A. C. McClelland 1947. New studies on Rhodesian Man. *Journal of the Royal Anthropological Institute* 77: 4–33.

Clark, J. G. D. 1977. *World prehistory: a new perspective*. Cambridge University Press, Cambridge.

Clottes, J. (ed.) 2001. *La Grotte Chauvet: l'art des origines*. Seuil, Paris.

Clottes, J. 2008. *Cave art*. Phaidon, London.

Clottes, J. 2012. U-series dating, evolution of art and Neandertal. *International Newsletter on Rock Art* 64: 1–6.

Clottes J., J.-M. Chauvet, E. Brunel-Deschamps, C. Hillaire, J.-P. Daugas, M. Arnold, H. Cachier, J. Evin, P. Fortin, C. Oberlin, N. Tisnerat and H. Valladas 1995. Les peintures paléolithiques de la Grotte Chauvet-Pont d'Arc, à Vallon-Pont-d'Arc (Ardèche, France): datations directes et indirectes par la méthode du radiocarbone. *Comptes Rendus de l'Académie des Sciences de Paris* 320, Ser. II: 1133–1140.

Clottes, J., B. Géli, C. Ghemis, É. Kaltnecker, V.-R. Lascu, C. Moreau, M. Philippe, F. Prud'homme and H. Valladas 2011. A very ancient art in Rumania: The Coliboaia dates. *International Newsletter on Rock Art* 61: 1–3.

Clowes, R. A. 2007. Self-regulation model of inner speech and its role in the organisation of human conscious experience. *Journal of Consciousness Studies* 14: 59–71.

Collado Giraldo, H. 2015. At the beginning: rock art in southwest Europe. In B. Půta and V. Soukup (eds), *The genesis of creativity and the origin of the human mind*, pp. 187–210. Karolinum Press, Prague.

Combier, J. and G. Jouve 2012. Chauvet Cave's art is not Aurignacian: a new examination of the archaeological evidence and dating procedures. *Quartär* 59: 131–152.

Compston, A. and A. Coles 2008. Multiple sclerosis. *Lancet* 372(9648): 1502–1517.

Conard, N. J. 2009. A female figurine from the basal Aurignacian of the Hohle Fels Cave in southwestern Germany. *Nature* 459: 248–252.

Conard, N. J., P. M. Grootes and F. H. Smith 2004. Unexpectedly recent dates for human remains from Vogelherd. *Nature* 430: 198–201.

Conard, N., K. Langguth and H.-P. Uerpmann 2003. Einmalige Funde aus dem Aurignacien und erste Belege für ein Mittelpaläolithikum im Hohle Fels bei Schelklingen, Alb-Donau-Kreis. In *Archäologische Ausgrabungen in Baden-Württemberg 2002*, pp. 21–27. Konrad Theiss Verlag, Stuttgart.

Corvinus, G. 2002. Arjun 3, a Middle Palaeolithic site, in the Deokhuri Valley, western Nepal. *Man and Environment* 27(2): 31–44.

Cosmides, L. and J. Tooby 1999. Toward an evolutionary taxonomy of treatable conditions. *Journal of Abnormal Psychology* 108(3): 453–464.

Craddock, N. and I. Jones 1999. Genetics of bipolar disorder: review article. *Journal of Medical Genetics* 26: 585–594.

Craddock, N., M. C. O'Donovan and M. J. Owen 2005. The genetics of schizophrenia and bipolar disorder: dissecting psychosis. *Journal of Medical Genetics* 42: 193–204.

Craig, J. and S. Baron-Cohen 1999. Creativity and imagination in autism and Asperger's Syndrome. *Journal of Autism and Developmental Disorders* 29, 319–326.

Crémades, M. 1996. L'expression graphique au paléolithique inférieur et moyen: L'exemple de l'Abri Suard (La Chaise-de-Vouthon, Charente). *Bulletin de la Société Préhistorique Française* 93: 494–501.

Crémades, M., H. Laville, N. Sirakov and J. K. Kozłowski 1995. Une pierre gravée de 50 000 ans B.P. dans les Balkans. *Palaeo* 7: 201–209.

Crochet, J.-Y., J.-L. Welcomme, J. Ivorra, G. Ruffet, N. Boulbes, R. Capdevila, J. Claude, C. Firmat, G. Métais, J. Michaux and M. Pickford 2009. Une nouvelle faune de vertébrés continentaux, associée à des artifacts dans le Pléistocène inférieur de l'Hérault (Sud de la France), ver 1,57 Ma. *Comptes Rendus Palevol* 8: 725–736.

Crockford, S. J. 2000. Dog evolution: a role for thyroid hormone physiology in domestication changes in dogs through time: an archaeological perspective. In S. J. Crockford (ed.), *Proceedings of the First ICAZ Symposium on the History of the Domestic Dog*, pp. 295–312. BAR International Series 889, Oxford.

Crockford, S. J. 2002. *Animal domestication and heterochronic speciation: the role of thyroid hormone.* Johns Hopkins University Press, Baltimore, MD.

Crockford, S. J. 2009. Evolutionary roots of iodine and thyroid hormones in cell–cell signaling. *Integrative and Comparative Biology* 49: 155–166.

Crow, T. J. 1995a. A Darwinian approach to the origins of psychosis. *The British Journal of Psychiatry* 167(1): 12–25.

Crow, T. J. 1995b. A theory of the evolutionary origins of psychosis. *European Neuropsychopharmacology* 5: 59–63.

Crow, T. J. 1997a. Is schizophrenia the price that *Homo sapiens* pays for language? *Schizophrenia Research* 28: 127–141.

Crow, T. J. 1997b. Schizophrenia as failure of hemispheric dominance for language. *Trends in Neurosciences* 20: 339–343.

Crow, T. J. 2000. Schizophrenia as the price that *Homo sapiens* pays for language: a resolution of the central paradox in the origin of the species. *Brain Research Reviews* 31: 118–129.

Crow, T. J. 2002. *The speciation of modern Homo sapiens.* Oxford University Press, Oxford.

Crow, T. J., D. J. Done and A. Sacker 1995. Childhood precursors of psychosis as clues to its evolutionary origins. *European Archives of Psychiatry and Clinical Neuroscience* 245: 61–69.

Cruciani, F., P. Santolamazza, P. Shen, V. Macaulay, P. Moral, A. Olckers, D. Modiano et al. 2002. A back migration from Asia to sub-Saharan Africa is supported by high-resolution analysis of human Y-chromosome haplotypes. *The American Journal of Human Genetics* 70: 1197–1214.

Cruz, F., C. Vilà and M. T. Webster 2008. The legacy of domestication: accumulation of deleterious mutations in the dog genome. *Molecular Biology and Evolution* 25(11): 2331–2336.

Cubuk, G. A. 1976. Altpaläolithische Funde von den Mittelmeerterassen bei Nea Skala auf Kephallinia (Griechenland). *Archäologisches Korrespondenzblatt* 6: 175–181.

Cummins, R. 1996. *Representations, targets, and attitudes.* MIT Press, Cambridge, MA.

Curnoe, D., X. Ji, A. I. R. Herries, B. Kanning, P. S. C. Taçon, B. Zhende et al. 2012. Human remains from the Pleistocene-Holocene transition of southwest China suggest a complex evolutionary history for East Asians. *PLoS One* 7: e31918.

Cynx, J. and S. J. Clark 1993. Ethological foxes and cognitive hedgehogs. *Behavioural and Brain Sciences* 16: 756–757.

Czarnetzki, A. 1983. Zur Entwicklung des Menschen in Südwestdeutschland. In H. Müller Beck (ed.), *Urgeschichte in Baden-Württemberg*, pp. 217–240. Konrad Theiss, Stuttgart.

Damasio, A. R., D. Tranel and H. Damasio 1990. Individuals with sociopathic behavior caused by frontal damage fail to respond autonomously to social stimuli. *Behavioral Brain Research* 41: 81–94.

Danto, A. C. 1986. *The philosophical disenfranchisement of art.* Columbia University Press, New York.

Danto, A. C. 1988. Artifact and art. In S. Vogel (ed.), *Exhibition catalogue for ART/artifact*, pp. 18–32. Center for African Art, New York.

Dart, R. A. 1925. *Australopithecus africanus*: the man-ape of South Africa. *Nature* 115: 195–199.

Dart, R. A. 1974. The waterworn australopithecine pebble of many faces from Makapansgat. *South African Journal of Science* 70: 167–169.

Darwin, C. [1859] 1959. *On the origin of species by means of natural selection, or the preservation of favoured races in the struggle for life*. John Murray, London.
Darwin, C. 1868. *The variation of animals and plants under domestication*, 2 vols. John Murray, London.
Darwin, C. 1871. *The descent of man, and selection in relation to sex*. John Murray, London.
Davidson, I. 2003. Comment on R. G. Bednarik, 'Seafaring in the Pleistocene'. *Cambridge Archaeological Journal* 13(1): 55–56.
Davidson, I. and W. Noble 1989. The archaeology of perception: traces of depiction and language. *Current Anthropology* 30(2): 125–155.
Davidson, I. and W. Noble 1990. Tools, humans and evolution: the relevance of the Upper Palaeolithic. Paper presented at the symposium 'Tools, language and intelligence: evolutionary implications', Cascais, Portugal.
Davidson, I. and W. Noble 1992. Why the first colonisation of the Australian region is the earliest evidence of modern human behaviour. *Archaeology in Oceania* 27: 113–119.
Dawkins, R. 1976. *The selfish gene*. Oxford University Press, Oxford.
Deacon, T. 1997. *The symbolic species: The co-evolution of language and the human brain*. Penguin Books, London.
Dean, D. and E. Delson 1995. *Homo* at the gates of Europe. *Nature* 373: 472–473.
De Beer, G. R. 1930. *Embryology and evolution*. Oxford University Press, Oxford.
De Beer, G. R. 1940. *Embryos and ancestors*. Oxford University Press, Oxford.
Decety, J. and J. A. Sommerville 2003. Shared representations between self and others: a social cognitive neuroscience view. *Trends in Cognitive Sciences* 7: 527–533.
Delluc, B. and G. Delluc 1978. Les manifestations graphiques aurignaciennes sur support rocheux des environs des Eyzies (Dordogne). *Gallia Préhistoire* 21: 213–438.
de Lumley, H. and A. Sonakia 1985. Contexte stratigraphique et archaéologique de l'homme de le Narmada, Hathnora, Madhya Pradesh, Inde. *L'Anthropologie* 89: 3–12.
de Lumley, M.-A. and A. Sonakia 1985. Première découverte d'un *Homo erectus* sur le continent indien, à Hathnora, dans le moyenne vallée de la Narmada. *L'Anthropologie* 89: 13–61.
De Miguel, C. and M. Henneberg 2001. Variation in hominid brain size: how much is due to method? *HOMO — Journal of Comparative Human Biology* 52(1): 3–58.
Dennell, R. W. 1995. The Early Stone Age of Pakistan: a methodological review. *Man and Environment* 20(1): 21–28.
Dennell, R. W. 2004. *Early hominin landscapes in northern Pakistan*. BAR International Series 1265, Archaeopress, Oxford.
Dennell, R. W., H. M. Rendell and E. Hailwood 1988. Early tool-making in Asia: two-million year-old artefacts in Pakistan. *Antiquity* 62: 98–106.
Dennett, D. C. 1991. *Consciousness explained*. Little, Brown, Boston.
de Panafieu, J.-B. with P. Gries 2007. *Evolution*. Seven Stories Press, New York.
d'Errico, F. 1991. Microscopic and statistical criteria for the identification of prehistoric systems of notation. *Rock Art Research* 8: 83–93.
d'Errico, F. 1994. *L'art gravé azilien. 31e supplément, Gallia Préhistoire*. CNRS Éditions, Paris.
d'Errico, F. 1995. Comment on R. G. Bednarik, 'Concept-mediated markings of the Lower Palaeolithic'. *Current Anthropology* 36: 618–620.
d'Errico, F., C. Gaillard and V. N. Misra 1989. Collection of non-utilitarian objects by *Homo erectus* in India. In *Hominidae. Proceedings of the 2nd International Congress of Human Paleontology*, pp. 237–239. Editoriale Jaca Book, Milan.

d'Errico, F., G. Lawson, C. Henshilwood, M. Vanhaeren, A.-M. Tillier, M. Soressi *et al.* 2003. Archaeological evidence for the emergence of language, symbolism, and music: an alternative multidisciplinary perspective. *Journal of World Prehistory* 17: 1–70.

d'Errico, F. and A. Nowell 2000. A new look at the Berekhat Ram figurine: implications for the origins of symbolism. *Cambridge Archaeological Journal* 10(1): 123–167.

d'Errico, F. and C. B. Stringer 2011. Evolution, revolution or saltation scenario for the emergence of modern cultures? *Philosophical Transactions of the Royal Society B* 366: 1060–1069.

d'Errico, F. and P. Villa 1997. Holes and grooves: the contribution of microscopy and taphonomy to the problem of art origins. *Journal of Human Evolution* 33: 1–31.

d'Errico, F., P. Villa, A. C. Pinto Llona and R. R. Idarraga 1998. A Middle Palaeolithic origin of music? Using cave-bear bone accumulations to assess the Divje babe I bone 'flute'. *Antiquity* 72: 65–79.

de Terra, H. and T. T. Paterson 1939. *Studies on the Ice Age in India and associated human cultures*. Carnegie Institution of Washington:Washington, DC.

Détroit, F., A. S. Mijares, J. Corny, G. Daver, C. Zanolli, E. Dizon, E. Robles, R. Grün and P. J. Piper 2019. A new species of *Homo* from the Late Pleistocene of the Philippines. *Nature* 568: 181–186.

De Veer, M. W. and R. Van Den Bos 1999. A critical review of methodology and interpretation of mirror self-recognition research in nonhuman primates. *Animal Behaviour* 58: 459–468.

Devièse, T., I. Karavanić, D. Comeskey, C. Kubiak, P. Korlević, M. Hajdinjak, S. Radović, N. Procopio, M. Buckley, S. Pääbo and T. Higham 2017. Direct dating of Neanderthal remains from the site of Vindija Cave and implications for the Middle to Upper Paleolithic transition. *Proceedings of the National Academy of Sciences of the U.S.A.* 114(40): 10606–10611.

Di Pellegrino, G., L. Fadiga, L. Fogassi, V. Gallese and G. Rizzolatti 1992. Understanding motor events: a neurophysiological study. *Experimental Brain Research* 91: 176–180.

Dobzhansky, T. 1962. *Mankind evolving: the evolution of the human species*. Yale University Press, New Haven, CT.

Dobzhansky, T. 1972. On the evolutionary uniqueness of man. In T. Dobzhansky, M. K. Hecht and W. C. Steere (eds), *Evolutionary biology*, pp. 415–430. Appleton-Century-Crofts, New York.

Donald, M. 1991. *Origins of the modern mind: three stages in the evolution of culture and cognition*. Harvard University Press, Cambridge, MA.

Donald, M. 2001. *A mind so rare: the evolution of human consciousness*. W. W. Norton, New York.

Douka, K., V. Slon, Z. Jacobs, C. Bronk Ramsey, M. V. Shunkov and A. P. Derevianko *et al.* 2019. Age estimates for hominin fossils and the onset of the Upper Palaeolithic at Denisova Cave. *Nature* 565: 640–644.

Drake, J. and E. Winner 2009. Precocious realists: perceptual and cognitive characteristics associated with drawing talent in non-autistic children. *Philosophical Transactions of the Royal Society of London, B, Biological Science* 364(1522): 1449–1458.

Draper, J. W. 1875. *History of the conflict between religion and science*. Henry S. King & Co., London.

Dretske, F. 1981. *Knowledge and the flow of information*. MIT Press, Cambridge, MA.

Drevets, W. C., J. L. Price, J. R. Thompson, R. D. Todd, R. Reich, T. M. Vannier and M. E. Raichle 1997. Subgenual prefrontal cortex abnormalities in mood disorders. *Nature* 386: 824–827.

Driscoll, C. A., D. W. Macdonald and S. J. O'Brien 2009. From wild animals to domestic pets, an evolutionary view of domestication. *Proceedings of the National Academy of Sciences of the U.S.A* 106(Suppl. 1): 9971–9978.
Dunbar, R. 1996. *Grooming, gossip and the evolution of language*. Faber and Faber, London.
Durham, W. H. 1991. *Coevolution: genes, culture, and human diversity*. Stanford University Press, Stanford, CA.
Dutton, D. 1993. Tribal art and artifact. *Journal of Aesthetics and Art Criticism* 51: 13–21.
Duvall, J. G. and W. T. Venner 1979. A statistical analysis of the lithics from the Calico site (SBCM 1500A), California. *Journal of Field Archaeology* 6: 455–462.
Eckhardt, R. B. 2000. *Human paleobiology*. Cambridge University Press, Cambridge.
Eckhardt, R. 2007. Palaeoanthropology, a science in need of a theoretical framework: the Liang Bua example. In E. Indriati (ed.), *Recent advances of Southeast Asia palaeoanthropology and archaeology*, pp. 30–36. Gadjah Mada University, Yogyakarta.
Eckhardt, R. B. and M. Henneberg 2010. LB1 from Liang Bua, Flores: craniofacial asymmetry confirmed, plagiocephaly diagnosis dubious. *American Journal of Physical Anthropology* 143: 331–334.
Eckhardt, R. B., M. Henneberg, A. S. Weller and K. J. Hsü 2014. Rare events in earth history include the LB1 human skeleton from Flores, Indonesia, as a developmental singularity, not a unique taxon. *Proceedings of the National Academy of Sciences of the U.S.A.* 111(33): 11961–11966. Available at: www.pnas.org/cgi/doi/10.1073/pnas.1407385111
Efremov, I. A. 1940. Taphonomy: a new branch of paleontology. *Pan-American Geologist* 74: 81–93.
Eibl-Eibesfeldt, I. 1970. *Ethology: the biology of behavior*. Holt, Rinehart and Winston, New York.
Eichmeier, J. and O. Höfer 1974. *Endogene Bildmuster*. Urban und Schwarzenberg, Munich.
Eitzman, W. I. 1958. Reminiscences of Makapansgat Limeworks and its bone-breccial layers. *South African Journal of Science* 54: 177–182.
Ellen, R. and A. Muthana 2013. An experimental approach to understanding the 'eolithic' problem: cultural cognition and the perception of plausibly anthropic artifacts. *Lithic Technology* 38: 109–123.
Ellmers, D. 1980. Ein Fellboot-Fragment der Ahrensburger Kultur aus Husum, Schleswig-Holstein? *Offa* 37: 19–24.
Enard, W., M. Przeworski, S. E. Fisher, C. S. Lai, V. Wiebe, T. Kitano, A. P. Monaco and S. Pääbo 2002. Molecular evolution of FOXP2, a gene involved in speech and language. *Nature* 418: 869–872.
Eswaran, V. 2002. A diffusion wave out of Africa. *Current Anthropology* 43(5): 749–774.
Evans, P. D., S. L. Gilbert, N. Mekel-Bobrov, E. J. Vallender, J. R. Anderson, L. M. Vaez-Azizi, S. A. Tishkoff, R. R. Hudson and B. T. Lahn 2005. Microcephalin, a gene regulating brain size, continues to evolve adaptively in humans. *Science* 309 (5741): 1717–1720.
Everett, D. L. 2005. Cultural constraints on grammar and cognition in Pirahã: another look at the design features of human language. *Current Anthropology* 76: 621–646.
Everett, D. L. 2008. *Don't sleep, there are snakes: life and language in the Amazonian jungle*. Pantheon, New York.
Ewens, W. J. 1983. The role of models in the analysis of molecular genetic data, with particular reference to restriction fragment data. In B. S. Weir (ed.), *Statistical analysis of DNA sequence data*, pp. 45–73. Marcel Dekker, New York.

Facorellis, Y., N. Kyparissi-Apostolika and Y. Maniatis 2001. The cave of Theopetra, Kalambaka: radiocarbon evidence for 50,000 years of human presence. *Radiocarbon* 43 (2B): 1029–1048.

Fahu C., F. Welker, S. Chuan-Chou, E. Bailey, I. Bergmann, S. Davis, X. Huan., W. Hui, R. Fischer, S. E. Freidline, Y. Tsai-Luen, M. M. Skinner, S. Stelzer, D. Guangrong, F. Qiaomei, D. Guanghui, W. Jian, Z. Dongju and J.-J. Hublin 2019. A late Middle Pleistocene Denisovan mandible from the Tibetan Plateau. *Nature* 569(7754), doi:10.1038/s41586-019-1139-x.

Fairservice, W. A., Jr. 1975. *The threshold of civilization*. Scribner, New York.

Fajardo, C., M. Escobar, E. Buritica, G. Arteoga, T. Umbarila, M. F. Casanova and H. Pimienta 2008. Von Economo neurons are present in the dorsolateral (dysgranular) prefrontal cortex in humans. *Neuroscience Letters* 435(3): 215–218.

Falconer, D. S. and T. F. C. Mackay 1996. *Introduction to quantitative genetics*, 4th edn. Longman, Burnt Mill, England.

Falk, D. 1975. Comparative anatomy of the larynx in man and chimpanzee: implications for language in Neanderthal. *American Journal of Physical Anthropology* 43: 123–132.

Falk, D. 1987. Hominid paleoneurology. *Annual Review of Anthropology* 16: 13–30.

Falk, D. 2009. *Finding our tongues: mothers, infants and the origins of language*. Basic Books, New York.

Fanelli, D. 2010. 'Positive' results increase down the hierarchy of the sciences. *PloS ONE* 5(4), e10068; doi:10.1371/journal.pone.0010068.

Farley, J. D. 1976. Phylogenetic adaptations and the genetics of psychosis. *Acta Psychiatry Scandinavia* 83(1): 173–192.

Farthing, G. W. 1992. *The psychology of consciousness*. Prentice Hall, Upper Saddle River, NJ.

Fedele, F. G., B. Giaccio and I. Hajdas 2008. Timescales and cultural process at 40,000 BP in the light of the Campanian Ignimbrite eruption, Western Eurasia. *Journal of Human Evolution* 55: 834–857.

Feldman, M. W. and L. L. Cavalli-Sforza 1989. On the theory of evolution under genetic and cultural transmission with application to the lactose absorption problem. In M. W. Feldman (ed.), *Mathematical evolutionary theory*, pp. 145–173. Princeton University Press, Princeton, NJ.

Felgenhauer, F. 1959. Das Paläolithikum von Willendorf in der Wachau, Niederösterreich. Vorbericht über die monographische Bearbeitung. *Forschungen und Fortschritte* 33(3): 152–155.

Ferentinos, G., M. Gkioni, M. Geraga and G. Papatheodorou 2012. Early seafaring activity in the southern Ionian islands, Mediterranean Sea. *Journal of Archaeological Science* 39: 2167–2176.

Fernández-Espejo, D., T. Bekinschtein, M. M. Monti, J. D. Pickard, C. Junque, M. R. Coleman and A. M. Owen 2011. Diffusion weighted imaging distinguishes the vegetative state from the minimally conscious state. *NeuroImage* 54: 103–112.

Fernández Rey, A., G. E. Adan Álvarez, M. Arbizu et al. 2005. Grafismo rupestre paleolítico de la Cueva del Conde (Tuñón, Santo Adriano, Asturias). *Zephyrus: Revista de prehistoria y arqueología* 58: 67–88.

Field, H. 1932. The cradle of *Homo sapiens*. *Journal of Archaeology* 36: 426–430.

Finlayson, C., K. Brown, R. Blasco, J. Rosell, J. J. Negro, G. R. Bortolotti, G. Finlayson, A. Sánchez Marco, F. G. Pacheco, J. Rodríguez Vidal, J. S. Carrión, D. A. Fa and J. M. Rodríguez Llanes 2012. Birds of a feather: Neanderthal exploitation of raptors and corvids. *Proceedings of the National Academy of Sciences of the U.S.A.* 7(9): e45927.

Fischer, E. 1914. Die Rassenmerkmale des Menschen als Domestikationerscheinungen. *Zeitschrift für Morphologie und Anthropologie* 18: 479–524.
Fisher, R. A. 1930. *The genetical theory of natural selection*. Clarendon Press, Oxford.
Fladerer,F. 1997. Drachenhöhle bei Mixnitz. In D. Döppes and G. Rabeder (eds), *Pliozäne und pleistozäne Faunen Österreichs*, pp. 295–304. Mitteilungen der Kommission für Quartärforschung der Österreichischen Akademie der Wissenschaften 10, Vienna.
Fodor, J. 1990. *A theory of content and other essays*. MIT Press, Cambridge, MA.
Foley, R. and M. M. Lahr 1997. Mode 3 technologies and the evolution of modern humans. *Cambridge Archaeological Journal* 7: 3–36.
Folstein, S. E. and B. Rosen-Sheidley 2001. Genetics of autism: complex aetiology for a heterogeneous disorder. *Nature Reviews Genetics* 2: 943–955.
Ford, C. S. and F. Beach 1951. *Patterns of sexual behavior*. Harper, New York.
Fortea Pérez, J. 1999. Abrigo de La Viña. Informe y primera valoración de las campañas 1995–1998. *Excavaciones arqueológicas en Asturias 1995–1998* 4: 31–41.
Fortea Pérez, J. 2007. 39 edades 14C AMS para el arte paleolítico rupestre en Asturias. *Excavaciones arqueológicas en Asturias 1999–2002* 5: 91–102.
Frankel, D. 1993. The excavator: creator or destroyer? *Antiquity* 67: 875–877.
Franklin, N. R. and P. J. Habgood 2015. Representation of scarification on the Venus of Hohle Fels. *Rock Art Research* 32(2): 231–233.
Frayer, D. W. 1986. Cranial variation at Mladeč and the relationship between Mousterian and Upper Palaeolithic hominids. *Anthropologie* 23: 243–256.
Frayer, D. W., M. H. Wolpoff, F. H. Smith, A. G. Thorne and G. G. Pope 1993. The fossil evidence for modern human origins. *American Anthropology* 95: 14–50.
Freeman, L. G. and J. Gonzalez Echegaray 1983. Tally-marked bone from Mousterian levels at Cueva Morín (Santander, Spain). In *Homenaje al Prof. M. Almagro Basch*, vol. I, pp. 143–147. Ministerio de Cultura, Madrid.
Frith, U., 1972. Cognitive mechanisms in autism: experiments with color and tone sequence production. *Journal of Autism and Childhood Schizophrenia* 2, 160–173.
Frith, U. 1989. *Autism: explaining the enigma*. Blackwell, Oxford.
Fu Q., M. Hajdinjak, O. T. Moldovan, S. Constantin, S. Mallick, P. Skoglund, N. Patterson, N. Rohland, I. Lazaridis, B. Nickel, B. Viola, K. Prüfer, M. Meyer, J. Kelso, D. Reich and S. Pääbo 2015. An early modern human from Romania with a recent Neanderthal ancestor. *Nature* 524: 216–219.
Fuentes, A. 2009. *Evolution of human behavior*. Oxford University Press, New York/Oxford.
Fuggazzola Delpino, M. A. and M. Mineo 1995. La piroga neolitica del Lago di Bracciano ('La Marmotta 1'). *Bolletino Paletnologia Italia (Rome)* 86: 197–266.
Fuhlrott, C. J. 1859. Menschliche Überreste aus einer Felsengrotte des Düsselthals. Ein Beitrag zur Frage über die Existenz fossiler Menschen. *Verhandlungen des Naturhistorischen Vereins Preussen und Rheinland Westphalen* 16: 131–153.
Fuss, J., N. Spassov, D. R. Begun and M. Böhme 2017. Potential hominin affinities of Graecopithecus from the Late Miocene of Europe. *PLoS ONE* 12(5): e0177127.
Gabora, L. 2005. Creative thought as a non-Darwinian evolutionary process. *Journal of Creative Behavior* 39(4): 65–87.
Gabora, L. 2011. Five clarifications about cultural evolution. *Journal of Cognition and Culture* 11: 61–83.
Gábori-Csánk, V. 1993. *Le Jankovichien: une civilisation paléolithiques en Hongrie*. ERAUL 53, Liège.

Gabunia, L. and A. Vekua 1995. A Plio-Pleistocene hominid from Dmanisi, East Georgia, Caucasus. *Nature* 373: 509–512.

Gaillard, C. 2006. Les premiers peuplements d'Asie du Sud: vestiges culturels. *Comptes Rendus Paleovol* 5: 359–369.

Gaillard, C., V. N. Misra and M. L. K. Murty 1990. Comparative study of three series of handaxes: one from Rajasthan and two from Andhra Pradesh. *Bulletin of the Deccan College Post-Graduate and Research Institute* 49: 137–143.

Gaillard, C., D. R. Raju, V. N. Misra and S. N. Rajaguru 1986. Handaxe assemblage from the Didwana region, the Thar Desert. India: a metrical analysis. *Proceedings of the Prehistoric Society* 52: 189–214.

Galanidou, N. 2013. Looking for the earliest occupants of the Aegean — Palaeolithic excavations at Rodafnidia, Lisvori, Lesvos. In M. Alvanou (ed.), *Island identities*, pp. 15–17. Mytilene, Athens.

Gallup, G. G., Jr. 1970. Chimpanzees: self recognition. *Science* 167: 86–87.

Gallup, G. G., Jr. 1998. Self-awareness and the evolution of social intelligence. *Behavioural Processes* 42: 239–247.

Gallup, G. G., Jr., J. L. Anderson and D. P. Shillito 2002. The mirror test. In M. Bekoff, C. Allen and G. M. Burghardt (eds), *The cognitive animal: empirical and theoretical perspectives on animal cognition*, pp. 325–333. University of Chicago Press, Chicago.

Gallup, G. G., Jr. and S. M. Platek 2002. Cognitive empathy presupposes self-awareness: evidence from phylogeny, ontogeny, neuropsychology, and mental illness. *Behavioral and Brain Sciences* 25: 36–37.

Gannon, P. J., R. L. Holloway, D. C. Broadfield and A. R. Braun 1998. Asymmetry of chimpanzee planum temporale: humanlike pattern of Wernicke's brain language area homolog. *Science* 279: 220–222.

Gannon, P. J., N. M. Kheck and P. R. Hof 2001. Language areas of the hominoid brain: a dynamic communicative shift on the upper east side planum. In D. Falk and K. R. Gibson (eds), *Evolutionary anatomy of the primate cerebral cortex*, pp. 216–240. Cambridge University Press, New York.

Gans, C. and R. G. Northcutt 1983. Neural crest and the origin of vertebrates: a new head. *Science* 220: 268–274.

Gao, X., W. Huang, Z. Xu, Z. Ma and J. W. Olsen 2004. 120–150 ka human tooth and ivory engravings from Xinglongdong Cave, Three Gorges region, south China. *Chinese Science Bulletin* 49(2): 175–180.

García-Diez, M., B. Ochoa Fraile and I. Barandiarán Maestu 2013. Neanderthal graphic behaviour. The pecked pebble from Axlor Rockshelter (northern Spain). *Journal of Anthropological Research* 69: 397–410.

Garrigan, D., Z. Mobasher, T. Severson, J. A. Wilder and M. F. Hammer 2005. Evidence for archaic Asian ancestry on the human X chromosome. *Molecular Biological Evolution* 22: 189–192.

Geay, P. 1957. Sur la découverte d'un squelette aurignacien en Charente-Maritime. *Bulletin de la Société Préhistoroque Française* 54: 193–197.

Geigl, E.-M. 2002. Why ancient DNA research needs taphonomy. Paper presented at Conférence ICAZ, 'Biosphere to Lithosphere'.

Germonpré, M., M. V. Sablin, R. E. Stevens, R. E. M. Hedges, M. Hofreiter, M. Stiller and V. R. Després 2009. Fossil dogs and wolves from Palaeolithic sites in Belgium, the Ukraine and Russia: osteometry, ancient DNA and stable isotopes. *Journal of Archaeological Science* 36(2): 473–490.

Geschwind, N. and W. Levitsky 1968. Human brain: left–right asymmetries in temporal speech region. *Science* 161: 186–187.
Ghika, J. 2008. Paleoneurology: neurodegenerative diseases are age-related diseases of specific brain regions recently developed by *Homo sapiens*. *Medical Hypotheses* 71(5): 788–801.
Gibbons, A. 1998. Calibrating the mitochondrial clock. *Science* 279: 28–29.
Gibbons, A. 2010. Close encounters of the prehistoric kind. *Science* 328: 680–684.
Gierliński, G. D., G. Niedźwiedzki, M. G. Lockley, A. Athanassiou, C. Fassoulas, Z. Dubicka, A. Boczarowski, M. R. Bennett and P. E. Ahlberg 2017. Possible hominin footprints from the late Miocene (c. 5.7 Ma) of Crete? *Proceedings of the Geologists' Association* 128: 697–710.
Gieseler, W. 1974. *Die Fossilgeschichte des Menschen*. Konrad Theiss, Stuttgart.
Gillespie, J. D., S. Tupakka and C. Cluney 2004. Distinguishing between naturally and culturally flaked cobbles: a test case from Alberta, Canada. *Geoarchaeology* 19: 615–633.
Ginesu, S., S. Sias and J. M. Cordy 2003. Morphological evolution of the Nurighe Cave (Logudoro, northern Sardinia, Italy) and the presence of man: first results. *Geografica Fisica e Dinamica Quaternaria* 26: 41–48.
Gloor, P. 1990. Experiential phenomena of temporal lobe epilepsy. *Brain* 113: 1673–1694.
Gloor, P. 1992. Role of the amygdala in temporal lobe epilepsy. In A. P. Aggleton (ed.), *The amygdala*, pp. 505–538. Wiley-Liss, New York.
Golenberg, E. M., A. Bickeland and P. Weihs 1996. Effect of highly fragmented DNA on PCR. *Nucleic Acids Research* 24: 5026–5033.
González Sainz C. and C. San Miguel 2001. *Las cuevas del desfiladero. Arte rupestre paleolítico en el valle del río Carranza (Cantabria-Vizcaya)*. Gobierno de Cantabria, Consejería de Cultura y Deporte, Santander.
Goodall, J. 1986. *The chimpanzees of Gombe: patterns of behavior*. Harvard University Press, Cambridge, MA.
Goodwin, F. K. and K. R. Jamison 1990. *Manic depressive illness*. Oxford University Press, Oxford.
Goody, J. 1977. *The domestication of the savage mind*. Cambridge University Press, Cambridge.
Goren-Inbar, N. 1986. A figurine from the Acheulian site of Berekhat Ram. *Mi'Tekufat Ha'Even* 19: 7–12.
Goren-Inbar, N., Z. Lewy and M. E. Kislev 1991. The taphonomy of a bead-like fossil from the Acheulian of Gesher Benot Ya'aqov, Israel. *Rock Art Research* 8: 83–87.
Goren-Inbar, N. and S. Peltz 1995. Additional remarks on the Berekhat Ram figurine. *Rock Art Research* 12: 131–132.
Gould, S. J. 1977. *Ontogeny and phylogeny*. Belknap Press, Cambridge, MA.
Gould, S. J. 1979. A biological homage to Mickey Mouse. *Natural History* 88(5): 30–36.
Gowaty, P. A. 1992. Evolutionary biology and feminism. *Human Nature* 3: 217–249.
Grammer, K. and R. Thornhill 1994. Human facial attractiveness and sexual selection: the role of symmetry and averageness. *Journal of Comparative Psychology* 108: 233–242.
Green, R. E., J. Krause, A. W. Briggs, T. Maricic, U. Stenzel, M. Kircher *et al.* 2010. A draft sequence of the Neandertal genome. *Science* 328: 710–722.
Green, R. E., J. Krause, S. E. Ptak, A. W. Briggs, M. T. Ronan, J. F. Simons, L. Du, M. Egholm, J. M. Rothberg, M. Paunovic and S. Pääbo 2006. Analysis of one million base pairs of Neanderthal DNA. *Nature* 444: 330–336.

Gregersen, G. 1983. *Sexual practices: the story of human sexuality*. F. Watts, New York.
Gregory, R. L. 1970. *The intelligent eye*. Weidenfeld and Nicolson, London.
Grine, F. E., R. M. Bailey, K. Harvati, R. P. Nathan, A. G. Morris, G. M. Henderson, I. Robot et al. 2007. Late Pleistocene human skull from Hofmeyr, South Africa, and modern human origins. *Science* 315: 226–229.
Grine, F. E., P. Gunz, L. Betti-Nash, S. Neubauer and A. G. Morris 2010. Reconstruction of the late Pleistocene human skull from Hofmeyr, South Africa. *Journal of Human Evolution* 59: 1–15.
Grinker, R. R. 2007. *Unstrange minds: remapping the world of autism*. Basic Books, New York.
Groucutt, H. S., M. D. Petraglia, G. Bailey, E. M. L. Scerri, A. Parton, L. Clark-Balzan et al. 2015. Rethinking the dispersal of *Homo sapiens* out of Africa. *Evolutionary Anthropology* 24: 149–164.
Groves, C. P. 1995. The origin of language, the use of language and other people's language. *Bulletin of the Archaeological and Anthropological Society of Victoria* 1995(4): 8–12.
Groves, C. P. 1999. The advantages and disadvantages of being domesticated. *Perspectives in Human Biology* 4: 1–12.
Guadelli, A. and J. L. Guadelli 2004. Une expression symbolique sur os dans le Paléolithique inférieur. Étude préliminaire de l'os incisé de la grotte Kozarnika, Bulgaire du nord-ouest. In M. Otte (ed.), *L'homme de Néanderthal: la spiritualité*, pp. 87–95. ERAUL, Liège.
Guérin, C., M. Faure, A. Argant, J. Argand, E. Crégut-Bonnoure, E. Debard, E. Delson, V. Eisenmann, M. Hugueney, N. Limondin-Louzouet, E. Martin-Suarez, P. Mein, C. Mourer-Chauviré, F. Parenti, J.-F. Pastre, S. Sen and A. Valli 2004. Le gisement Pliocène supérieur de Saint-Vallier (Drôme, France): synthèse biostratigraphique et paléoécologique. *Geobios* 37: 349–360.
Guidon, N. 1984. Les premières occupations humaines de l'aire archéologique de São Raimundo Nonato – Piauí – Brésil. *L'Anthropologie* 88: 263–271.
Guidon, N. and G. Delibrias 1986. Carbon-14 dates point to man in the Americas 32,000 years ago. *Nature* 321: 769–771.
Guillo, D. 2012. Does culture evolve by means of Darwinian selection? The lessons of Candide's travels. *Social Science Information* 51: 364–388
Gunz, P., S. Neubauer, L. Golovanova, V. Doronichev, B. Maureille and J. J. Hublin 2012. A uniquely modern human pattern of endocranial development. Insights from a new cranial reconstruction of the Neandertal newborn from Mezmaiskaya. *Journal of Human Evolution* 62: 300–313.
Gunz, P., S. Neubauer, B. Maureille and J. J. Hublin 2010. Brain development after birth differs between Neanderthals and modern humans. *Current Biology* 20: R921–R922.
Gusnard, D. A. 2005. Being a self: considerations from functional imaging. *Consciousness and Cognition* 14: 679–697. Gutierrez, G. and A. Marin 1998. The most ancient DNA recovered from amber-preserved specimen may not be as ancient as it seems. *Molecular Biological Evolution* 15: 926–929.
Gutierrez, G., D. Sanchez and A. Marin 2002. A reanalysis of the ancient mitochondrial DNA sequences recovered from Neandertal bones. *Molecular Biological Evolution* 19: 1359–1366.
Guzder, S. 1980. *Quaternary environments and Stone Age cultures of the Konkan, coastal Maharashtra, India*. Deccan College, Pune.
Gvozdover, M. 1989. The typology of female figurines of the Kostenki Paleolithic culture. *Soviet Anthropology and Archaeology* 27(4): 32–94.

Gvozdover, M. 1995. *Art of the mammoth hunters: the finds from Avdeevo*. Oxbow Books, Oxford.

Gyllensten, U., D. Wharton, A. Josefsson and A. C. Wilson 1991. Paternal inheritance of mitochondrial DNA in mice. *Nature* 352: 255–257.

Hahn, E. 1896. *Die Haustiere und ihre Beziehungen zur Wirtschaft des Menschen. Eine geographische Studie. Mit einer chromolithischen Karte: Die Wirtschaftsformen der Erde*. Verlag Duncker & Humblot, Leipzig.

Hahn, J. 1986. *Kraft und Aggression: Die Botschaft der Eiszeitkunst im Aurignacien Süddeutschlands?*Archaeologica Venatoria, Institut für Urgeschichte der Universität Tübingen, Tübingen.

Hahn, J. 1988a. *Das Geißenklösterle 1; Forschungen und Berichte zur Vor- und Frühgeschichte in Baden-Württemberg 26*. Konrad Theiss Verlag, Stuttgart.

Hahn, J. 1988b. *Die Geißenklösterle-Höhle im Achtal bei Blaubeuren 1: Fundhorizontbildung im Mittelpaläolithikum und Aurignacien*. Konrad Theiss, Stuttgart.

Hahn, J. 1991. Höhlenkunst aus dem Hohlen Fels bei Schelklingen, Alb-Donau-Kreis. *Fundberichte aus Baden-Württemberg* 16: 19–22.

Hahn, M. W. and R. A. Bentley 2003. Drift as a mechanism for cultural change: an example from baby names. *Proceedings of the Royal Society B* 270: S120–S123.

Hakeem, A. Y., C. C. Sherwood, C. J. Bonar, C. Butti, P. R. Hof and J. M. Allman 2009. Von Economo neurons in the elephant brain. *The Anatomical Record* 292: 242–248.

Haldane, J. B. S. 1932. *The causes of evolution*. Longmans, Green & Co., London.

Hammer, K. 1984. Das Domestikationssyndrom. *Kulturpflanze* 32: 11–34.

Hammer, M. F. 1995. A recent common ancestry for human Y chromosomes. *Nature* 378: 376–378.

Handwerker, W. P. 1989. The origins and evolution of culture. *American Anthropologist* 91: 313–326.

Happé, F. and U. Frith (eds) 2010. *Autism and talent*. Oxford University Press, Oxford.

Happé, F. and P. Vital 2009. What aspects of autism predispose to talent? *Philosophical Transactions of the Royal Society of London, B Biological Science* 364(1522): 1369–1375.

Hardy, J., A. Pittman, A. Myers, K. Gwinn-Hardy, H. C. Fung, R. de Silva, M. Hutton and J. Duckworth 2005. Evidence suggesting that *Homo neanderthalensis* contributed the H2 MAPT haplotype to *Homo sapiens*. *Biochemical Society Transactions* 33: 582–585.

Hare, E. H. 1988. Schizophrenia as a recent disease. *British Journal of Psychiatry* 153: 521–531.

Hare, B. and M. Tomasello 2005. Human-like social skills in dogs? *Trends in Cognitive Sciences* 9(9): 439–444.

Hare, B., V. Wobber and R. Wrangham 2012. The self-domestication hypothesis: evolution of bonobo psychology is due to selection against aggression. *Animal Behaviour* 83(3): 573–585.

Harpending, H. C., M. A. Batzer, M. Gurven, L. B. Jorde, A. R. Rogers and S. T. Sherry 1998. Genetic traces of ancient demography. *Proceedings of the National Academy of Sciences of the U.S.A.* 95: 1961–1967.

Harrison, P. J. 1999. The neuropathology of schizophrenia: a critical review of the data and their interpretation. *Brain* 122: 593–624.

Hartfield, M. and P. D. Keightley 2012. Current hypotheses for the evolution of sex and recombination. *Integrative Zoology* 7: 192–209.

Hartl, D. and A. Clark 1997. *Principles of population genetics*. Sinauer, Sunderland, MA.

Hawks, J. and M. H. Wolpoff 2001. The accretion model of Neandertal evolution. *Evolution* 55: 1474–1485.

Haworth, K. 2006. Upper Paleolithic art, autism, and cognitive style: implications for the evolution of language. *Semiotica* 162: 127–174.

Hayashi, M. 2006. Spindle neurons in the anterior cingulate cortex of humans and great apes. In T. Matsuzawa, M. Tomonaga and M. Tanaka (eds), *Cognitive development in chimps*, pp. 64–74. Springer-Verlag, New York.

Haynes, C. V. 1973. The Calico site: artifacts or geofacts? *Science* 181: 305–310.

Hedges, R., P. B. Pettitt and C. Bronk Ramsey 1994. Radiocarbon dates from the Oxford AMS system: archeometry datelist 18. *Archeometry* 36(2): 337–374.

Hedges, S. B., S. Kumar, K. Tamura and M. Stoneking 1992. Human origins and analysis of mitochondrial DNA sequences. *Science* 255: 737–739.

Heimer, L., G. W. Van Hoesen, M. Trimble and D. S. Zahm 2008. *Anatomy of neuropsychiatry: the new anatomy of the basal forebrain and its implications for neuropsychiatric illness*. Academic Press, Burlington, MA.

Hellenthal, G., A. Auton and D. Falush 2008. Inferring human colonization history using a copying model. *PLOS Genetics* 4: e1000078.

Helm, C. W., H. C. Cawthra, J. C. De Vynck, C. J. Z. Helm, R. Rust and W. Stear in press. Drawing a line in the sand. *Rock Art Research* 37(1): 95-99.

Helvenston, P. A. 1999. Concepts regarding melancholia and mania in medieval Europe with an emphasis on England. M.A. in History, University of Northern Arizona, Flagstaff, AZ.

Helvenston, P. A. 2013. Differences between oral and literate cultures: what we can know about Upper Paleolithic minds. In R. G. Bednarik (ed.), *The psychology of human behaviour*, pp. 59–110. Nova Press, New York.

Helvenston, P. A. and R. G. Bednarik 2011. Evolutionary origins of brain disorders in *Homo sapiens sapiens*. *Brain Research Journal* 3(2): 113–139.

Henke, W. and R. Protsch 1978. Die Paderborner Calvaria—ein diluvialer *Homo sapiens*. *Anthropologischer Anzeiger* 36: 85–108.

Henke, W. and H. Rothe 1994. *Paläoanthropologie*. WILEY-VCH Verlag GmbH, Berlin.

Henneberg, M. 1988. Decrease of human skull size in the Holocene. *Human Biology* 60: 395–405.

Henneberg, M. 1990. Brain size/body weight variability in *Homo sapiens*: consequences for interpreting hominid evolution. *HOMO — Journal of Comparative Human Biology* 39(3–4): 121–130.

Henneberg, M. 1998. Evolution of the human brain: is bigger better? *Clinical and Experimental Pharmacology and Physiology* 25(9): 745–749.

Henneberg, M. 2004. The rate of human morphological microevolution and taxonomic diversity of hominids. *Studies in Historical Anthropology* 4: 49–59.

Henneberg, M., R. B. Eckhardt, S. Chavanaves and K. J. Hsü 2014. Evolved developmental homeostasis disturbed in LB1 from Flores, Indonesia, denotes Down syndrome and not diagnostic traits of the invalid species *Homo floresiensis*. *Proceedings of the National Academy of Sciences of the U.S.A.* Available at: www.pnas.org/cgi/doi/10.1073/pnas.1407382111

Henneberg, M. and A. Saniotis 2009. Evolutionary origins of human brain and spirituality. *Anthropologischer Anzeiger* 67(4): 427–438.

Henneberg, M. and J. Schofield 2008. *The Hobbit trap: money, fame, science and the discovery of a 'new species'*. Wakefield Press, Kent Town, South Australia.

Henneberg, M. and M. Steyn 1993. Trends in cranial capacity and cranial index in Sub-Saharan Africa during the Holocene. *American Journal of Human Biology* 5: 473–479.

Henneberg, M. and J. F. Thackeray 1995. A single-lineage hypothesis of hominid evolution. *Evolutionary Theory* 11: 31–38.

Henneberg, M. and A. Thorne 2014. Flores human may be pathological *Homo sapiens*. *Before Farming* 2004: 2–3.

Henrich, J. and R. Boyd 2002. On modelling cognition and culture: why replicators are not necessary for cultural evolution. *Journal of Cognition and Culture* 2: 87–112.

Henrich, J., R. Boyd and P. J. Richerson 2008. Five misunderstandings about cultural evolution. *Human Nature* 19: 119–137.

Henriksen, R., M. Johnsson, L. Andersson, P. Jensen and D. Wright 2016. The domesticated brain: genetics of brain mass and brain structure in an avian species. *Scientific Reports* 6: 34031.

Henry-Gambier, D. 2002. Les fossiles de Cro-Magnon (Les-Eyzies-de-Tayac, Dordogne): Nouvelles données sur leur position chronologique et leur attribution culturelle. *Bulletin et Mémoires de la Société d'Anthropologie de Paris* 14(1–2): 89–112.

Hermelin, B. and N. O'Connor 1970. *Psychological experiments with autistic children*. Pergamon Press, Oxford.

Hershfield, A., D. Yurgelun-Todd, R. Kikinis, F. A. Jolesz and R. W. McCarley 2006. Middle and inferior temporal gyrus gray matter volume abnormalities in first-episode schizophrenia: an MRI study. *American Journal of Psychiatry* 163(12): 2103–2110.

Heston, J. J. 1966. Psychiatric disorders in foster home reared children of schizophrenic mothers. *The British Journal of Psychiatry* 112: 819–825.

Heyes, C. M. 1998. Theory of mind in nonhuman primates. *Behavioral and Brain Sciences* 21(1): 101–134.

Hobsbawm, E. J. 1992. Ethnicity and nationalism in Europe today. *Anthropology Today* 8(1): 3–13.

Hodder, I. 1990. *The domestication of Europe: structure and contingency in Neolithic societies*. Basil Blackwell, Oxford.

Hoffmann, D. L., C. D. Standish, M. García-Diez, P. B. Pettitt, J. A. Milton, J. Zilhão, J. J. Alcolea-González, P. Cantalejo-Duarte, H. Collado, R. de Balbín, M. Lorblanchet, J. Ramos-Muñoz, G.-C. Weniger and A. W. G. Pike 2018. U-Th dating of carbonate crusts reveals Neandertal origin of Iberian cave art. *Science* 359(6378): 912–915.

Hofstadter, D. 2007. *I am a strange loop*. Basic Books, New York.

Hölldobler, B. and E. O. Wilson 1990. *The ants*. Belknap Press, Cambridge, MA.

Holloway, R. L. 1995. Toward a synthetic theory of human brain evolution. In J. P. Changeaux and J. Chavillon (eds), *Origins of the human brain*, pp. 42–55. Clarendon Press, Oxford.

Holloway, R. L. 1996. Evolution of the human brain. In A. Lock and C. Peters (eds), *Handbook of human symbolic evolution*, pp. 74–116. Clarendon Press, Oxford.

Holloway, R. L. 2001. Revisiting australopithecine visual striate cortex: newer data from chimpanzee and human brains suggest it could have been reduced during australopithecine times. In D. Falk and K. R. Gibson (eds), *Evolutionary anatomy of the primate cerebral cortex*, pp. 177–186. Cambridge University Press, New York.

Hooijer, D. A. 1957. A stegodon from Flores. *Treubia* 24: 119–129.

Hooper, J. and D. Teresi 1992. *The determinism problem*. Oxford University Press, Oxford.

Horrobin, D. F. 1998. Schizophrenia: the illness that made us human. *Medical Hypotheses* 50: 269–288.

Horrobin, D. 2002. *The madness of Adam and Eve: how schizophrenia shaped humanity*. Bantam, London.

Hovatta, I., T. Varilo, J. Suvisaari, J. D. Terwilliger, V. Ollikainen, R. Arajärvi et al. 1999. A genomewide screen for schizophrenia genes in an isolated Finnish sub-population, suggesting multiple susceptibility loci. *American Journal of Human Genetics* 65: 1114–1124.

Howells, W. 1959. *Mankind in the making: the story of human evolution.* Doubleday & Co., Garden City, NY.

Huang, W., R. Ciochon, Y. Gu, R. Larick, Q. Fang, H. Schwarcz, C. Yonge, J. De Vos and W. Rink 1995. Early *Homo* and associated artefacts from Asia. *Nature* 378: 275–278.

Huang, W. and Fang, Q. 1991. *Wushan hominid site.* Ocean Press, Beijing.

Hublin, J.-J., F. Spoor, M. Braun, F. Zonneveld and S. Condemi 1996. A late Neanderthal associated with Upper Palaeolithic artefacts. *Nature* 381: 224–226.

Hughes, C., I. Soares-Boucaud, J. Hochmann and U. Frith 1997. Social behaviour in pervasive developmental disorders: effects of informant, group and 'theory-of-mind'. *European Child and Adolescent Psychiatry* 6(4): 191–198.

Humphrey, N. 1998. Cave art, autism, and the evolution of the human mind. *Cambridge Archaeological Journal* 8(2): 165–191.

Hurcombe, L. 2004. The stone artefacts from the Pabbi Hills. In R. W. Dennell (ed.), *Early hominin landscapes in northern Pakistan: investigations in the Pabbi Hills*, pp. 222–292. International Series 1265, British Archaeological Reports, Oxford.

Huyge, D. 1990. Mousterian skiffle? Note on a Middle Palaeolithic engraved bone from Schulen, Belgium. *Rock Art Research* 7(2): 125–132.

Hwang, J., I. K. Lyoo, S. R. Dager, S. D. Friedman, J. S. Oh, J. Y. Lee, S. J. Kim, D. L. Dunner and P. F. Renshaw 2006. Basal ganglia shape alterations in bipolar disorder. *American Journal of Psychiatry* 163(2): 276–285.

Hyman, B. T., G. W. Van Hoesen, A. R. Damasio and C. L. Barnes 1984. Alzheimer's disease: cell-specific pathology isolates the hippocampal formation. *Science* 225(9): 1168–1170.

Imber, C. E. 2003. Robusticity and rugosity in the modern human skeleton. PhD thesis, University College London.

International Chronographic Chart 2014. Available at: www.stratigraphy.org/ICSchart/ChronostratChart2014-02.jpg

Jablensky, A., N. Sartorius, G. Ernberg, M. Anker, A. Korten, J. E. Cooper, R. Day and A. Bertelsen 1992. Schizophrenia: manifestations, incidence and course in different cultures. A World Health Organization ten-country study. *Psychological Medicine Monographs* 20: 1–97.

Jablonka, E. and M. Lamb 2005. *Evolution in four dimensions: genetic, epigenetic, behavioral, and symbolic variation in the history of life.* MIT Press, Cambridge, MA.

Jacob, T., E. Indriati, R. P. Soejono, K. Hsü, D. W. Frayer, R. B. Eckhardt, A. J. Kuperavage, A. Thorne and M. Henneberg 2006. Pygmoid Australomelanesian *Homo sapiens* skeletal remains from Liang Bua, Flores: population affinities and pathological abnormalities. *Proceedings of the National Academy of the Sciences of the U.S.A.* 103(36): 13421–13426.

Jelínek, J. 1987. Historie, identifikace a význam mladečských antropologických nálezů z počátku mladého paleolitu. *Anthropologie* 25: 51–69.

Jelínek, J., M. H. Wolpoff and D. W. Frayer 2005. Evolutionary significance of the Quarry Cave specimens from Mladeč. *Anthropologie* 43: 215–228.

Jensen, E. 2005. *Teaching with the brain in mind*, 2nd edn. Association for Supervision and Curriculum Development, Alexandria, VA.

Jerison, H. J. 1973. *Evolution of the brain and intelligence.* Academic Press, New York.

Jeste, D. V., R. Del Carmen, J. B. Lohr and R. J. Wyatt 1985. Did schizophrenia exist before the eighteenth century? *Comprehensive Psychiatry* 26: 493–503.

Jin C., W. Dong, J. Liu, G. Wei, Q. Xu, J. Zheng, L. Zheng, L. Han and F. Wang 2000. A preliminary study on the Early Pleistocene deposits and the mammalian fauna from the Renzi Cave, Fanchang, Anhui, China. *Acta Anthropologica Sinica* 19: 235–245.

Joffe, T. H. 1997. Social pressures have selected for an extended juvenile period in primates. *Journal of Human Evolution* 32: 593–605.

Johnson, D. L. 1980. Problems in the land vertebrate zoogeography of certain islands and the swimming powers of elephants. *Journal of Biogeography* 7: 383–398.

Jones, D. M. 1995. Sexual selection, physical attractiveness and facial neoteny: cross-cultural evidence and implications. *Current Anthropology* 36(5): 723–748.

Jones, D. M. 1996. An evolutionary perspective on physical attractiveness. *Evolutionary Anthropology* 5(3): 97–109.

Joordens, J. C. A., F. d'Errico, F. P. Wesselingh, S. Munro, J. de Vos, J. Wallinga et al. 2014. Homo erectus at Trinil in Java used shells for tool production and engraving. *Nature* 518: 228–231.

Jost, P. 1966. *Lubrication (tribology): a report on the present position and industry's needs.* H.M. Stationery Office, London.

Kandel, E. R. and C. Pittenger 1999. The past, the future and the biology of memory storage. *Philosophical Transactions of the Royal Society B, Biological Sciences* 354(1392): 2027–2052.

Karavanic, I. 1998. The early Upper Paleolithic of Croatia. In *Proceedings of the XIII International Congress of Prehistoric and Protohistoric Sciences (Forlì, Italy), 8–14 September 1996*, vol. 2, pp. 659–665.

Karlsson, J. L. 1970. Genetic association of giftedness and creativity with schizophrenia. *Hereditas* 66: 177–181.

Karlsson, J. L. 1974. Inheritance of schizophrenia. *Acta Psychiatrica Scandinavica* 247: 77–88.

Karlsson, J. L. 1984. Creative intelligence in relatives of mental patients. *Hereditas* 100: 83–86.

Kaskan, P. M. and B. L. Finley 2001. Encephalization and its developmental structure: how many ways can a brain get big? In D. Falk and K. R. Gibson (eds), *Evolutionary anatomy of the primate cerebral cortex*, pp. 14–29. Cambridge University Press, New York.

Kavvadias, G. 1984. *Palaeolithic Kephalonia.* Fytrakik, Athens.

Keenan, J. P., D. Falk and G. C. Gallup, Jr. 2003. *The face in the mirror: the search for the origins of consciousness.* HarperCollins, New York.

Keller, M. C. and G. Miller 2006. Resolving the paradox of common, harmful, heritable mental disorders: which evolutionary genetic models work best? *Behavioral and Brain Sciences* 29: 385–452.

Kellman, J. 1998. Ice Age art, autism and vision: how we see/how we draw. *Studies in Art Education* 39(2): 117–131.

Kellman, J. 1999. Drawing with Peter: autobiography, narrative, and the art of a child with autism. *Studies in Art Education* 40(3): 258–274.

Kellogg, R. M., M. Knoll and J. Kugler 1965. Form-similarity between phosphenes of adults and preschool children's scribblings. *Nature* 208: 1129–1130.

Kennedy, D. P., K. Semendeferi and E. Courchesne 2007. No reduction of spindle neuron number in the frontoinsular cortex in autism. *Brain and Cognition* 64(2): 124–129.

Kennedy, J. L., L. A. Farrer, N. C. Andreasen, R. Mayeux and P. St George-Hyslop 2003. The genetics of adult-onset neuropsychiatric disease: complexities and conundra? *Science* 302: 822–826.

Kennedy, K. A. R., A. Sonakia, J. Chiment and K. K. Verma 1991. Is the Narmada hominid an Indian *Homo erectus*? *American Journal of Physical Anthropology* 86: 475–496.

Khatri, A. P. 1963. Mahadevian: an Oldowan pebble culture of India. *Asian Perspectives* 6: 186–197.

Kim, Y. S., B. Leventhal, Y.-J. Koh, E. Fombonne, E. Laska, E.-C. Lim, K.-A. Chun, S.-J. Kim, Y.-K. Kim, H.-J. Lee, D.-H. Song and R. R. Grinker, 2011. Prevalence of autism spectrum disorders in a total population sample. *American Journal of Psychiatry* 168(90): 904–912.

Kind, C.-J., N. Ebinger-Rist, S. Wolf, T. Beutelspacher and K. Wehrberger 2014. The smile of the lion man. Recent excavations in Stadel Cave (Baden-Württemberg, south-western Germany) and the restoration of the famous Upper Palaeolithic figurine. *Quartär* 61: 129–145.

King, H. D. and H. H. Donaldson 1929. Life processes and size of the body and organs in gray Norway rats during ten generations in captivity. *American Anatomical Memoirs* 14: 5–72.

Klamer, A., R. M. Solow and D. N. McCloskey 1989. *The consequences of economic rhetoric*. Cambridge University Press, New York.

Klein, R. G. 2009. *The human career: human biological and cultural origins*. Chicago University Press, Chicago.

Klyosov, A. A. and I. L. Rozhanskii 2012a. Re-examining the out of Africa theory and the origin of Europeoids (Caucasoids) in light of DNA genealogy. *Advances in Anthropology* 2: 80–86.

Klyosov, A. A. and I. L. Rozhanskii 2012b. Haplogroup R1a as the proto Indo-Europeans and the legendary Aryans as witnessed by the DNA of their current descendants. *Advances in Anthropology* 2: 1–13.

Klyosov, A. A., I. L. Rozhanskii and L. E. Ryanbchenko 2012. Re-examining the Out-of-Africa theory and the origin of Europeoids (Caucasoids). Part 2. SNPs, haplogroups and haplotypes in the Y-chromosome of chimpanzee and humans. *Advances in Anthropology* 2: 198–213.

Klyosov, A. A. and G. T. Tomezzoli 2013. DNA genealogy and linguistics: Ancient Europe. *Advances in Anthropology* 3: 101–111.

Koch, C. 2004. *The quest for consciousness*. Roberts & Company Publishers, Greenwood Village, CO.

Kohl, P. L. and C. Fawcett 2000. Archaeology in the service of the state: theoretical considerations. In P. L. Kohl and C. Fawcett (eds), *Nationalism, politics, and the practice of archaeology*, pp. 3–18. Cambridge University Press, Cambridge.

Kooij, P., M. Schiott, J. Boomsma and H. De Fine Licht 2011. Rapid shifts in Atta cephalotes fungus-garden enzyme activity after a change in fungal substrate (*Attini, Formicidae*). *Insectes Sociaux* 58(2): 145–151.

Kopaka, K., P. Drossinou and Y. Christodoulakos 1994–1995. Surface survey on Gavdos. *Kritiki Estia* 5: 242–244.

Kopaka, K. and C. Matzanas 2009. Palaeolithic industries from the island of Gavdos, near neighbour to Crete in Greece. *Antiquity* 83(321). Available at: http://antiquity.ac.uk/antiquityNew/projgall/kopaka321/index.html#author

Korisettar, R. 2002. The archaeology of the south Asian Lower Palaeolithic: history and current studies. In S. Settar and R. Korisettar (eds), *Prehistory. Archaeology of south Asia*, pp. 1–65. Indian Archaeology in Retrospect, vol. 1, Indian Council of Historical Research, Manohar.

Kourtessi-Philippaki, G. 1999. The Lower and Middle Palaeolithic in the Ionian islands: new finds. In G. N. Bailey, E. Adam, E. Panagopoulou, C. Perlès and K. Zachos (eds), *The Palaeolithic archaeology of Greece and adjacent areas*, pp. 282–287. British School at Athens, London.

Kozłowski, J. K. 1990. A multiaspectual approach to the origins of the Upper Palaeolithic in Europe. In P. Mellars (ed.), *The emergence of modern humans. An archaeological perspective*, pp. 419–438. Edinburgh University Press, Edinburgh.

Krause, J., C. Lalueza-Fox, L. Orlando, W. Enard, R. E. Green, H. A. Burbano *et al.* 2007. The derived FOXP2 variant of modern humans was shared with Neandertals. *Current Biology* 17(21): 1908–1912.

Krings, M., A. Stone, R. W. Schmitz, H. Krainitzki, M. Stoneking and S. Pääbo 1997. Neandertal DNA sequences and the origin of modern humans. *Cell* 90: 19–30.

Kroeber, A. L. 1900. The Eskimo of Smith Sound. *Bulletin of the American Museum of Natural History* 12: 265–327.

Kruska, D. 1988a. Mammalian domestication and its effect on brain structure and behaviour. In H. J. Jerison and I. Jerison (eds), *Intelligence and evolutionary biology*, pp. 211–250. Springer, New York.

Kruska, D. 1988b. Effects of domestication on brain structure and behavior in mammals. *Human Evolution* 3(6): 473–485.

Kshirsagar, A. 1993. The role of fluorine in the chronometric dating of Indian Stone Age cultures. *Man and Environment* 18(2): 23–32.

Kshirsagar, A. and G. L. Badam 1990. Biochronology and fluorine analysis of some Pleistocene fossils from central and western India. *Bulletin of the Deccan College Post-Graduate and Research Institute* 49: 199–211.

Kshirsagar, A. and V. D. Gogte 1990. Fluorine determinations in fossil bones with ion-selective electrode. *Bulletin of the Deccan College Post-Graduate and Research Institute* 49: 213–215.

Kshirsagar, A. and K. Paddayya 1988–1989. Relative chronology of Stone Age cultures of Hunsgi-Baichbal valley, northern Karnataka, by fluorine analysis. *Bulletin of the Deccan College Post-Graduate and Research Institute* 47/48: 143–145.

Kuhlwilm, M., I. Gronau, M. J. Hubisz, C. De Filippo, J. Prado-Martinez, M. Kircher *et al.* 2016. Ancient gene flow from early modern humans into eastern Neanderthals. *Nature* 530(7591): 429–433.

Kuhn, S. L. 1995. *Mousterian lithic technology. An ecological perspective*. Princeton University Press, Princeton, NJ.

Kuhn, S. L. and A. Bietti 2000. The Late Middle and Early Upper Paleolithic in Italy. In O. Bar-Yosef and D. Pilbeam (eds), *The geography of Neandertals and modern humans in Europe and the Greater Mediterranean*, pp. 49–76. Peabody Museum of Archaeology and Ethnology, Cambridge, MA.

Kuhn, S. L. and M. C. Stiner 2001. The antiquity of hunter-gatherers. In C. Panter-Brick, R. H. Layton and P. Rowley-Conwy (eds), *Hunter-gatherers: an interdisciplinary perspective*, pp. 99–142. Cambridge University Press, Cambridge.

Kuhn, T. 1970. *The structure of scientific revolutions*, 2nd edn. University of Chicago Press, Chicago.

Kumar, G., R. G. Bednarik, A. Pradhan and R. Krishna 2016. The stone tools from Daraki-Chattan Cave – part 1. *Purakala* 26: 25–64.

Kumar, G. and R. Krishna 2014. Understanding the technology of the Daraki-Chattan cupules: the cupule replication project. *Rock Art Research* 31(2): 177–186.

Künzl, C. and N. Sachser 1999. The behavioral endocrinology of domestication: a comparison between the domestic guinea pig (*Cavia aperea f. porcellus*) and its wild ancestor, the cavy (*Cavia aperea*). *Hormones and Behavior* 35: 28–37.

Kuroki, N., M. E. Shenton, D. F. Salisbury, Y. Hirayasu, T. Onitsuka, H. Ersner-Hershfield, D. Yurgelun-Todd, R. Kikinis, F. A. Jolesz and R. W. McCarley 2006. Middle and inferior temporal gyrus gray matter volume abnormalities in first-episode schizophrenia: an MRI study. *American Journal of Psychiatry* 163(12): 2103–2110.

Kyparissi-Apostolika, N. (ed.) 2000. *Theopetra Cave: Twelve years of excavation and research, 1987–1998*. Institute for Aegean Prehistory, Athens.

Kyrle, G. 1931. Die Höhlenbärenjägerstation. In O. Abel and G. Kyrle (eds), *Die Drachenhöhle bei Mixnitz*, pp. 804–962. Band 7–9, Speläologische Monographien, Vienna.

Lahr, M. M. and R. A. Foley 1998. Towards a theory of modern human origins: geography, demography and diversity in recent human evolution. *Yearbook of Physical Anthropology* 41: 137–176.

Lahr, M. M. and R. V. S. Wright 1996. The question of robusticity and the relationship between cranial size and shape in *Homo sapiens*. *Journal of Human Evolution* 31(2): 157–191.

Laidler, P. W. 1933. Dating evidence concerning the Middle Stone Ages and a Capsio-Wilton culture, in the south-east Cape. *South African Journal of Science* 30: 530–542.

Laidler, P. W. 1934. The archaeological and geological sequence in the Transkei and Ciskei. *South African Journal of Science* 31: 535–546.

Laland, K. N. 1994. Sexual selection with a culturally transmitted mating preference. *Theoretical Population Biology* 45: 1–15.

Laland, K., J. Odling-Smee and M. W. Feldman 2000. Niche construction, biological evolution and cultural change. *Behavioral and Brain Sciences* 23: 131–175.

Langley, M. C., C. Clarkson and S. Ulm 2008. Behavioural complexity in Eurasian Neanderthal populations: a chronological examination of the archaeological evidence. *Cambridge Archaeological Journal* 18(3): 289–307.

Lashley, K. S. 1923a. The behavioristic interpretation of consciousness. *Psychological Review* 30, Part I: 237–272; Part II: 329–353.

Lashley, K. S. 1923b. Temporal variation in the function of the gyrus precentralis in primates. *American Journal of Physiology* 65: 585–602.

Lashley, K. S. 1924. The theory that synaptic resistance is reduced by the passage of the nerve impulse. *Psychological Review* 31: 369–375.

Lashley, K. S. 1930. Brain mechanisms and intelligence. *Psychological Review* 37: 1–24.

Lashley, K. S. 1932. *Studies in the dynamics of behavior*. University of Chicago Press, Chicago.

Lashley, K. S. 1935. The mechanism of vision, Part 12: Nervous structures concerned in the acquisition and retention of habits based on reactions to light. *Comparative Psychology Monographs* 11: 43–79.

Lashley, K. S. 1943. Studies of cerebral function in learning: loss of the maze habit after occipital lesions in blind rats. *Journal of Comparative Neurology* 79(3): 431–462.

Lashley, K. S. 1950. In search of the engram. *Society of Experimental Biology, Symposium* 4: 454–482.

Latour, B. 1993. *We have never been modern*. Harvard University Press, Cambridge, MA.

Latour, B. 2000. When things strike back: a possible contribution of 'science studies' to the social sciences. *British Journal of Sociology* 51: 107–123.

Leach, H. M. 2003. Human domestication reconsidered. *Current Anthropology* 44(3): 349–368.
Leary, M. R. and N. R. Buttermore 2003. The evolution of the human self: tracing the natural history of self-awareness. *Journal for the Theory of Social Behaviour* 33: 365–404.
Lee, S.-H. and M. H. Wolpoff 2003. The pattern of evolution in Pleistocene human brain size. *Paleobiology* 29: 186–196.
Leigh, S. R. 1992. Cranial capacity evolution in *Homo erectus* and early *Homo sapiens*. *American Journal of Physical Anthropology* 87: 1–14.
Leiner, H. C., A. L. Leiner and R. S. Dow 1995. Human brain map. In J. D. Schmahmann (ed.), *The cerebellum and cognition*, p. 244. Academic Press, San Diego, CA.
Leonard, W. R. 2002. Food for thought: dietary change was a driving force in human evolution. *Scientific American* 287(6): 106–115.
Leonard, W. R. and M. L. Robertson 1992. Nutritional requirements and human evolution: a bioenergetics model. *American Journal of Human Biology* 4: 179–195.
Leonard, W. R. and M. L. Robertson 1994. Evolutionary perspectives on human nutrition: the influence of brain and body size on diet and metabolism. *American Journal of Human Biology* 6: 77–88.
Leonard, W. R. and M. L. Robertson 1997. Comparative primate energetics and hominid evolution. *American Journal of Physical Anthropology* 102(2): 265–281.
Leonardi, P. 1988. Art Paléolithique mobilier et pariétalen Italie. *L'Anthropologie* 92: 139–202.
Lewis-Williams, J. D. 1993. Southern African archaeology in the 1990s. *South African Archaeological Bulletin* 48: 45–50.
Lewis-Williams, J. D. and T. A. Dowson 1988. The signs of all times: entoptic phenomena in Upper Palaeolithic art. *Current Anthropology* 29: 201–245.
Li, D., D. A. Collier and L. He 2006. Meta-analysis shows strong positive association of the neuregulin 1 (NRG1) gene with schizophrenia. *Human Molecular Genetics* 15: 1995–2002.
Lieberman, P. 1991. *Uniquely human: the evolution of speech, thought and selfless behavior*. Harvard University Press, Cambridge, MA.
Lieberman, P. 2007. The evolution of human speech: its anatomical and neural bases. *Current Anthropology* 48(1): 39–66.
Lindhal, T. and B. Nyberg 1972. Rate depurination of native deoxyribonucleic acid. *Biochemistry* 11(19): 3610–3618.
Lisiecki, L. E. 2005. *Ages of MIS boundaries*. Boston University, Boston.
Liu, Q., Y. Zhou, P. L. Morrell and B. S. Gaut 2017. Deleterious variants in Asian rice and the potential cost of domestication. *Molecular Biology and Evolution* 34(4): 908–924.
Loganovsky, K. N. and T. K. Loganovskaja 2000. Schizophrenia spectrum disorders in persons exposed to ionizing radiation as a result of the Chernobyl accident. *Schizophrenia Bulletin* 26: 751–753.
Longley, A. J. 2001. Depression is an adaptation. *Archives of General Psychiatry* 58: 1085–1086.
Lourandos, H. 1997. *Continent of hunter-gatherers: new perspectives in Australian prehistory*. Cambridge University Press, Cambridge.
Lovejoy, C. O. 1981. The origin of man. *Science* 211(4480): 341–350.
Lu, J., T Tang., H. Tang, J. Huang, S. Shi and C Wu. 2006. The accumulation of deleterious mutations in rice genomes: a hypothesis on the cost of domestication. *Trends in Genetics* 22(3): 126–131.

Lubinski, P. M., K. Terry and P. T. McCutcheon 2014. Comparative methods for distinguishing flakes from geofacts: a case study from the Wenas Creek Mammoth site. *Journal of Archaeological Science* 52: 308–320.

Luedtke, B. E. 1986. An experiment in natural fracture. *Lithic Technology* 15: 55–60.

Lumsden, C. E. 1970. The neuropathology of multiple sclerosis. In P. J. Vinken and G. W. Bruyn (eds), *Handbook of clinical neurology*, vol. 9, pp. 217–309. North-Holland, Amsterdam.

Lutz, H., T. Engel, B. Lischewsky and A. von Berg 2017. A new great ape with startling resemblances to African members of the hominin tribe, excavated from the Mid-Vallesian Dinotheriensande of Eppelsheim. First report (Hominoidea, Miocene, MN 9, Proto-Rhine River, Germany). *Mainzer Naturwissenschaftliches Archiv* 54.

MacAndrew, F. T. 2002. New evolutionary perspective on altruism: multilevel-selection and costly-signaling theories. *Current Directions in Psychological Science* 11: 79–82.

Maddison, D. R. 1991. African origin of human MtDNA re-examined. *Systematic Zoology* 40: 355.

Maddison, D. R., M. Ruvolo and D. L. Swofford 1992. Geographic origins of human mitochondrial DNA: phylogenetic evidence from control region sequences. *Systematic Biology* 41: 111–124.

Maguire, B. 1980. Further observations on the nature and provenance of the lithic artefacts from the Makapansgat Limeworks. *Palaeontologia Africana* 23: 127–151.

Makino, T., C. J. Rubin, M. Carneiro, E. Axelsson, L. Andersson and M. T. Webster 2018. Elevated proportions of deleterious genetic variation in domestic animals and plants. *Genome Biology and Evolution* 10(1): 276–290.

Malafouris, L. 2004. The cognitive basis of material engagement: where brain, body and culture conflate. In E. DeMarrais, C. Gosden and C. Renfrew (eds), *Rethinking materiality: the engagement of mind with the material world*, pp. 53–62. McDonald Institute for Archaeological Research, Cambridge.

Malez, M. 1959. Das Paläolithikum der Veternicahöhle und der Bärenkult. *Quartär* 11: 171–188.

Mania, D. and U. Mania 1988. Deliberate engravings on bone artefacts of *Homo erectus*. *Rock Art Research* 5(2): 91–107.

Maravita, A., C. Spence and J. Driver 2003. Multisensory integration and the body schema: close to hand and within reach. *Current Biology* 13: R531–R539.

Marchetti, M. P. and G. A. Nevitt 2003. Effects of hatchery rearing on brain structures of rainbow trout, *Oncorhynchus mykiss*. *Environmental Biology of Fishes* 66(1): 9–14.

Maringer, J. and T. Verhoeven 1972. Steingeräte aus dem Waiklau-Trockenbett bei Maumere auf Flores, Indonesien. Eine Patjitanian-artige Industrie auf der Insel Flores. *Anthropos* 67: 129–137.

Maringer, J. and T. Verhoeven 1975. Die Oberflächenfunde von Marokoak auf Flores, Indonesien. Ein weiterer altpaläolithischer Fundkomplex von Flores. *Anthropos* 70: 97–104.

Markó, A. 2015. Istállóskö revisited: lithic artefacts and assemblages, sixty years after. *Acta Archaeologica Academiae Scientiarum Hungaricae* 66: 5–38.

Marsden, C. D., D. Ortega-Del Vecchyo, D. P. O'Brien, J. F. Taylor, O. Ramirez, C. Vilà, T. Marques-Bonet, R. D. Schnabel, R. K. Wayne and K. E. Lohmueller 2016. Bottlenecks and selective sweeps during domestication have increased deleterious genetic variation in dogs. *Proceedings of the National Academy of Sciences of the U.S.A.* 113(1): 152–157.

Marshack, A. 1985. Theoretical concepts that lead to new analytic methods, modes of enquiry and classes of data. *Rock Art Research* 2: 95–111.

Marshack, A. 1989. Methodology in the analysis and interpretation of Upper Palaeolithic image: theory versus contextual analysis. *Rock Art Research* 6: 17–38.

Marshack, A. 1991. The female image: a 'time-factored' symbol. (A study in style and aspects of image use in the Upper Palaeolithic). *Proceedings of the Prehistoric Society* 57 (1): 17–31.

Marshall, J. C. 1989. Reply to P. Lieberman, J. T. Laitman, J. S. Reidenberg, K. Landahl and P. J. Gannon, 'Folk physiology and talking hyoids'. *Nature* 342: 486–487.

Marshall-Pescini, S., S. Cafazzo, Z. Virányi and F. Range 2017. Integrating social ecology in explanations of wolf–dog behavioral differences. *Current Opinion in Behavioral Sciences* 16: 80–86.

Martin, R. D., A. M.MacLarnon, J. L.Phillips and W. B. Dobyns 2006. Flores hominid: new species or microcephalic dwarf? *The Anatomical Record Part A: Discoveries in Molecular, Cellular, and Evolutionary Biology* 288(11): 1123–1145.

Martini, F. 1992. Early human settlement in Sardinia: the Palaeolithic industries. In R. Tyko and T. Andrews (eds), *Sardinia in the Mediterranean: a footprint in the sea*, pp. 40–48. Sheffield Academic Press, Sheffield.

Marvanová, M., J. Ménager, E. Bezard, R.E. Bontrop, L. Pradier and G. Wong 2003. Microarray analysis of nonhuman primates: validation of experimental models in neurological disorders. *The FASEB Journal* 17: 929–931.

Mathalon, D. H. and J. M. Ford 2008. Divergent approaches converge on frontal lobe dysfunction in schizophrenia. *American Journal of Psychiatry* 154(8): 944–948.

Maturana, H. and F. Varela 1980. Autopoiesis and cognition: the realization of the living. In R. S. Cohen and M. W. Wartofsky (eds), *Boston Studies in the Philosophy of Science*, vol. 42, Kluwer Academic Publishers Group, Dordrecht.

Mayr, E. 2001. *What evolution is*. Orion Books Ltd, London.

McBrearty, S. and A. S. Brooks 2000. The revolution that wasn't: a new interpretation of the origin of modern human behaviour. *Journal of Human Evolution* 39: 453–563.

McDougall, I., F. H. Brown and J. G. Fleagle 2005. Stratigraphic placement and age of modern humans from Kibish, Ethiopia. *Nature* 433: 733–736.

McGann, J. 1991. *The textual condition*. Princeton University Press, Princeton, NJ.

McGrath, J. J., S. Saha, J. Welham, O. El Saadi, C. MacCauley and D. Chant 2004. A systematic review of the incidence of schizophrenia: the distribution of rates and the influence of sex, urbanicity, migrant status and methodology. *BMC Medicine* 2(1): 13.

Mealey, L. 1995. The sociobiology of sociopathy: an integrated evolutionary model. *Behavioral and Brain Sciences* 18: 523–599.

Mega, M. S. and J. Cummings 1997. The cingulate and cingulate syndromes. In M. R. Trimble and J. L. Cummings (eds), *Contemporary behavioural neurology*, pp. 189–214. Butterworths, Oxford.

Mekel-Bobrov, N., S. L. Gilbert, P. D. Evans, E. J. Vallender, J. R. Anderson, S. A. Tishkoff and B. T. Lahn 2005. Ongoing adaptive evolution of ASPM, a brain size determinant in *Homo sapiens*. *Science* 309(5741): 1720–1722.

Mellars, P., K. C. Gori, M. Carr, P. A. Soares and M. B. Richards 2013. Genetic and archaeological perspectives on the initial modern human colonization of southern Asia. *Proceedings of the National Academy of Sciences of the U.S.A.* 110: 10699–10704.

Mellars, P. and C. Stringer 1989. Introduction. In P. Mellars and C. Stringer (eds), *The human revolution: behavioural and biological perspectives on the origins of modern humans*, pp. 1–14. Edinburgh University Press, Edinburgh.

Menary, R. 2007. *Cognitive integration: mind and cognition unbounded*. Palgrave Macmillan, Hampshire.
Mesoudi, A., A. Whiten and K. Laland 2006. Towards a unified science of cultural evolution. *Behavioral and Brain Sciences* 29: 329–383.
Meulemans, D. and M. Bronner-Fraser 2004. Gene-regulatory interactions in neural crest evolution and development. *Developmental Cell* 7(3): 291–299.
Miklósi, Á. 2007. *Dog behaviour, evolution, and cognition*. Oxford University Press, New York
Miller, G. F. 2000. *The mating mind: how sexual choice shaped the evolution of human nature*. Doubleday, New York.
Mishra, S. 1982. On the effects of basalt weathering on the distribution of Lower Palaeoliths sites in the Deccan. *Bulletin of the Deccan College Post-Graduate and Research Institute* 41: 107–151.
Mishra, S. 1992. The age of the Acheulian in India: new evidence. *Current Anthropology* 33: 325–328.
Mishra, S. 1994. The south Asian Lower Palaeolithic. *Man and Environment* 19(1/2): 57–71.
Mishra, S. and S. N. Rajaguru 1994. Comment on 'Toba ash on the Indian sub-continent and its implications for the correlation of Late Pleistocene alluvium' by S. K. Acharyya and P. K. Basu. *Quaternary Research* 41: 396–397.
Misra, V. N. 1989. Stone Age India: an ecological perspective. *Man and Environment* 14: 17–64.
Misra, V. N. 1995. Geoarchaeology of the Thar Desert, northwest India. In S. Wadia, R. Korisettar and V. S. Kale (eds), *Quaternary environments and geoarchaeology of India*, pp. 210–224. Geological Society of India, Bangalore.
Misra, V. N., S. N. Rajaguru and H. Raghvan 1988. Late Middle Pleistocene environment and Acheulian culture around Didwana, Rajasthan. *Proceedings of the Indian National Science Academy* 54A(3): 425–438.
Misra, V. N., S. N. Rajaguru, D. R. Raju and H. Raghvan 1982. Acheulian occupation and evolving landscape around Didwana, in the Thar Desert. *Man and Environment* 7: 112–131.
Mitchell, R. W. 1993. Mental models of mirror-self-recognition: two theories. *New Ideas in Psychology* 11: 295–325.
Mitchell, R. W. 1997. Kinesthetic-visual matching and the self-concept as explanations of mirror-self-recognition. *Journal for the Theory of Social Behaviour* 27: 18–39.
Mitchell, R. W. 2002. Subjectivity and self-recognition in animals. In M. R. Leary and J. P. Tangney (eds), *Handbook of self and identity*, pp. 567–595. Guilford Press, New York.
Mol, D. and F. Lacombat 2009. *Mammuthus trogontherii* (Pohlig, 1885), the steppe mammoth of Nolhac. *Quaternaire* 20(4): 569–574.
Moncel, M. H. 2003. Tata (Hongrie). Un assemblage microlithique du début du Pléistocène supérieur en Europe Centrale. *L'Anthropologie* 107: 117–151.
Moog, F. 1939. Paläolithische Freilandstation im Älteren Löß von Wyhlen (Amt Lörrach). *Badische Fundberichte* 15: 36–52.
Moreno-Estrada A., J. C. Fernandez-Lopez, F. Zakharia, M. Sikora, A. V. Contreras, V. Acuna-Alonzo, K. Sandoval, C. Eng, S. Romero-Hidalgo and P. Ortiz-Tello 2014. The genetics of Mexico recapitulates Native American substructure and affects biomedical traits. *Science* 344(6189): 1280–1285.
Morey, D. F. 1994. The early evolution of the domestic dog. *American Scientist* 82(4): 336–347.
Morikawa, M. and S. Hashimoto 1994. *Torihama kaizuka*. Yomiuri Shinbin-sha, Tokyo.

Morin, A. 2002. Right hemispheric self-awareness: a critical assessment. *Consciousness and Cognition* 11: 396–401.

Morin, A. 2003. Let's face it. *Evolutionary Psychology* 1: 177–187.

Morin, A. 2004. A neurocognitive and socioecological model of self-awareness. *Genetic, Social and General Psychological Monographs* 130: 197–222.

Morin, E. and V. Laroulandie 2012. Presumed symbolic use of diurnal raptors by Neanderthals. *PLoS One* 7(3): e32856.

Morin, A. and J. Michaud 2007. Self-awareness and the left inferior frontal gyrus: inner speech use during in self-related processing. *Brain Research Bulletin* 74: 387–396.

Morris, A. A. M. and R. N. Lightowlers 2000. Can paternal mtDNA be inherited? *The Lancet* 355: 1290–1291.

Morris, D. 1967. *The naked ape: a zoologist's study of the human animal*. Jonathan Cape, London.

Mortensen, P. 2008. Lower to Middle Palaeolithic artefacts from Loutró on the south coast of Crete. *Antiquity* 82(317): 1–6.

Morwood, M. J., P. B. O'Sullivan, F. Aziz and A. Raza 1998. Fission-track ages of stone tools and fossils on the east Indonesian island of Flores. *Nature* 392: 173–179.

Morwood, M. J., R. P. Soejono, R. G. Roberts, T. Suktina, C. S. M. Turnkey, K. E. Westaway et al. 2004. Archaeology and age of a new hominin from Flores in eastern Indonesia. *Nature* 431: 1087–1091.

Mottl, M. 1951. Die Repolust-Höhle bei Peggau (Steiermark) und ihre eiszeitlichen Bewohner. *Archaeologica Austriaca* 8: 1–78.

Mottron, L. and S. Belleville 1993. A study of perceptual analysis in a high-level autistic subject with exceptional graphic abilities. *Brain and Cognition* 23: 279–309.

Mottron, L. and S. Belleville 1995. Perspective production in a savant autistic draughtsman. *Psychological Medicine* 25: 639–648.

Mottron, L., S. Belleville and E. Ménard 1999. Local bias in autistic subjects as evidenced by graphic tasks: perceptual hierarchization or working memory deficit. *Journal of Child Psychology and Psychiatry* 40: 743–756.

Mueller, R.-A. and E. Courchesne 1998. The cerebellum: so much more. *Science* 181 (5390): 879–880.

Mueller, U. G., N. M. Gerardo, D. K. Aanen, D. L. Six and T. R. Schultz 2005. The evolution of agriculture in insects. *Annual Review of Ecology, Evolution, and Systematics* 36: 563–595.

Munkacsi, A. B., J. J. Pan, P. Villesen, U. G. Mueller, M. Blackwell and D. J. McLaughlin 2004. Convergent coevolution in the domestication of coral mushrooms by fungus-growing ants. *Proceedings of the Royal Society B, Biological Sciences* 271(1550): 1777–1782.

Napierala, H. and H.-P. Uerpmann, 2010. A 'new' Palaeolithic dog from central Europe. *International Journal of Osteoarchaeology*. doi:10.1002/oa.1182.

Nash, D. T. 1993. Distinguishing stone artifacts from naturefacts created by rockfall processes. In P. Goldberg, D. T. Nash and M. D. Petraglia (eds), *Formation processes in archaeological context*, pp. 125–138. Prehistory Press, Madison, WI.

Natsoulas, T. 1998. Consciousness and self-awareness. In M. D. Ferrari and R. J. Sternberg (eds), *Self-awareness: its nature and development*, pp. 12–33. The Guilford Press, New York.

Nelson, E. C., J. T. Manning and A. G. M. Sinclair 2006. Using the length of the 2nd and 4th digit ratio (2D:4D) to sex cave art hand stencils: factors to consider. *Before Farming* 2000(1): 1–7.

Nelson, E., C. Rolian, L. Cashmore and S. Shultz 2011. Digit ratios predict polygyny in early apes, Ardipithecus, Neanderthals and early modern humans but not in Australopithecus. *Proceedings of the Royal Society of London B, Biological Sciences* 278(1711): 1556–1563.
Nelson, E. W. 1899. The Eskimo about Bering Strait. In J. W. Powell (ed.), *Eighteenth Annual Report of the Bureau of American Ethnology*. Government Printing Office, Washington, DC.
Nelson, K. 2005. Emerging levels of consciousness in early human development. In H. S. Terrace and J. Metcalfe (eds), *The missing link in cognition: origins of self-reflective consciousness*, pp. 116–141. Oxford University Press, Oxford.
Nestler, H. 1961. Spongien aus der weißen Schreibkreide (Unt. Maastricht) der Insel Rügen (Ostsee). *Paläontologische Abhandlungen* 1(1): 1–70.
Nettle, D. 2001. *Strong imagination: madness, creativity and human nature*. Oxford University Press, Oxford.
Nettle, D. 2004. Evolutionary origins of depression: a review and reformulation. *Journal of Affective Disorders* 81: 91–102.
Nettle, D. 2006. Reconciling the mutation-selection balance model with the schizotypy-creativity connection. *Behavioral and Brain Sciences* 29: 418.
Neubauer, S., P. Gunz and J. J. Hublin 2010. Endocranial shape changes during growth in chimpanzees and humans: a morphometric analysis of unique and shared aspects. *Journal of Human Evolution* 59: 555–566.
Niego, A. and A. Benítez-Burraco 2019. Williams syndrome, human self-domestication, and language evolution. *Frontiers in Psychology*. doi:10.3389/fpsyg.2019.00521.
Nimchimsky, E. A., E. Gilissen, J. M. Allman, D. P. Perl, J. M. Erwin and P. Hof 1999. A neuronal morphologic type unique to humans and great apes. *Proceedings of the National Academy of Sciences of the U.S.A.* 96: 5268–5273.
Nimchinsky, E. A., B. A. Vogt, J. H. Morrison and P. R. Hof 1995. Spindle neurons of the human anterior cingulate cortex. *Journal of Comparative Neurology* 355: 27–37.
Noble, W. and I. Davidson 1996. *Human evolution, language and mind: a psychological and archaeological inquiry*. Cambridge University Press, Cambridge.
Oakley, K. P. 1981. Emergence of higher thought, 3.0–0.2 Ma B.P. *Philosophical Transactions of the Royal Society of London B* 292: 205–211.
O'Connell, J. F., K. Hawkes and N. G. B. Jones 1999. Grandmothering and the evolution of *Homo erectus*. *Journal of Human Evolution* 36: 461–485.
O'Connor, N. and B. Hermelin, 1987. Visual and graphic abilities of the idiot savant artist. *Psychological Medicine* 17, 79–90.
O'Connor, N. and B. Hermelin, 1990. The recognition failure and graphic success of idiot savant artists. *Journal of Child Psychology and Psychiatry* 31, 203–215.
O'Connor, T. P. 1997. Working at relationships: another look at animal domestication. *Antiquity* 71(271): 149–156.
Odling-Smee, F. J., K. N. Laland and M. W. Feldman 2003. *Niche construction: the neglected process in evolution*. Princeton University Press, Princeton, NJ.
Ogden, C. K. and I. A. Richards 1923. *The meaning of meaning: A study of the influence of language upon thought and of the science of symbolism*. Harcourt, Brace & World, New York.
Olson, M. V. and A. Varki 2003. Sequencing the chimpanzee genome: insights into human evolution and disease. *Nature Reviews Genetics* 4: 20–28.
Ongur, D., W. C. Drevets and J. L. Price 1998. Glial reduction in the subgenual prefrontal cortex in mood disorders. *Proceedings of the National Academy of Science of the U.S.A.* 95: 13290–13295.

Oppenheimer, S. 2012. A single southern exit of modern humans from Africa: before or after Toba? *Quaternary International* 258: 88–99.

Osadschuk, L. V. 1997. Effects of domestication on the adrenal cortisol production of silver foxes during embryonic development. In L. N. Trut and L. V. Osadschuk (eds), *Evolutionary-genetic and genetic-physiological aspects of fur animal domestication*, pp. 336–344. Scientifur, Oslo.

Oskina, I. N. 1997. Analysis of the functional state of the pituitaryadrenal axis during postnatal development of domesticated silver foxes (*Vulpes vulpes*). In L. N. Trut and L. V. Osadschuk (eds), *Evolutionary-genetic and genetic-physiological aspects of fur animal domestication*, pp. 55–63. Scientifur, Oslo.

Otake, M. 1996. Threshold for radiation-related severe mental retardation in prenatally exposed A-bomb survivors: a reanalysis. *International Journal of Radiation Biology* 70: 755–763.

Oyama, S. 2000. *Evolution's eye: a systems view of the biology-culture divide*. Duke University Press, Durham, NC.

Oyama, S., P. E. Griffiths and R. D. Gray 2001. *Cycles of contingency: developmental systems and evolution*. MIT Press, Cambridge, MA.

Pacher, M. and A. J. Stuart 2009. Extinction chronology and palaeobiology of the cave bear (*Ursus spelaeus*). *Boreas* 38(2): 189–206.

Paddayya, K. 1991. The Acheulian culture of the Hunsgi and Baichbal valleys, peninsular India: a processual study. *Quartär* 41/42: 111–138.

Paddaya, K., B. A. B. Blackwell, R. Jhaldiyal, M. D. Petraglia, S. Ferrier, D. A. Chaderton, J. I. B. Blickstein and A. R. Skinner 2002. Recent findings on the Acheulian of Hunsgi and Baichbal valleys, Karnataka, with special reference to Isampur excavation and its dating. *Current Science* 83(5): 641–647.

Palma di Cesnola, A. 1976. Le leptolithique archaïque en Italie. In B. Klíma (ed.), *Périgordien et Gravettien en Europe*, pp. 66–99. Congrès IX, Colloque XV, UISPP, Nice.

Palma di Cesnola, A. 1989. L'Uluzzien: faciès italien du Leptolithique archaïque. *L'Anthropologie* 93: 783–811.

Panagopolou, E., E. Kotjaboulou and P. Karkanas 2001. Geoarchaeological research in Alonnisos: new evidence for the Palaeolithic and Mesolithic in the Aegean region. In A. Sampson (ed.), *Archaeological investigations in the northern Sporades*, pp. 121–151. Community of Alonnisos, Alonnisos.

Pappu, S., Y. Gunnell, K. Akhilesh, R. Braucher, M. Taieb, F. Demory and N. Thouveny 2011. Early Pleistocene presence of Acheulian hominins in south India. *Science* 331(6024): 1596–1599.

Patnaik, R., P. R. Chauhan, M. R. Rao, B. A. B. Blackwell, A. R. Skinner, A. Sahni, M. S. Chauhan and H. S. Khan 2009. New geochronological, paleoclimatological, and archaeological data from the Narmada Valley hominin locality, central India. *Journal of Human Evolution* 56(2): 114–133.

Patterson, L. W. 1983. Criteria for determining the attributes of man-made lithics. *Journal of Field Archaeology* 10: 297–307.

Pavlov, P., J. I. Svendsen and S. Indrelid 2001. Human presence in the European Arctic nearly 40,000 years ago. *Nature* 413: 64–67.

Payen, L. 1982. Artifacts or geofacts at Calico: application of the Barnes test. In J. Ericson, R. Taylor and R. Berger (eds), *Peopling of the New World*, pp. 193–201. Ballena Press, Los Altos, CA.

Peacock, E. 1991. Distinguishing between artifacts and geofacts: a test case from eastern England. *Journal of Field Archaeology* 18: 345–361.

Pearlson, G. D. and B. S. Folley 2008. Schizophrenia, psychiatric genetics, and Darwinian psychiatry: an evolutionary framework. *Schizophrenia Bulletin* 34(4): 722–733.

Peled, A., A. Pressman, A. B. Geva and I. Modai 2003. Somatosensory evoked potentials during a rubber-hand illusion in schizophrenia. *Schizophrenia Research* 64(2–3): 157–163.

Penfield, W. 1952. Memory mechanisms. *AMA Archives of Neurology and Psychiatry* 67: 178–198.

Penfield, W. 1954. The permanent records of the stream of consciousness. *Acta Physiologica* 11: 47–69.

Penfield, W. 1958. *The excitable cortex in conscious man*. The Sherrington Lectures V. Liverpool University Press, Liverpool.

Pennisi, E. 1999. Genetic study shakes up Out of Africa Theory. *Science* 283: 1828.

Perpère, M. 1971. L'aurignacien en Poitou-Charentes (étude des collections d'industries lithiques). PhD thesis, University of Paris.

Perpère, M. 1973. Les grands gisements aurignaciens du Poitou. *L'Anthropologie* 77: 683–716.

Pettitt, P. and P. Bahn 2003. Current problems in dating Palaeolithic cave art: Candamo and Chauvet. *Antiquity* 77: 134–141.

Pettitt, P. B., P. Bahn and C. Züchner 2009. The Chauvet conundrum: are claims for the 'birthplace of art' premature? In P. Bahn (ed.), *An enquiring mind: studies in honor of Alexander Marshack*, pp. 239–262. Oxbow Books, Oxford, and American School of Prehistoric Research Monograph Series, Cambridge, MA.

Petraglia, M. D. 1998. The Lower Palaeolithic of India and its bearing on the Asian record. In M. D. Petraglia and R. Korisettar (eds), *Early human behaviour in global context: the rise and diversity of the Lower Palaeolithic record*, pp. 343–390. Routledge, London.

Petraglia, M. D., M. Haslam, D. Q. Fuller, N. Boivin and C. Clarkson 2010. Out of Africa: new hypotheses and evidence for the dispersal of *Homo sapiens* along the Indian Ocean rim. *Annals of Human Biology* 37: 288–311.

Petrie, F. 1899. Sequences in prehistoric remains. *The Journal of the Anthropological Institute of Great Britain and Ireland* 29(3/4): 295–301.

Peyrégne, S., M. J. Boyle, M. Dannemann and K. Prüfer 2017. Detecting ancient positive selection in humans using extended lineage sorting. *Genome Research* 27(9): 1563–1572.

Peyrony, D. 1934. La Ferrassie. *Préhistoire* 3: 1–92.

Pickard, C., B. Pickard and C. Bonsall, 2011. Autistic spectrum disorder in prehistory. *Cambridge Archaeological Journal* 21(3), 357–364.

Pike, A. W. G., D. L. Hoffmann, M. García-Diez, P. B. Pettitt, J. Alcolea, R. de Balbin et al. 2012. U-series dating of Paleolithic art in 11 caves in Spain. *Science* 336: 1409–1413.

Plagnes, V., C. Causse, M. Fontugne, H. Valladas, J.-M. Chazine and L.-H. Fage 2003. Cross dating (Th/U-14C) of calcite covering prehistoric paintings in Borneo. *Quaternary Research* 60(2): 172–179.

Plogmann, D. and D. Kruska 1990. Volumetric comparison of auditory structures in the brains of European wild boars (*Sus scrofa*) and domestic pigs (*Sus scrofa f. dom.*). *Brain, Behavior and Evolution* 35: 146–155.

Plotkin, H. 2002. *The imagined world made real: towards a natural science of culture*. Penguin Books, London.

Polimeni, J. 2006. Mental disorders are not a homogeneous construct. *Behavioral and Brain Sciences* 29: 418–419.

Praslov, N. D. and A. N. Rogachev (eds) 1982. *Palaeolithic of the Kostenki-Borschevo area on the River Don, 1879–1979. Results of the field investigations*. Nauka, Leningrad.

Press, J. S. and J. M. Tanur 2001. *The subjectivity of scientists and the Bayesian approach*. John Wiley & Sons, New York.

Prestwich, J. 1859. On the occurrence of flint-implements, associated with the remains of extinct mammalia, on undisturbed beds of a late geological period. *Proceedings of the Royal Society of London* 10: 50–59.

Preti, A. and P. Miotto 2006. Mental disorders, evolution, and inclusive fitness. *Behavioral and Brain Sciences* 29: 419–420.

Preuss, T. M. 2001. The discovery of cerebral diversity: an unwelcome scientific revolution. In D. Falk and K. R. Gibson (eds.), *Evolutionary anatomy of the primate cerebral cortex*, pp. 138–164. Cambridge University Press, Cambridge.

Preuss, T. M. and J. H. Kaas 1999. Human brain evolution. In F. E. Bloom, S. C. Landes, J. L. Robert, L. R. Squire and M. J. Zigmond (eds), *Fundamental neuroscience*, pp. 1283–1311. Academic Press, San Diego, CA.

Price, E. O. 2002. *Animal domestication and behavior*. CABI Publishing, New York.

Pring, L. and B. Hermelin 1993. Bottle, tulip and wineglass: semantic and structural picture processing by savant artists. *Journal of Child Psychology and Psychiatry* 34: 1365–1385.

Prinz, J. 2006. Putting the brakes on enactive perception. *Psyche* 12: 1–19.

Protsch, R. 1975. The absolute dating of Upper Pleistocene sub-Saharan fossil hominids and their place in human evolution. *Journal of Human Evolution* 4: 297–322.

Protsch, R. and H. Glowatzki 1974. Das absolute Alter des paläolithischen Skeletts aus der Mittleren Klause bei Neuessing, Kreis Kelheim, Bayern. *Anthropologischer Anzeiger* 34: 140–144.

Protsch, R. and A. Semmel 1978. Zur Chronologie des Kelsterbach-Hominiden. *Eiszeitalter und Gegenwart* 28: 200–210.

Protsch von Zieten, R. R. R. 1973. The dating of Upper-Pleistocene Sub-Saharan fossil hominids and their place in human evolution: with morphological and archaeological implications. PhD thesis, University of California, Los Angeles.

Prüfer, K., F. Racimo, N. Patterson, F. Jay, S. Sankararaman et al. 2014. The complete genome sequence of a Neanderthal from the Altai Mountains. *Nature* 505 (7481): 43–49.

Qian F., Q. Li, P Wu., S. Yuan, R. Xing, H. Chen and H. Zhang 1991. *Lower Pleistocene, Yuanmou formation: Quaternary geology and paleoanthropology of Yuanmou, Yunnan, China*. Science Press, Beijing.

Rabeder, G. 1985. Die Grabungen des Oberösterreichischen Landesmuseums in der Ramesch-Knochenhöhle (Totes Gebirge, Warschenegg Gruppe). *Jahrbuch des OÖ. Musealvereines Gesellschaft für Landeskunde* 130: 161–181.

Rabeder, G. and N. Kavcik 2013. 'Drachenhöhle' Mixnitz. In G. Rabeder and H. Kavcik (eds), *Abstracts & excursions*, pp. 27–28. 19th International Cave Bear Symposium, Semriach (Styria, Austria).

Racimo, F. 2016. Testing for ancient selection using cross-population allele frequency differentiation. *Genetics* 202(2): 733–750.

Rada-Iglesias, A., R. Bajpai, S. Prescott, S. A. Brugmann, T. Swigut and J. Wysocka 2012. Epigenomic annotation of enhancers predicts transcriptional regulators of human neural crest. *Cell Stem Cell* 11(5): 633–648.

Raghvan, H., S. N. Rajaguru and V. N. Misra 1989. Radiometric dating of a Quaternary dune section, Didwana, Rajasthan. *Man and Environment* 13: 19–22.

Rajaguru, S. N. 1985. The problem of Acheulian chronology in western and southern India. In V. N. Misra and P. Bellwood (eds), *Recent advances in Indo-Pacific prehistory*, pp. 13–18. Oxford-IBH, New Delhi.

Rajkowska, G. 2009. Reductions in neuronal and glial density characterize the dorsolateral prefrontal cortex in bipolar disorder. *Biological Psychiatry* 49(9): 741–752.

Ramachandran, V. S. 2009. Mirror neurons and imitation learning as the driving force behind 'the great leap forward' in human evolution. *Edge*. Available at: www.edge.org/3rd_culture/ramachandran/ramachandran_index.html (accessed 20 March 2019).

Randall, P. L. 1998. Schizophrenia as a consequence of brain evolution. *Schizophrenia Research* 30: 143–148.

Rathelot, R.-A. and P. L. Strick 2009. Subdivisions of primary motor cortex based on cortico-motoneuronal cells. *Proceedings of the National Academy of Sciences of the U.S.A.* 106(3): 918–923.

Raynal, J.-P. and R. Séguy 1986. Os incisé acheuléen de Sainte-Anne 1 (Polignac, Haute-Loire). *Revue archéologique du Centre de la France* 25: 79–80.

Reich, D., R. E. Green, M. Kircher, J. Krause, N. Patterson, E. Y. Durand et al. 2010. Genetic history of an archaic hominin group from Denisova Cave in Siberia. *Nature* 468: 1053–1060.

Reich, D., N. Patterson, M. Kircher, F. Delfin, M. R. Nandineni, I. Pugach et al. 2011. Denisova admixture and the first modern human dispersals into southeast Asia and Oceania. *American Journal of Human Genetics* 89: 516–528.

Relethford, J. H. 2001. *Genetics and the search for modern human origins*. Wiley-Liss, New York.

Relethford, J. H. 2012. *Human population genetics*. Wiley-Blackwell, Hoboken, NJ.

Relethford, J. H. and L. B. Jorde 1999. Genetic evidence for larger African population size during recent human evolution. *Journal of Physical Anthropology* 108(3): 251–260.

Rendell, H. M., R. W. Dennell and M. A. Halim 1989. *Pleistocene and Palaeolithic investigations in the Soan valley, northern Pakistan*. British Archaeological Mission to Pakistan, Series 2. International Series 544, British Archaeological Reports, Oxford.

Reyes-Centeno, H., S. Ghirotto, F. Détroit, D. Grimaud-Hervé, G. Barbujani and K. Harvati 2014. Genomic and cranial phenotype data support multiple modern human dispersals from Africa and a southern route into Asia. *Proceedings of the National Academy of Sciences of the U.S.A.* 111: 7248–7253.

Richerson, P. J. and R. Boyd 2005. *Not by genes alone: how culture transformed human evolution*. University of Chicago Press, Chicago.

Ridley, M. 1996. *Evolution*, 2nd edn. Blackwell-Wiley, Hoboken, NJ.

Ridley, M. 2000. *Mendel's demon: gene justice and the complexity of life*. Orion Books, London.

Riel-Salvatore, J. and G. A. Clark 2001. Grave markers. Middle and Early Upper Paleolithic burials and the use of chronotypology in contemporary Paleolithic research. *Current Anthropology* 42: 449–479.

Rigaud, S. 2006–2007. Révision critique des *Porosphaera globularis* interprétées comme éléments de parure acheuléens. Unpublished MA thesis, Université Bordeaux 1, France.

Rigaud, S., F. d'Errico, M. Vanhaeren and C. Neumann 2009. Critical reassessment of putative Acheulean [sic] *Prosphaere globularis* beads. *Journal of Archaeological Science* 36: 25–34.

Rightmire, G. P. 2004. Brain size and encephalization in early to mid-Pleistocene *Homo*. *American Journal of Physical Anthropology* 124: 109–123.

Rigollot , M.-J. 1854. *Mémoire sur les instruments en silex trouvés à Saint-Acheul, près Amiens*. Rapports géologiques et archéologiques, Amiens.

Riley, B. and K. S. Kendler 2006. Molecular genetic studies of schizophrenia. *European Journal of Human Genetics* 14: 669–680.

Rindos, D. 1984. *The origins of agriculture: An evolutionary perspective*. Academic Press, Orlando, FL.

Rizzolatti, G., L. Fadiga, V. Gallese and L. Fogassi 1996. Premotor cortex and the recognition of motor actions. *Cognitive Brain Research* 3: 131–141.

Rodman, K. E. 2003. *Asperger's syndrome and adults… Is anyone listening?* Jessica Kingsley Publishers, London.

Rodriguez-Trelles, F., R. Tarrio and F. J. Ayala 2001. Erratic overdispersion of three molecular clocks: GPDH, SOD, and XDH. *Proceedings of the National Academy of Sciences of the U.S.A.* 98: 11405–11410.

Rodriguez-Trelles, F., R. Tarrio and F. J. Ayala 2002. A methodological bias toward overestimation of molecular evolutionary time scales. *Proceedings of the National Academy of Sciences of the U.S.A.* 99: 8112–8115.

Rodríguez-Vidal, J., F. d'Errico, F. G. Pacheco, R. Blasco, J. Rosell, R. P. Jennings et al. 2014. A rock engraving made by Neanderthals in Gibraltar: reconstruction of human evolution: bringing together genetic, archaeological, and linguistic data. *Proceedings of the National Academy of Sciences of the U.S.A.* 111(37): 13301–13306.

Rogachev, A. N., N. D. Praslov, M. V. Anikovich, V. I. Belyaeva and T. N. Dmitrieva 1982. Kostenki 1. In N. D. Praslov and A. N. Rogachev (eds), *Paleolit Kostenkovsko-Borshevskogo rajona na Donu 1879–1979*, pp. 42–66. Nauka, Leningrad.

Rogers, A. R., R. J. Bohlender and C. D. Huff 2017. Early history of Neanderthals and Denisovans. *Proceedings of the National Academy of Sciences of the U.S.A.* 114(37): 9859–9863.

Ronen, A. 1991. The Yiron-gravel lithic assemblage; artefacts older than 2.4 Ma in Israel. *Archäologisches Korrespondenzblatt* 21: 159–164.

Rose, M. R. 2016. Darwinian evolution of free will and spiritual experience. In J. Carroll, D. P. McAdams and E. O. Wilson (eds), *Darwin's bridge: uniting the humanities and sciences*, pp. 69–85. Oxford University Press, New York.

Rougier, H., Ş. Milota, R. Rodrigo, M. Gherase, L. Sarcină et al. 2007. Peştera cu Oase 2 and the cranial morphology of early modern Europeans. *Proceedings of the National Academy of Sciences of the U.S.A.* 104(4): 1165–1170.

Rowland, M. 1999. *The body in mind: understanding cognitive processes*. Cambridge University Press, Cambridge.

Rowland, M. 2003. Comment on R. G. Bednarik, 'Seafaring in the Pleistocene'. *Cambridge Archaeological Journal* 13(1): 52–54.

Rowley-Conwy, P. 2007. *From genesis to prehistory: the archaeological three age system and its contested reception in Denmark, Britain, and Ireland*. Oxford University Press, Oxford.

Rubinsztein, D. C., W. Amos, J. Leggo, S. Goodburn, R. S. Ramesar, J. Old, R. Dontrop, R. McMahon, D. E. Barton and M. A. Ferguson-Smith 1994. Mutational bias provides a model for the evolution of Huntington's disease and predicts a general increase in disease prevalence. *Nature Genetics* 7(7): 525–530.

Ruff, C. B., E. Trinkaus and T.W. Holliday 1997. Body mass and encephalization in Pleistocene *Homo*. *Nature* 387: 173–176.

Rupert, R. 2004. Challenges to the hypothesis of extended cognition. *The Journal of Philosophy* 101: 1–40.

Sackett, J. R. 1981. From de Mortillet to Bordes: a century of French Palaeolithic research. In G. Daniel (ed.), *Towards a history of archaeology*, pp. 85–99. Thames and Hudson, London.

Sackett, J. R. 1988. The Mousterian and its aftermath: a view from the Upper Paleolithic. In H. L. Dibble and A. Montet-White (eds), *The Upper Pleistocene prehistory of western Eurasia*, pp. 413–426. University of Pennsylvania Press, Philadelphia, PA.

Sadier, B., J.-J. Delannoy, L. Benedetti, D. L. Bourlès, S. Jaillet, J.-M. Geneste, A.-E. Lebatard and M. Arnold 2012. Further constraints on the Chauvet Cave artwork elaboration. *Proceedings of the National Academy of Sciences of the U.S.A.* 109(21): 8002–8006.

Sahasrabudhe, Y. S. and S. N. Rajaguru 1990. The laterites of the Maharashtra State. *Bulletin of the Deccan College Postgraduate and Research Institute* 49: 257–270.

Sailor, M., B. Fischl, D. Salat, C. Tempelmann, E. Busa, N. Bodammer et al. 2003. Focal cortical thinning of the cerebral cortex in multiple sclerosis. *Brain* 126: 1734–1744.

Säll, T. and B. O. Bengtsson 2017. *Understanding population genetics*. Wiley, Oxford.

Sampson, A. 2006. *The prehistory of the Aegean basin: Palaeolithic-Mesolithic-Neolithic*. Atrapos, Athens.

Sánchez-Villagra, M. R., M. Geiger and R. A. Schneider 2016. The taming of the neural crest: a developmental perspective on the origins of morphological covariation in domesticated mammals. *Open Science* 3(6): 160107.

Sanes, D. 2012. *Development of the nervous system*, 3rd edn. Elsevier, Oxford.

Saniotis, A. and M. Henneberg 2013. Evolutionary medicine and future of humanity: will evolution have the final word? *Humanities* 2(2): 278–291.

Sanjuan, J., A. Tolosa, J. C. Gonzalez, E. J. Aguilar, J. Perez-Tur, C. Najera, M. D. Molto and R. Frutos 2006. Association between FOXP2 polymorphisms and schizophrenia with auditory hallucinations. *Psychiatric Genetics* 16: 67–72.

Sankalia, H. D. 1974. *Prehistory and protohistory of India and Pakistan*. Deccan College, Pune.

Sankaranarayanan, K. 2001. Estimation of the hereditary risks of exposure to ionizing radiation: history, current status, and emerging perspectives. *Health Physics* 80: 363–369.

Sankararaman, S., S. Mallick, M. Dannemann, K. Prüfer, J. Kelso, S. Pääbo, N. Patterson and D. Reich 2014. The genomic landscape of Neanderthal ancestry in present-day humans. *Nature* 507: 354–357.

Sankararaman, S., N. Patterson, H. Li, S. Pääbo and D. Reich 2012. The date of interbreeding between Neandertals and modern humans. *PloS Genetics* 8(10): e1002947.

Sankhyan, A. R. 1997. Fossil clavicle of a Middle Pleistocene hominid from the central Narmada valley, India. *Journal of Human Evolution* 32: 3–16.

Sankhyan, A. R. 1999. The place of the Narmada hominid in the jigsaw puzzle of human origins. *Gondwana Geological Magazine* 4: 335–345.

Sapolsky, R. M. 1992. Neuroendocrinology of the stress response. In J. B. Becker, S. M. Breedlove and D. Crews (eds), *Behavioral endocrinology*, pp. 287–324. MIT Press, Cambridge, MA.

Schekman, R. 2013. How journals like *Nature*, *Cell* and *Science* are damaging science. *The Guardian*, 9 December. Available at: www.theguardian.com/commentisfree/2013/dec/09/how-journals-nature-science-cell-damage-science

Schildkraut, J. J. 1965. The catecholamine hypothesis of affective disorders: a review of supporting evidence. *American Journal of Psychiatry* 122: 509–528.

Schmid, E., J. Hahn and U. Wolf 1989. Die altsteinzeitliche Elfenbeinstatuette aus der Höhle Stadel im Hohlenstein bei Asselfingen, Alb-Donau-Kreis. *Fundberichte aus Baden-Württemberg* 14: 33–118.

Schuldberg, D. 1988. Creativity and schizotypal traits: creativity test scores and perceptual aberration, magical ideation, and impulsive nonconformity. *Journal of Nervous and Mental Disease* 176: 648–657.

Schuldberg, D. 2000. Six subclinical spectrum traits in normal creativity. *Creativity Research Journal* 13: 5–16.

Schultz, W. 1969. Zur Kenntnis des Hallstromhundes (Canis hallstromi, Troughton 1957). *Zoologischer Anzeiger* 183: 42–72.

Schulz, H.-P. 2002. The lithic industry from layers IV–V, Susiluola Cave, western Finland, dated to the Eemian interglacial. *Préhistoire Européenne* 16–17: 7–23.

Schulz, H.-P., B. Eriksson, H. Hirvas, P. Huhta, H. Jungner, P. Purhonen, P. Ukkonen and T. Rankama 2002. Excavations at Susiluola Cave. *Suomen Museo* 2002: 5–45.

Schulz, M. 2004. Die Regeln mache ich. *Der Spiegel* 34(18 August): 128–131.

Schütz, K. E., B. Forkman and P. Jensen 2001. Domestication effects on foraging strategy, social behaviour and different fear responses: a comparison between the red junglefowl (*Gallus gallus*) and a modern layer strain. *Applied Animal Behaviour Science* 74: 1–14.

Schwalbe, G. 1901. Der Neanderthalschädel. *Bonner Jahrbücher* 206: 1–72.

Schwalbe, G. and E. Fischer (eds) 1923. *Anthropologie*. B.G. Teubner, Leipzig.

Schwartz, J. M. and S. Begley 2002. *The mind and the brain: neuroplasticity and the power of mental force*. HarperCollins, New York.

Schwartz, M. and J. Vissing 2002. Paternal inheritance of mitochondrial DNA. *New England Journal of Medicine* 347: 576–580.

Scott, N., S. Neubauer, J. J. Hublin and P. Gunz 2014. A shared pattern of postnatal endocranial development in extant hominoids. *Evolutionary Biology* 41: 572–594.

Searle, J. R. 1995. *The construction of social reality*. Allen Lane, London.

Seeley, W. W., J. M. Allman, D. A. Carlin, R. K. Crawford, M. N. Macedo et al. 2007. Divergent social functioning in behavioral variant frontotemporal dementia and Alzheimer disease: reciprocal networks and neuronal evolution. *Alzheimer Disease and Associated Disorders* 21(94): S50–S57.

Seeley, W. W., D. A. Carlin and J. M. Allman 2006. Early frontotemporal dementia targets neurons unique to apes and humans. *Annals of Neurology* 60(6): 660–667.

Seferiadis, M. 1983. Un centre industriel préhistorique dans les Cyclades: les ateliers de débitage di silex de Stelida (Naxos). In G. Rougemont and M. Bousac (eds), *Les Cyclades: matériaux pour une étude de géographie historique*, pp. 67–73. CNRS, Paris.

Selfe, L. 1977. *Nadia: a case of extraordinary ability in an autistic child*. Harcourt Brace Jovanovich, New York.

Selfe, L. 1983. *Normal and anomalous representational drawing ability in children*. Academic Press, London.

Semendeferi, K. 1997. The evolution of the frontal lobes: a volumetric analysis based on three-dimensional reconstructions of magnetic resonance scans of human and ape brains. *Journal of Human Evolution* 32: 375–388.

Semendeferi, K. 2001a. Advances in the study of hominoid brain evolution: magnetic resonance imaging (MRI) and a 3-D reconstruction. In D. Falk and K. R. Gibson (eds), *Evolutionary anatomy of the primate cerebral cortex*, pp. 257–289. Cambridge University Press, New York.

Semendeferi, K. 2001b. Before or after the split: hominid neural specializations. In A. Nowell (ed.), *In the mind's eye: multidisciplinary approaches to the evolution of human cognition*, pp. 107–120. International Monographs in Prehistory, Ann Arbor, MI.

Semendeferi, K. and H. Damasio 2000. The brain and its main anatomical subdivisions in living hominoids using magnetic resonance imaging. *Journal of Human Evolution* 38: 317–332.

Semenov, S. A. 1964. *Prehistoric technology. An experimental study of the oldest tools and artefacts from traces of manufacture and wear* (trans. M. W. Thompson). Cory, Adams and Mackay, London.

Semon, R. 1904. *Die Mneme*. W. Engelmann, Leipzig.

Semon, R. 1921. *The mneme*. George Allen & Unwin, London.

Sestieri, C., M. Corbetta, G. L. Romani and G. L. Shulman 2011. Episodic memory retrieval, parietal cortex, and the default mode network: functional and topographic analyses. *Journal of Neuroscience* 31: 4407–4420.

Seyfarth, R. M. and D. L. Cheney 2000. Social awareness in monkeys. *American Zoologist* 40: 902–909.

Shackelford, T. K. and R. J. Larsen 1997. Facial asymmetry as an indicator of psychological, emotional, and physiological distress. *Journal of Personality and Social Psychology* 72(1): 456–466.

Sharma, H. C. and S. K. Roy 1985. On the discovery of a pebble-tool industry in the Garo Hills, Meghalaya. In V. N. Misra and P. Bellwood (eds), *Recent advances in Indo-Pacific prehistory*, pp. 89–91. Oxford and IBH Publishing, New Delhi.

Shea, B. T. 1989. Heterochrony in human evolution: the case for neoteny reconsidered. *American Journal of Physical Anthropology* 32(S10): 69–101.

Shea, J. J. 2010. Stone Age visiting cards revisited: a strategic perspective on the lithic technology of early hominin dispersal. In J. G. Fleagle, J. J. Shea, F. E. Grine, A. L. Baden and R. E. Leakey (eds), *Out of Africa I: the first hominin colonization of Eurasia*, pp. 47–64. Springer, New York.

Shea, J. J. 2011. The archaeology of an illusion: the Middle-Upper Paleolithic transition in the Levant. In J. M. Le Tenesor, R. Jagher and M. Otte (eds), *The Lower and Middle Paleolithic in the Middle East and neighbouring regions*, pp. 169–182. ERAUL, Liège.

Sherwood, C. C., A. D. Gordon, J. S. Allen, K. A. Phillips, J. M. Erwin, P. R. Hof and W. D. Hopkins 2011. Aging of the cerebral cortex differs between humans and chimpanzees. *Proceedings of the National Academy of Sciences of the U.S.A.* doi:10.1073/pnas.1016709108.

Shipman, M. D. 1988. *The limitations of social research*. Longman, London.

Siegal, M. and R. Varley 2002. Neural systems involved in 'theory of mind'. *Nature Reviews Neuroscience* 3(6): 463–471.

Silberman, N. A. 1982. *Digging for God and country*. Knopf, New York.

Silberman, N. A. 1989. *Between past and present: archaeology, ideology and nationalism in the modern Middle East*. Henry Holt, New York.

Silberman, N. A. 1995. Promised lands and chosen peoples: the politics and poetics of archaeological narrative. In P. L. Kohl and C. Fawcett (eds), *Nationalism, politics, and the practice of archaeology*, pp. 249–262. Cambridge University Press, Cambridge.

Silk, J. B. 2007. Social component of fitness in primate groups. *Science* 317: 1347–1351.

Simmons, A. H. 2014. *Stone Age sailors: Paleolithic seafaring in the Mediterranean*. Left Coast Press Inc., Walnut Creek, CA

Simões-Costa, M. and M. E. Bronner 2015. Establishing neural crest identity: a gene regulatory recipe. *Development* 142(2): 242–257.

Simonton, D. K. 2004. Psychology's status as a scientific discipline: its empirical placement within an implicit hierarchy of the sciences. *Review of General Psychology* 8: 59–67.

Singh, D. 1993. Body shape and women's attractiveness: the critical role of waist-to-hip ratio. *Human Nature* 4: 297–321.

Smith, D. A. 2002. Imaging the progression of Alzheimer pathology through the brain. *Proceedings of the National Academy of Sciences of the U.S.A.* 99(7): 4135–4137.

Smith, F. H. 1982. Upper Pleistocene hominid evolution in south-central Europe: a review of the evidence and analysis of trends. *Current Anthropology* 23: 667–686.

Smith, F. H. 1985. Continuity and change in the origin of modern *Homo sapiens*. *Zeitschrift für Morphologie und Anthropologie* 75: 197–222.

Smith, F. H. 1992. The role of continuity in modern human origins. In G. Bräuer and F. H. Smith (eds), *Continuity or replacement? Controversies in Homo sapiens evolution*, pp. 145–156. Balkema, Rotterdam.

Smith, F. H., I. Janković and I. Karavanić 2005. The assimilation model, modern human origins in Europe, and the extinction of Neandertals. *Quaternary International* 137: 7–19.

Smith, F. H. and G. Ranyard 1980. Evolution of the supraorbital region in Upper Pleistocene fossil hominids from south-central Europe. *American Journal of Physical Anthropology* 53: 589–610.

Smith, F. H., E. Trinkaus, P. B. Pettitt, I. Karavanić and M. Paunović 1999. Direct radiocarbon dates for Vindija G1 and Velika Pećina Late Pleistocene hominid remains. *Proceedings of the National Academy of Sciences of the U.S.A.* 96(22): 12281–12286.

Smith, W. G. 1894. *Man, the primeval savage*. Edward Stanford, London.

Soffer, O., J. M. Adovasio and D. C. Hyland 2000. The 'Venus' figurines. *Current Anthropology* 41(4): 511–536.

Soficaru, A., A. Doboş and E. Trinkaus 2006. Early modern humans from the Peştera Muierii, Baia de Fier, Romania. *Proceedings of the National Academy of Sciences of the U.S.A.* 103(46): 17196–171201.

Solomon, S. 1990. What is this thing taphonomy? In S. Solomon, I. Davidson and D. Watson (eds), *Problem solving in taphonomy: archaeological and palaeontological studies from Europe, Africa and Oceania*, pp. 25–33. Tempus 2, University of Queensland, St Lucia.

Somel, M., H. Franz, Y. Zheng, A. Lorenc, G. Song, T. Giger, J. Kelso, B. Nickel, M. Dannemann, S. Bahn, M. J. Webster, C. S. Weickert, M. Lachmann, S. Pääbo and P. Khaitovich 2009. Transcriptional neoteny in the human brain. *Proceedings of the National Academy of Sciences of the U.S.A.* 106(14): 5743–5748.

Sonakia, A. 1984. The skull cap of early man and associated mammalian fauna from Narmada valley alluvium, Hoshangabad area, M.P. (India). *Records of the Geological Survey of India* 113: 159–172.

Sonakia, A. 1997. The Narmada *Homo erectus* — its morphology and analogues. In *Quaternary geology and the Narmada valley*, pp. 123–125. Geological Survey of India, Kolkata.

Sondaar, P. Y., G. D. van den Bergh, B. Mubroto, F. Aziz, J. de Vos and U. L. Batu 1994. Middle Pleistocene faunal turnover and colonization of Flores (Indonesia) by *Homo erectus*. *Comptes Rendus de l'Académie des Sciences Paris* 319: 1255–1262.

Sperber, D. 1996. *Explaining culture: a naturalistic approach*. Blackwell, Malden, MA.

Spikins, P. 2009. Autism, the integrations of 'difference' and the origins of modern human behaviour. *Cambridge Archaeological Journal* 19(2): 179–201.

Spinks, R., H. K. Sandhu, N. C. Andreasen and R. A. Philibert 2004. Association of the HOPA12bp allele with a large X-chromosome haplotype and positive symptom schizophrenia. *American Journal of Medical Genetics, Part B: Neuropsychiatric Genetics* 127: 20–27.

Spriggs, M. 2003. Comment on R. G. Bednarik, 'Seafaring in the Pleistocene'. *Cambridge Archaeological Journal* 13(1): 54–55.
Squire, L. R. and E. Kandel 1999. *Memory: from mind to molecules*. Scientific American Library, New York.
Squire, L. R. and S. Zola-Morgan 1991. The medial temporal lobe memory system. *Science* 253(5026): 1380–1386.
Sridharan, D., D. J. Levitin and V. Menon 2008. A critical role for the right fronto-insular cortex in switching between central-executive and default-mode networks. *Proceedings of the National Academy of Sciences of the U.S.A.* 105(34): 12569–12574.
Srinivasan, S., F. Bettella, S. Hassani, Y. Wang, A. Witoelar, A. J. Schork *et al.* 2017. Probing the association between early evolutionary markers and schizophrenia. *PLoS One* 12: e0169227.
Srinivasan, S., F. Bettella, M. Mattingsdal, Y. Wang, A. Witoelar, A. J. Schork *et al.* 2016. Genetic markers of human evolution are enriched in schizophrenia. *Biological Psychiatry* 80: 284–292.
Steguweit, L. 1999. Intentionelle Schnittmarken auf Tierknochen von Bilzingsleben — Neue lasermikroskopische Untersuchungen. *Praehistoria Thuringica* 3: 64–79.
Steinmetz, J. E., D. G. Lavond, D. Ivkovich, C. G. Logan and R. F. Thompson 1992. Disruption of classical eyelid conditioning after cerebellar lesions: damage to a memory trace system or a simple performance deficit? *Journal of Neuroscience* 12: 4403–4426.
Steinmetz, J. E., C. G. Logan, D. J. Rosen, J. K. Thompson, D. G. Lavond and R. F. Thompson 1987. Initial localization of the acoustic conditioned stimulus projection system to the cerebellum essential for classical eyelid conditioning. *Proceedings of the National Academy of the Sciences of the U.S.A.* 84: 3531–3535.
Steinmetz, J. E. and R. F. Thompson 1991. Brain substrates of aversive classical conditioning. In J. Madden (ed.), *Neurobiology of learning, emotion and affect*, pp. 97–120. Raven Press, New York.
Stepanchuk, V. N. 1993. Prolom II, a Middle Palaeolithic cave site in the eastern Crimea with non-utilitarian bone artefacts. *Proceedings of the Prehistoric Society* 59: 17–37.
Stepniewska, I., T. M. Preuss and J. H. Kaas 1993. Architectonics, somatotopic organization and ipsilateral cortical connections of the primary motor area (M1) of owl monkeys. *Journal of Comparative Neurology* 330: 238–271.
Sterling, T. D. 1959. Publication decisions and their possible effects on inferences drawn from tests of significance – or vice versa. *Journal of American Statistical Association* 54 (285): 30–34.
Stern, D. N. 1985. *The interpersonal world of the infant*. Basic Books, New York.
Stiner, M. C. 1994. *Honor among thieves. A zooarchaeological study of Neandertal ecology*. Princeton University Press, Princeton, NJ.
Strasser, T. 2012. The Damnoni excavation: Ákoue. *Newsletter of the American School of Classical Studies at Athens* 66: 11.
Strasser, T. F., E. Panagopoulou, C. N. Runnels, P. M. Murray, N. Thompson, P. Karkanas, F. W. McCoy and K. W. Wegmann 2010. Stone Age seafaring in the Mediterranean. Evidence from the Plakias region for Lower Paleolithic and Mesolithic habitation of Crete. *Hesperia* 79: 145–190.
Strasser, T. F., C. Runnels, K. Wegmann, E. Panagopoulou, F. W. McCoy, C. Digregorio, P. Karkanas and N. Thompson 2011. Dating Palaeolithic sites in southwestern Crete, Greece. *Journal of Quaternary Science* 26: 553–560.
Stringer, C. 2011. *The origin of our species*. Penguin, London.

Stringer, C. 2014. Why we are not all multiregionalists now. *Trends in Ecology and Evolution* 29(5): 248–251.
Stringer, C. B. and P. Andrews 1988. Genetic and fossil evidence for the origin of modern humans. *Science* 239: 1263–1268.
Stuss, D. T., T. W. Picton and M. P. Alexander 2001. Consciousness, self-awareness and the frontal lobes. In S. Salloway, P. Malloy and J. Duffy (eds), *The frontal lobes and neuropsychiatric illness*, pp. 101–109. American Psychiatric Press, Washington, DC.
Sugita, Y. 2009. Innate face processing. *Current Opinion in Neurobiology* 19(1): 39–44.
Sutikna, T., M. W. Tocheri, M. J. Morwood, E. W., J. Saptomo, Jatmiko, R. D. Awe et al. 2016. Revised stratigraphy and chronology for *Homo floresiensis* at Liang Bua in Indonesia. *Nature* 532: 366–369.
Sutton, J. 2008. Material agency, skills, and history: distributed cognition and the archaeology of memory. In L. Malafouris and C. Knappett (eds), *Material agency: towards a non-anthropocentric approach*, pp. 37–55. Springer, Berlin.
Sutton, J. 2009. Remembering. In P. Robbins and M. Aydede (eds), *The Cambridge handbook of situated cognition*, pp. 217–235. Cambridge University Press, Cambridge.
Suzuki, K., H. Yamada, T. Kobayashi and K. Okanoya 2012. Decreased fecal corticosterone levels due to domestication: a comparison between the white-backed munia (*Lonchura striata*) and its domesticated strain, the Bengalese finch (*Lonchura striata var. domestica*) with a suggestion for complex song evolution. *Journal of Experimental Zoology* 317: 561–570.
Svoboda, J. 1990. The Bohunician. In J. K. Kozłowski (ed.), *La mutation*, pp. 169–192. ERAUL, Liège.
Svoboda, J. 1993. The complex origin of the Upper Paleolithic in the Czech and Slovak Republics. In H. Knecht, A. Pike-Tay and R. White (eds), *Before Lascaux: the complete record of the early Upper Paleolithic*, pp. 23–36. CRC Press, Boca Raton, FL.
Swartz, K. B. 1997. What is mirror self-recognition in nonhuman primates, and what is it not? In J. G. Snodgrass and R. L. Thompson (eds), *The self across psychology: self-recognition, self-awareness, and the self-concept*, pp. 65–71. New York Academy of Sciences, New York.
Swisher, C. C., G. H. Curtis, T. Jacob, A. G. Getty, A. Suprijo and P. Widiasmoro 1994. The age of the earliest hominids in Indonesia. *Science* 263: 1118–1121.
Szabo, B. J., C. McKinney, T. S. Dalbey and K. Paddayya 1990. On the age of the Acheulian culture of the Hunsgi-Baichbal valleys, peninsular India. *Bulletin of the Deccan College Postgraduate and Research Institute* 50: 317–321.
Tang H., A. Jin and R. G. Bednarik 2016. The earliest known logboats of China. *The International Journal of Nautical Archaeology* 45(2): 441–466.
Tassabehji, M., K. Metcalfe, A. Karmiloff-Smith, M. J. Carette, J. Grant, N. Dennis et al. 1999. Williams syndrome: use of chromosomal microdeletions as a tool to dissect cognitive and physical phenotypes. *American Journal of Human Genetics* 64(1): 118–125.
Tattersall, I. 1995. *The fossil trail: how we know what we think we know about human evolution*. Oxford University Press, Oxford.
Taubert, J., S. G. Wardle, M. Flessert, D. A. Leopold and L. G. Ungerleider 2017. Face pareidolia in the rhesus monkey. *Current Biology* 27(16): 2505–2509.
Templeton, A. R. 1992. Human origins and analysis of mitochondrial DNA sequences. *Science* 255: 737.
Templeton, A. R. 1993. The 'Eve' hypothesis: a genetic critique and re-analysis. *American Anthropologist* 95: 51–72.
Templeton, A. R. 1996. Gene lineages and human evolution. *Science* 272: 1363.

Templeton, A. R. 2002. Out of Africa again and again. *Nature* 416: 45–51.
Templeton, A. R. 2005. Haplotype trees and modern human origins. *Yearbook of Physical Anthropology* 48: 33–59.
Terberger, T. 1998. Endmesolithische Funde von Drigge, Lkr. Rügen — Kannibalen auf Rügen? *Jahrbuch für Bodendenkmalpflege Mecklenburg-Vorpommern* 46: 7–44.
Terberger, T. and M. Street 2003. Jungpaläolithische Menschenreste im westlichen Mitteleuropa und ihr Kontext. In J. M. Burdukiewicz, L. Fiedler, W.-D. Heinrich, A. Justus and E. Brühl (eds), *Erkenntnisjäger: Kultur und Umwelt des frühen Menschen*, pp. 579–591, Landesmuseum für Vorgeschichte, Halle.
Thakkar, K. N., H. S. Nichols, L. G. McIntosh and S. Park 2011. Disturbances in body ownership in schizophrenia: Evidence from the rubber hand illusion and case study of a spontaneous out-of-body experience. *PLoS ONE* 6(10), e27089; doi:10.1371/journal.pone.0027089.
Theofanopoulou, C., S. Gastaldon, T. O'Rourke, B. D. Samuels, A. Messner, P. T. Martins *et al.* 2017. Self-domestication in *Homo sapiens*: insights from comparative genomics. *PLoS ONE* 12(10): e0185306.
Theveneau, E. 2012. Neural crest delamination and migration: from epithelium-to-mesenchyme transition to collective cell migration. *Developmental Biology* 366(1): 34–54.
Thiessen, D. D. 1997. *Bittersweet destiny: the stormy evolution of human behavior*. Transaction Publishers, Piscataway, NJ.
Thioux, M., D. E. Stark, C. Klaiman and R. T. Schultz 2006. The day of the week when you were born in 700 ms: calendar computation in an autistic savant. *Journal of Experimental Psychology: Human Perception and Performance* 32(5): 1155–1168.
Thomas, J. G. 2013. Self-domestication and language evolution. PhD thesis, Linguistics and English Language School of Philosophy, Psychology and Language Sciences, University of Edinburgh.
Thompson, J. R. 2012. The allegory of the hammer and the nail gun and other unstable orthodoxies of 'modernity': possible pitfalls of 'behavioural modernity'. *AURA Newsletter* 29: 3–12.
Thompson, J. R. 2014. Archaic modernity vs the high priesthood: on the nature of unstable archaeological/palaeoanthropological orthodoxies. *Rock Art Research* 31(2): 131–156.
Thompson, R. F. 1967. *Foundations of physiological psychology*. Harper & Row, New York.
Thompson, R. F. 1986. The neurobiology of learning and memory. *Science* 233: 941–947.
Thompson, R. F. 1990. Neural mechanisms of classical conditioning in mammals. *Philosophical Transactions, Royal Society of London B* 329: 161–170.
Thompson, R. F., T. W. Berger, C. F. Cegavske, M. M. Patterson, R. A. Roemer, T. J. Teyler and R. A. Young 1976. The search for the engram. *American Psychologist* 31: 209–227.
Thorne, A. G. and M. H. Wolpoff 2003. The multiregional evolution of humans. *Scientific American* 13: 46–53.
Tindale, N. 1962. Some population changes among the Kaiadilt of Bentinck Island, Queensland. *Records of the South Australian Museum* 14(2): 297–336.
Tobias, P. V. 1995. The bearing of fossils and mitochondrial DNA on the evolution of modern humans, with a critique of the 'mitochondrial Eve' hypothesis. *South African Archaeological Bulletin* 50: 155–167.
Tooby, J. and L. Cosmides 1992. The psychological foundations of culture. In J. H. Barkow, L. Cosmides and J. Tooby (eds), *The adapted mind: evolutionary psychology and the generation of culture*, pp. 19–36. Oxford University Press, New York.

Trehub, A. 2009. Comment on Ramachandran. *Edge*. Available at: www.edge.org/3rd_culture/rama08/rama08_index.html (accessed 25 February 2019).
Trigger, B. G. 1984. Alternative archaeologies: nationalist, colonialist, imperialist. *Man* 19: 355–370.
Trigger, B. G. 1985. The past as power: anthropology and the North American Indian. In I. McBryde (ed.), *Who owns the past?*, pp. 11–40. Oxford University Press, Oxford.
Trigger, B. G. 1989. *A history of archaeological thought*. Cambridge University Press, Cambridge.
Trinkaus, E. 2005. Early modern humans. *Annual Review of Anthropology* 34: 207–230.
Trinkaus, E. and M. Le May 1982. Occipital bunning among Later Pleistocene hominids. *American Journal of Physical Anthropology* 57: 27–35.
Trinkaus, E., O. Moldovan, Ş. Milota, A. Bîlgar, L. Sarcina, S. Athreya, S. E. Bailey, R. Rodrigo, G. Mircea, T. Higham, C. Bronk Ramsey and J. van der Plicht 2003. An early modern human from the Peştera cu Oase, Romania. *Proceedings of the National Academy of Sciences of the United States of America of the U.S.A.* 100(20): 11231–11236.
Trut, L. N. 1999. Early canid domestication: the farm-fox experiment. *American Scientist* 87: 160–169.
Trut, L., I. Oskina and A. Kharlamova 2009. Animal evolution during domestication: the domesticated fox as a model. *BioEssays* 31(3): 349–360.
Tsuang, M. T., W. S. Stone and S. V. Faraone 2001. Genes, environment and schizophrenia. *The British Journal of Psychiatry* 178: s18–s24.
Turk, I. (ed.) 1997. *Mousterienska koscena piscal in druge najdbe iz Divjih Bab I v Sloveniji* [Mousterian bone flute and other finds from Divje babe I Cave site in Slovenia]. Znanstvenoraziskovalni Center Sazu, Ljubljana, Slovenia.
Turk, I., J. Dirjec and B. Kavur 1995. Ali so v Sloveniji našli najstarejše glasbilo v Evropi? [The oldest musical instrument in Europe discovered in Slovenia?]. *Razprave IV, razreda SAZU* 36: 287–293.
Turk, M. and L. Dimkaroski 2011. Neanderthal flute from Divje babe I: old and new findings. In B. Toškan (ed.), *Fragments of Ice Age environments: Proceedings in honour of Ivan Turk's jubilee*, pp. 251–265. ZRC SAZU, Ljubljana, Slovenia.
Turner, M. A. 1999. Generating novel ideas: fluency performance in high-functioning and learning disabled individuals with autism. *Journal of Child Psychology and Psychiatry and Allied Diseases* 40: 189–201.
Usik, V. I., J. I. Rose, Y. H. Hilbert, P. Van Peer and A. E. Marks 2013. Nubian complex reduction strategies in Dhofar, southern Oman. *Quaternary International* 300: 244–266.
Valoch, K. 1968. Evolution of the Paleolithic in central and eastern Europe. *Current Anthropology* 9(5): 351–390.
Valoch, K. 1987. The early Palaeolithic site Stránská skála I near Brno (Czechoslovakia). *Anthropologie* 25: 125–142.
Van Gerven, D. P., G. J. Armelagos and A. Rohr 1977. Continuity and change in cranial morphology of three Nubian archaeological populations. *Man* 12: 270–277.
van Heekeren, H. R. 1957. *The Stone Age of Indonesia*. Martinus Nijhoff, 's-Gravenhage.
van Os, J. and S. Kapur 2009. Schizophrenia. *The Lancet* 374(9690): 635–645.
Van Peer, P., R. Fullager, S. Stokes, R. M. Bailey, J. Moeyersons, F. Steenhoudt, A. Geerts, T. Vanderbeken, N. De Dapper and F. Geus 2003. The Early to Middle Stone Age transition and the emergence of modern behaviour at site 8-B-11, Sai Island, Sudan. *Journal of Human Evolution* 45(2): 187–193.
Van Sommers, P. 1984. *Drawing and cognition*. Cambridge University Press, Cambridge.

Van Zeist, W. 1957. De mesolitische Boot van Pesse. *Nieuwe Drentse Volksalmanak* 75: 4–11.
Vawter, M. P., W. J. Freed and J. E. Kleinman 2000. Neuropathology of bipolar disorder. *Society of Biological Psychiatry* 48: 486–504.
Vaz, A. P., S. K. Inati, N. Brunl and K. Zaghloul 2019. Coupled ripple oscillations between the medial temporal lobe and neocortex retrieve human memory. *Science* 363(6430): 975–978.
Velly, L., M. F. Rey, N. J. Bruder, F. A. Gouvitsos, T. Witjas, J. M. Regis, J. C. Peragut and F. M. Gouin 2007. Differential dynamic of action on cortical and subcortical structures of anesthetic agents during induction of anesthesia. *Anesthesiology* 107: 202–212.
Verhoeven, T. 1958. Pleistozäne Funde in Flores. *Anthropos* 53: 264–265.
Verhoeven, T. 1968. Vorgeschichtliche Forschungen auf Flores, Timor und Sumba. In *Anthropica: Gedenkschrift zum 100. Geburtstag von P. W. Schmidt*, pp. 393–403. Studia Instituti Anthropos 21, St Augustin.
Vernot, B. and J. Akey 2014. Resurrecting surviving Neandertal lineages from modern human genomes. *Science* 343(6174): 1017–1021.
Vernot, B., S. Tucci, J. Kelso, J. G. Schraiber, A. B. Wolf, R. M. Gittelman et al. 2016. Excavating Neandertal and Denisovan DNA from the genomes of Melanesian individuals. *Science* doi:10.1126/science.aad9416.
Vértes, L. 1959. Die Rolle des Höhlenbären im ungarischen Paläolithikum. *Quartär* 11: 151–170.
Vértes, L. 1964. *Tata: eine mittelpaläolithische Travertin Siedlung in Ungarn.* Akadémiai Kiadó, Budapest.
Vértes, L. 1965. *Az öskökor és az átmeneti kökor emlékei Magyarországon.* Akadémiai Kiadó, Budapest.
Viegas, J. 2015. Ancient human with 10% Neandertal genes found. Available at: http://news.discovery.com/human/genetics/ancient-human-human-with-10-percent-neanderthal-genes-found-150622.htm
Vigilant, L., M. Stoneking, H. Harpending, K. Hawkes and A. C. Wilson 1991. African populations and the evolution of human mitochondrial DNA. *Science* 253: 1503–1507.
Vishnyatsky, L. B. 1994. 'Running ahead of time' in the development of Palaeolithic industries. *Antiquity* 68: 134–140.
Voight, B. F., S. Kudaravalli, X. Wen and J. K. Pritchard 2006. A map of recent positive selection in the human genome. *PLoS Biology* 4(3): e72.
von Koenigswald, G. H. R. and A. K. Ghosh 1973. Stone implements from the Trinil Beds of Sangiran, central Java. *Koninklijk Nederlands Akademie Wetenschappen, Proc. Ser. B* 76(1): 1–34.
Vorstman, J. A. S., E. J. Breetvelt, K. I. Thode, E. W. C. Chow and A. S. Bassett 2013. Expression of autism spectrum and schizophrenia in patients with a 22q11.2 deletion. *Schizophrenia Research* 143(1): 55–59.
Wade, N. 2014. *A troublesome inheritance: genes, race and human history.* Penguin Books, New York.
Wakankar, V. S. 1975. Bhimbetka: the prehistoric paradise. *Prachya Pratibha* 3(2): 7–29.
Walberg, M. W. and D. A. Clayton 1981. Sequence and properties of the human KB cell and mouse L cell D-loop regions of mitochondrial DNA. *Nucleic Acids Research* 9: 5411–5421.
Walker, L. C. and L. C. Cork 1999. The neurobiology of aging in nonhuman primates. In R. D. Terry, R. Katzman, K. L. Bick and S. S. Sisodia (eds), *Alzheimer's disease*, 2nd edn., pp. 233–243. Lippincott Williams and Wilkins, Philadelphia, PA.

Wang S., C. M. Lewis, M. Jakobsson, S. Ramachandran, N. Ray, G. Bedoya et al. 2007. Genetic variation and population structure in Native Americans. *PLoS Genetics* 3(11): 2049–2067.

Wang Z., T. Yonezawa, B. Liu., T. Ma, X. Shen, J. Su et al. 2011. Domestication relaxed selective constraints on the yak mitochondrial genome. *Molecular Biology and Evolution* 28(5): 1553–1556.

Ward, R. H., B. L. Frazier, K. Dew-Jager and S. Pääbo 1991. Extensive mitochondrial diversity within a single Amerindian tribe. *Proceedings of the National Academy of Sciences of the U.S.A.* 88: 8720–8724.

Warner, C. and R. G. Bednarik 1996. Pleistocene knotting. In J. C. Turner and P. van de Griend (eds), *History and science of knots*, pp. 3–18. World Scientific, Singapore.

Waterhouse, L. 1988. Extraordinary visual memory and pattern perception by an autistic boy. In L. K. Obler and D. Fein (eds), *The exceptional brain*, pp. 325–338. Guilford Press, New York.

Watson, E., K. Bauer, R. Aman, G. Weiss, A. von Haeseler and S. Pääbo 1996. MtDNA sequence diversity in Africa. *American Journal of Human Genetics* 59: 437–444.

Watson, K. K., T. K. Kones and J. M. Allman 2006. Dendritic architecture of the von Economo neurons. *Neuroscience* 141: 1107–1112.

Wegner, C. and P. M. Mathews 2003. A new view of the cortex, new insights into multiple sclerosis. *Brain* 126: 1810–1821.

Weidenreich, F. 1946. *Apes, giants, and man*. University of Chicago Press, Chicago.

Weiner, W. S., K. P. Oakley and W. E. Le Gros Clark 1953. The solution of the Piltdown problem. *Bulletin of the British Museum (Natural History) Geology* 2(3): 141–146.

Weintraub, K. 2011. Autism counts. *Nature* 479(7371): 22–24.

Weiss, M. L. and A. E. Mann 1978. *Human biology and behavior: an anthropological perspective*. Little, Brown and Co., Boston.

Wendt, W. E. 1974. 'Art mobilier' aus der Apollo 11-Grotte in Südwest-Afrika. Die ältesten datierten Kunstwerke Afrikas. *Acta Praehistorica et Archaeologia* 5: 1–42.

Wendt, W. E. 1976. 'Art mobilier' from the Apollo 11 Cave, South West Africa: Africa's oldest dated works of art. *South African Archaeological Bulletin* 31: 5–11.

Werry, E. and B. Kazenwadel 1999. Garten Eden in der Sahara. *Bild der Wissenschaft* 4/1999: 18–23.

White, R. 1993. Technological and social dimensions of Aurignacian-age body ornaments across Europe. In H. Knecht, A. Pike-Tay and R. White (eds), *Before Lascaux: the complex record of the early Upper Palaeolithic*, pp. 277–299. CRC Press, Boca Raton, FL.

White, T. D., B. Asfaw, D. DeGusta, H. Gilbert, G. D. Richards, G. Suwa and F. C. Howell 2003. Pleistocene *Homo sapiens* from Middle Awash, Ethiopia. *Nature* 425: 742–747.

Whitley, D. 2009. *Cave paintings and the human spirit*. Prometheus Books, Amherst, NY.

Wickler, S. 2001. *The prehistory of Buka: a stepping stone island in the northern Solomons*. Terra Australia 16, Australian National University, Canberra.

Wickler, S. and M. J. T. Spriggs 1988. Pleistocene human occupation of the Solomon Islands, Melanesia. *Antiquity* 62: 703–706.

Wild, E. M., M. Teschler-Nicola, W. Kutschera, P. Steier, E. Trinkaus and W. Wanek 2005. Direct dating of Early Upper Palaeolithic human remains from Mladeč. *Nature* 435: 332–335.

Wilkins, A. S., R. W. Wrangham and W. T. Fitch 2014. The 'domestication syndrome' in mammals: a unified explanation based on neural crest cell behavior and genetics. *Genetics* 197(3): 795–808.

Williams, R. S. 2002. Another surprise from the mitochondrial genome. *New England Journal of Medicine* 347: 609–611.
Wilson, D. R. 1998. Evolutionary epidemiology and manic depression. *British Journal of Medical Psychiatry* 71: 367–395.
Wilson, D. R. 2006. The evolution of evolutionary epidemiology: a defense of pluralistic epigenetic modes of transmission. *Behavioral and Brain Sciences* 29: 427–429.
Wilson, P. 1988. *The domestication of the human species*. Yale University Press, New Haven, CT.
Wolpoff, M. 1999. *Paleoanthropology*, 2nd edn. McGraw-Hill, New York.
Wolpoff, M. and R. Caspari 1996. *Race and human evolution: a fatal attraction*. Simon & Schuster, New York.
Wolpoff, M., F. H. Smith, M. Malez, J. Radovčić and D. Rukavina 1981. Upper Pleistocene hominid remains from Vindija Cave, Croatia, Yugoslavia. *American Journal of Physical Anthropology* 54: 499–545.
Wood, B. A. and A. Turner 1995. Out of Africa and into Asia. *Nature* 378: 239–240.
Wrangham, R. W. 2009. *Catching fire: how cooking made us human*. Basic Books, New York.
Wrangham, R. and N. Conklin-Brittain 2003. Cooking as a biological trait. *Comparative Biochemistry and Physiology Part A: Molecular & Integrative Physiology* 136(1): 35–46.
Wrangham, R. W., J. H. Jones, G. Laden, D. Pilbeam and N. L. Conklin-Brittain 1999. The raw and the stolen. *Current Anthropology* 40(5): 567–594.
Wright, A. F., B. Charlesworth, I. Rudan, A. Carothers and H. Campbell 2003. A polygenic basis for late-onset disease. *Trends in Genetics* 19: 97–106.
Wyatt, J. 1862. On some further discoveries of flint implements in the gravel near Bedford. *Quarterly Journal of the Geological Society* 18: 113–114.
Xiao H. and J. P. Saint-Jeannet 2004. Induction of the neural crest and the opportunities of life on the edge. *Developmental Biology* 275(1): 1–11.
Xu M., D. St Clair and L. He 2006. Meta-analysis of association between ApoE epsilon4 allele and schizophrenia. *Schizophrenia Research* 84: 228–235.
Yoshikawa, T., M. Kikuchi, K. Saito, A. Watanabe, K. Yamada, H. Shibuya *et al.* 2001. Evidence for association of the myo-inositol monophosphatase 2 (IMPA2) gene with schizophrenia in Japanese samples. *Molecular Psychiatry* 6: 202–210.
Zahavi, A. and A. Zahavi 1997. *The handicap principle: a missing piece of Darwin's puzzle*. Oxford University Press, New York.
Zeder, M. A. 2006. Central questions in the domestication of plants and animals. *Evolutionary Anthropology: Issues, News, and Reviews* 15(3): 105–117.
Zeder, M. A. 2012a. The domestication of animals. *Journal of Anthropological Research* 68 (2): 161–190.
Zeder, M. A. 2012b. Pathways to animal domestication. In P. Gepts, T. R. Famula, R. L. Bettinger *et al.* (eds), *Biodiversity in agriculture: domestication, evolution and sustainability*, pp. 227–259. Cambridge University Press, Cambridge.
Zeder, M. 2014. Domestication: definition and overview. In C. Smith (ed.), *Encyclopedia of global archaeology*, pp. 2184–2194. Springer Science, New York.
Zeder, M. A., E. Emshwiller, B. D. Smith and D. G. Bradley 2006. Documenting domestication: the intersection of genetics and archaeology. *Trends in Genetics* 22(3): 139–155.
Zeuner, F. E. 1950. *Stone Age and Pleistocene chronology of Gujarat*. Deccan College, Pune.
Zhang, J., D. M. Webb and O. Podlaha 2002. Accelerated protein evolution and origins of human-specific features: FOXP2 as an example. *Genetics* 162: 1825–1835.

Zhou, Z.-Y. 2014. New dating of the *Homo erectus* cranium from Lantian (Gongwangling), China. *Journal of Human Evolution* 78: 144–157.

Zhu, R. X., K. A. Hoffman, R. Potts, C. L. Deng, X. Y. Pan, B. Guo, C. D. Shi, Z. T. Guo, B. Y. Yuan, Y. M. Hou and W. W. Hunags 2001. Earliest presence of humans in northeast Asia. *Nature* 413–417.

Zhu, R. X., R. Potts, F. Xie, K. A. Hoffman, C. L. Deng, C. D. Shi, Y. X. Pan, H. Q. Wang, R. P. Shi, Y. C. Wang, G. H. Shi and N. Q. Wu 2004. New evidence on the earliest human presence at high northern latitudes in northeast Asia. *Nature* 431: 559–562.

Ziegert, H. 2007. A new dawn for humanity: Lower Palaeolithic village life in Libya and Ethiopia. *Minerva* 18(4): 8–9.

Zirkle, C. 1941. Natural selection before the 'Origin of species'. *Proceedings of the American Philosophical Society* 84(1): 71–123.

Zischler, H., H. Geisert, A. von Haeseler and S. Pääbo 1995. A nuclear 'fossil' of the mitochondrial D-loop and the origin of modern humans. *Nature* 378: 489–492.

Zotz, L. F. 1951. *Altsteinzeitkunde Mitteleuropas*. F. Enke, Stuttgart.

Zuechner, C. 1996. The Chauvet Cave: radiocarbon versus archaeology. *International Newsletter on Rock Art* 13: 25–27.

Index

This index contains no geographical terms because many hundreds of site names are listed in the text. Names of individuals are limited to a few that are of historical relevance.

abolition of oestrus 111–12, 116, 185
Aborigines 49, 61, 89, 114
academic cliques 184
Acheulian 12–13, 43–4, 46, 48, 54, 70, 100, 131, 149–50, 153–61
aetiology 8–9, 25, 85, 88, 93, 106–7, 124, 128, 142, 157
African Eve 4–5, 8, 19–20, 26–8, 30, 32–7, 56, 59–60, 62–3, 71, 78, 105, 112–13, 116, 122, 127, 146, 159, 162, 182–4
allele 29, 86, 93, 95–7, 106, 112, 12–16
Alpine Palaeolithic 62
Altamira 3, 39, 77,
Alzheimer's 82, 85, 91–2, 94
AMH 32–8, 47, 55–6, 60–1
amygdala 80, 87–8, 90–2, 140
anatomically modern 8, 15, 20, 31–2, 66, 83, 93, 97, 104–6, 112–13, 122–3, 138, 180, 183
anthropocentric 101–2, 104
anthropogenic 43, 142, 151, 155, 158–9
Ardipithecus 15, 68
armchair archaeologists 50
artefacts 2, 9–10, 12, 14, 16, 18, 34, 42, 47, 52, 89, 141, 149, 152, 161, 177, 179–80
Asperger's 89–90
asymmetry 80, 86
attractiveness 108–9, 115, 119, 129, 133
Aurignacian 31–5, 38–42, 105–8, 110, 131–2
Australopithecus 4, 15, 47, 67–8, 82
autism 84, 86–90, 94, 119, 165

auto-domestication 99, 104, 115–16, 119–20, 122, 124, 126, 129, 136, 139, 144, 165, 168, 182, 184
autopoiesis 166–7

basal ganglia 87, 93, 164
bead 34, 42, 44–5, 70, 77–8, 131, 134, 142, 146, 148, 152–9
behaviour 8, 17, 58, 62, 74–6, 79, 81, 89–91, 95, 98, 100, 102–3, 107–8, 113–14, 120, 128, 130, 137, 142, 164, 170, 174, 177, 180–1
biological species 175
bipolar 82, 84, 87–9, 98
body decoration 122, 130, 134, 142–3, 169
bottleneck 125–6
Boucher de Perthes, J. 1–3, 5, 7, 19, 44, 153–4, 177, 179
brain atrophy 67, 82–3, 107, 118, 139, 168–9, 182
brain size 65–7, 70, 82–3, 102–3, 106, 109, 118, 123, 136, 184
brain volume 10, 24, 82–3, 103, 110, 112, 118, 136, 145, 168, 170
Broca's area 80, 164
Brunhes 11, 54

catastrophist 30, 60, 104, 162
cave art 3, 6–8, 38–40, 77, 110, 143,
cave bear 42, 62
cerebellum 80, 88, 92, 104
Châtelperronian 31–2, 34, 40–1
chimpanzee 22, 64–5, 68, 83, 98, 111, 121, 164

Index

cingulate cortex 79, 81, 85, 90, 92
cleidocranial dysplasia 93, 124
cognition 80, 84, 96, 107, 109, 111, 113, 130, 142, 166–7, 177, 180
cognitive evolution 24, 141, 152
cognitive sciences 170
confirmation bias 1
consciousness 25, 60, 78, 129–30, 133, 167, 169, 171–2
constructivism 166
constructs of reality 25, 111, 138, 144, 166–9, 180
cooking hypothesis 120
cranial volume 24, 66–7, 71, 82, 112, 118, 170
Crô-Magnon 31, 60, 138
cultural construct 115, 119, 133
cultural evolution 17, 37–8, 74, 105, 142, 145
culture 6, 17–18, 20, 36–7, 46, 55, 59, 61–2, 72–5, 78, 86, 94, 101, 105, 130, 135, 145, 160, 168–71, 173, 174–5, 179
cupule 39, 43, 47, 150, 160–1
cytogenesis 83

Dart, R.A. 3, 36, 147
Darwin, C. 2, 72–3, 75–6, 99, 101, 104, 108, 117, 123, 163–4
Darwinian 63–4, 72–3, 107, 109, 136
dating 9, 13–14, 16, 38, 40, 42, 46, 52, 68–9, 78, 131, 152, 159, 163
datings 4, 12, 27, 32, 38, 40
dementia 82, 85, 87, 91–2, 94
demyelination 84, 91
Denisovan 16, 23, 29, 109–10, 112, 122, 141, 146, 175, 183
depression 82
dimorphism 61, 69, 76, 107, 112
DNA 27–9, 36, 58, 68, 73–4, 76, 125
docility 102–3, 119, 123
domesticate 99, 101–2, 110, 114, 117–18, 123–5, 129
domestication hypothesis 97, 99, 106–7, 112–16, 121–5, 127, 173, 182
domestication syndrome 102, 104, 117–19, 123, 128–9, 168, 184–5
domesticator 99–102, 117, 123–4
domesticity 101–2, 117
Dubois, E. 3–4, 10
dwarfism 10, 48, 51, 69, 175
dysteleological 17, 28, 38, 74, 96, 105–6, 115, 181

electroencephalographic oscillations 165
emic 18, 143, 175, 177
encephalisation 24, 65–7, 82–3, 85, 94–5, 107, 109, 118, 136–7, 168–70
encephalisation quotient 83
endocranial 66, 82, 118
engram 139–40, 145, 165
epigenetic 73–4
episodic memory 130, 165
etic 18, 145, 179–80
exogram 111, 143–4, 146, 148, 151, 159, 161, 168–70
exogrammic 41, 46, 141, 144–5, 148–50, 161
extended mind studies 141
external storage 139, 170
external world 168, 171

facial symmetry 108–9
Falconer, H. 2, 153–4
falsifiability 20, 173, 180, 184
fecundity 109, 116
figurine 41, 46, 108, 110, 131–5, 137, 142–3
finger flutings 39–40
flute 41–2
foetal 64–5, 94, 111, 137
foetalisation 64, 110–11
FOXP2 85–6, 117
frontal lobe 79–80, 84–5, 104
frontoinsular 81
frontotemporal dementia 85, 92
Fuhlrott, J.C. 2–3

gene-culture co-evolution 74, 101
genetics 16, 22, 28, 34, 59, 71, 73, 75–6, 86, 97–8, 100–1, 119, 122–3, 125–7, 165, 183
genitalia 64, 132–3,
genocide 8, 114
genome 29, 83–4, 95, 115, 122–6, 129, 175
geofact 177–9
globularisation 104, 112
Graciles 5, 24, 31, 34, 59, 64, 86, 105, 107–8, 110, 112, 182–3
gracilisation 27, 32, 60, 62, 64, 104, 107, 112, 131
gracility 20, 56, 58, 60–1, 105, 109, 164
Graecopithecus 9, 15, 68
great apes 78–80, 82, 86, 129, 144

hand-axe 2, 12, 44, 52, 54–5, 69, 149–51, 153
haplotype 27–9, 84

Index

heterozygosity 127
hippocampus 80–1, 85, 88, 91, 140
Hobbit 10, 48, 71, 175–6
Holocene 4, 6, 14, 30, 33, 41, 51, 55, 60, 66, 68, 94, 100–1, 103, 107, 112–13, 118, 136, 168, 173
hominin evolution 16, 57, 65, 75, 78, 111, 142, 182
hominin history 23, 101, 162, 170, 174
Homo antecessor 14, 23, 55, 70, 122
Homo erectus 3, 10, 43, 46, 52, 54, 69, 77–8, 120, 122, 152, 159, 169
Homo floresiensis 10, 48, 71, 127, 175
Homo habilis 69, 122, 162–3
Homo luzonensis 127, 175
Homo sapiens 19, 23, 29, 31, 40, 55–6, 59, 79, 82–3, 94, 107–9, 111–12, 122, 125–6, 162, 169, 175, 183–4
homology 80, 169
homosexuality 24, 111, 113, 185
human condition 8, 24, 26, 77, 98, 105, 185
human domestication 99, 100–2, 117
human modernity 5, 107, 115, 142, 162, 169, 181
humanistic 30, 38, 102, 105, 170
humanities 1, 17, 19, 21, 55, 72, 101, 104–5, 170
Huntington's 85, 92, 94
Huxley, T.H. 2, 21–3, 176
hybridisation 16, 123
hyoid bone 117, 163

Ice Age 1, 6, 9, 38, 68, 177
iconicity 148
imagined world 145, 167, 171
incommensurabilities 15–17, 38, 55, 127
intelligence 63, 67, 82, 84, 89
interbreeding 15, 29, 56
interfertile 122, 126–7
introgression 16, 30, 35, 58–9, 108, 115

Jarawas 114, 162
jaspilite cobble 46, 49, 147

Keller and Miller paradox 90, 94, 169
Kenyanthropus 68, 147

Lamarckian 73, 137
language origins 63, 86, 117, 141, 162–5
Late Pleistocene 8, 11, 24, 31, 60–1, 105, 111, 122, 128, 134, 137, 148, 169
limbic 79–81, 91–2, 103
lithocentric 145, 179

Lower Palaeolithic 10, 12, 17, 34, 43–5, 52, 54–5, 70, 78, 100, 122, 131, 149, 151–2, 155, 159, 161–2, 181
lumpers 15, 56, 70–1
luxury journals 21, 176

maladaptive 86, 95–7, 106
malformed clavicles 93, 124
manic-depressive 84, 87
manuport 46, 48–9, 146, 149, 159
maritime 47–8, 53, 55, 78, 117, 174, 181
mathematics 130, 167
Matuyama 9, 13–14
meme 19–20, 64, 75, 78, 115, 143, 146
memory traces 79, 139–40, 143, 165, 167, 169–71
Mendel, G. 71, 101
Mendelian disorder 24, 85, 92–3, 97, 106
menopause 109, 112, 116, 128
mental disorder 95–7, 106
mental illness 24–5, 87, 89–90, 124
Mesolithic 30, 38, 58, 82, 100, 118
Micoquian 33, 42
microcephalin 93, 124
Middle Palaeolithic 6, 11–12, 33–4, 39–43, 54–5, 59, 61, 89, 114, 137, 147, 149, 151–2, 174, 181
Middle Stone Age 6, 33, 61, 138, 151–2
Miocene 9, 15, 54
mirror neurons 168
mirror test 129
Mode 1 13–14, 16, 46, 122
Mode 2 14, 46, 149, 151
Mode 3 6, 33–4, 45, 55, 59, 113, 149–51, 174, 181
Mode 4 33–4, 36, 113
modern humans 3, 4, 8, 11, 15, 19–20, 22, 24, 26–7, 29, 31–2, 35, 38, 41, 56, 58, 61–2, 64, 66–7, 69–70, 82, 86, 93, 97, 101, 104, 105–7, 109, 111–13, 115, 122–4, 145, 162, 171–2, 180, 182–3, 185
modernity 5, 90, 98, 107, 115, 142, 162, 169, 180–1
Moderns 23, 29, 31–4, 59, 162, 183
Mousterian 33–4, 41–3, 45, 131
multiple sclerosis 85, 91
multiregional 5, 56, 183
multiregionalists 56, 183
mutation 16, 28, 73–4, 76, 92–8, 106, 111, 116–17, 119, 125–6, 136, 141, 181, 185

Index

naked ape 64, 171
nationalism 7, 21
natural selection 24, 63–4, 72–7, 83–5, 88, 93–8, 106, 108–10, 114–16, 118, 128–9, 155, 167–9, 181
Neanderthal 2, 4, 16, 19–20, 22–3, 26, 29–32, 34, 35, 37–41, 55, 58, 60–3, 66–7, 70, 82, 106–7, 110, 112, 117, 122, 124–5, 134, 138, 141, 162–3, 175, 183
Neanderthaloid 29, 31–2, 37–9, 60, 62, 110, 138
Neolithic 30, 70, 82, 100–1, 109–10, 117, 146–7, 174–5
neotenous ape 64–5, 171
neoteny 63–6, 102, 106–11, 113–16, 119, 123, 128, 131, 133, 171
neural crest cell 103, 123
neural structures 140, 163–4, 180
neurodegenerative 24, 84–5, 92, 94, 107, 185
neuropathology 88, 93–4, 106–7, 111, 167–9
neuroplasticity 60, 114, 167
neuroscience 142, 165, 170
niche construction 73, 75, 84

obsessive-compulsive 85, 90, 94, 98
obstetric 65–6, 83, 94, 137
oestrus 24, 102, 106, 111–2, 116, 123, 185
Oldowan 10, 13, 47, 54, 69, 120, 122, 160
Olschewian 31–3, 40, 62
onomatopoeic 148, 162
orbitofrontal 79, 88, 90

palaeoanthropologists 4, 23, 56, 59, 67–8, 70–1, 94, 111, 126, 175
palaeoanthropology 1, 15, 58, 72, 127, 176, 180, 184
palaeoart 6, 8, 18, 20, 34–6, 38, 40–2, 44–5, 47, 56, 62–3, 87, 89, 105, 110, 113–15, 137–9, 145, 162, 164, 169–70, 180–1, 183
palaeontologists 154, 175
Paranthropus 15, 68–9,
pareidolia 147–8
pareidolic 46–7, 147–8, 159
parietal lobe 88, 168
Parkinson's 85, 92, 94
pathology 10, 68, 84, 94, 98, 106, 185
pendant 34–5, 42, 78, 131, 134, 142, 146, 152–3, 159

petroglyph 13, 37, 39, 42, 46, 122, 138, 146, 159–60
phenotype 86, 93, 126
phosphene 149–51
phylogenetic 29, 79, 84, 106–7, 112, 115, 150–1
pictogram 39, 42
pigment 39, 46, 103, 131, 146, 155, 161, 181
Piltdown 3–6, 36, 126, 173
Pitt Rivers Museum 154
Planck, M. 5, 22
planum temporale 80, 86
pleiotropic 103, 118–9, 119, 125
Pleistocene 2–6, 8–11, 13–14, 16–24, 26, 31, 34–8, 45–6, 50–4, 58, 60–1, 66, 68, 71, 78, 82, 89–90, 94, 96, 98, 100, 105, 107, 111–13, 117–18, 120, 122, 128, 134, 136–7, 143, 145–6, 148, 153, 162–3, 167–70, 173, 175–7, 180–5
Pliocene 9, 23, 68, 173
political 6–7, 37, 100, 173, 180
popular science 4, 10, 19, 21–2, 37, 56, 71, 105, 173, 176, 180
Porosphaera globularis 44, 154–5, 158
precocious realism 89
prefrontal 79, 81, 87–8, 90, 130, 164, 168
pre-History 19, 99
Prestwich, J. 2, 44, 153–4
primate 23–4, 47, 51, 54, 64, 66, 77, 79, 81, 83–4, 90, 94–5, 104, 106–9, 120, 147–8, 164, 172
primitiveness 78, 161–2
prion disease 92, 94
prognathism 32, 64, 112
proprioceptors 130, 171
proto-sculptures 146, 159
Protsch, R. 4, 19, 27, 29–30, 34, 56, 113, 182
psychiatry 1, 25
psychology 1, 25, 98, 133
psychosis 84, 124

raft 49–50, 52, 106, 129
random access memory 141, 143, 145
reality constructs 7, 142
recognition 81, 88, 146
referent 139, 149, 170
referrer 149, 170
religion 7, 72–3, 105, 162
religious 37, 50, 108, 173
replacement hypothesis 5, 8, 16, 19, 26, 28, 45, 61, 63, 86, 105–7, 112–13,

122, 127, 142, 145, 148, 152, 163, 182, 184
reproduction 76, 95, 102, 111, 171
reproductive success 76, 107
Rindos, D. 3, 101
robusticity 24, 33, 56, 58–61, 76, 106, 109, 113, 184
Robusts 5, 19, 22, 24, 29, 31, 40, 58, 61, 63, 86, 106–8, 112–13, 115, 122, 135, 148, 183
Rose, M.R. 21, 105
rubber hand illusion 87
running ahead 78, 146, 181

Sautuola, M. de 3, 77
Schekman, R. 21–2, 56, 176
schizophrenia 82, 84–9, 94, 97, 119, 124, 165
sciences 1, 17, 55, 72, 77, 101, 170, 173, 176–7, 184
seafaring 26, 34, 36, 52, 122, 162,
self-awareness 25, 47, 129–31, 138, 155 169, 171
self-domestication 99–102, 116–7, 164, 170,
sentience 129, 171
sexual selection 75–7, 106, 116, 118, 128, 133, 169, 172, 184
sexuality 107, 131
sexy son hypothesis 76, 133
shaman 87, 89, 151
sickle cell anaemia 119
simile of the cave 171
single-nucleotide polymorphisms 29, 126
skeletal 23–4, 28, 32, 50, 58–60, 63, 69, 71, 105, 107–9, 112, 127–8, 175, 181
Société d'Émulation 2
Société de Linguistique 162
Somme 2, 153
speciation 5, 16, 19, 59–60, 64, 83, 86, 104, 122, 126, 183
splitters 15, 56, 69–71
superior parietal lobule 167
susceptibility alleles 86, 95–7, 106

symbol 78, 131–3, 135, 143, 148, 151, 169–70, 180
symbolic 34, 73–5, 113, 135, 143, 146, 148–9, 170
symbolling 34, 46, 78, 142, 171

tameness 102, 119, 123–4
taphonomy 12, 23, 51, 137, 145–6, 148, 151, 157, 159–60, 174, 181
Tasmanians 162, 181
technology 14, 33–4, 45, 55, 59, 61–3, 75, 77, 109–10, 113–14, 122, 128, 138, 146–7, 160, 168–9, 174–5, 180–1
teleological 17, 37–8, 66, 74–5, 105, 115, 145, 181
teleology 17, 74, 106, 181
temporal lobe 80–1, 85, 88, 91–2, 104, 139, 165–6
testability 1, 7, 180, 184
Theory of Mind 17, 25, 130–1
therianthrope 40, 132
Thomsen, C.J. 2
tool use 78, 147, 171
transition 32, 34, 62, 113, 152,
tribology 17, 155, 159, 174
typology 47, 54, 141

undiscovered primate 79, 172
Upper Palaeolithic 6, 27, 31–8, 40, 44–5, 60–3, 89, 106–7, 110, 113–14, 131, 136–8, 142, 159, 168–9, 180–1
uranium-thorium 40, 176

VENs 79, 81–2, 88, 92
Verhoeven, T. 52
von Economo neurons 79, 81, 129–30
vulva 39, 108, 131, 133

watercraft 49, 51
Wernicke's area 81, 163
Williams syndrome 164–5
World Heritage List 6, 36, 173

zoomorph 39–40

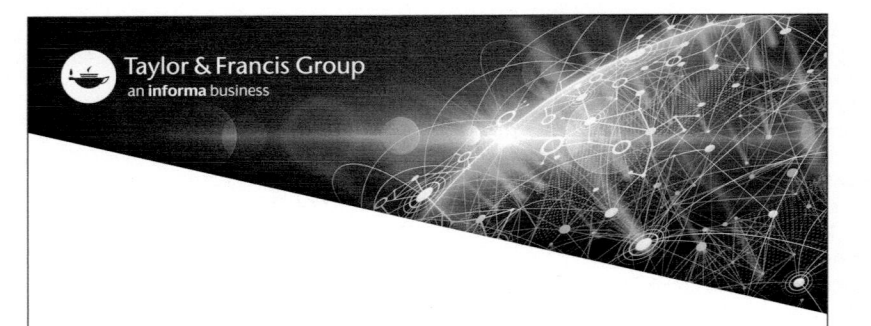